The Anthropocene Judgments Project

This book is a collection of speculative judgments that, along with accompanying commentaries, pursue a novel enquiry into how judges might respond to the formidable and planetary-scaled challenges of the Anthropocene.

The book's contributors – from Australia, Asia, Europe, and the United Kingdom – take up a range of issues: including multispecies justice, the challenges of intergenerational justice, dimensions of postcolonial justice, the potential contribution of AI platforms to the judgment process, and the future of judging and law in and beyond the Anthropocene. The project takes its inspiration from existing critical judgment projects. It is, however, thoroughly interdisciplinary. In anticipating future scenarios, and designing or adapting legal principles to respond to them, the book's contributors have been assisted by climate scientists with expertise in future modelling; they have benefitted from the experience of fiction writers in future world-building; and they have incorporated elements of the future worlds depicted in various texts of speculative fiction and artworks. The judgments are, of necessity, speculative and hypothetical in their subject matter. Thus, taken together, they constitute a collaborative experiment in creating the inclusive and radical imaginaries of the future common law.

The Anthropocene Judgments Project will appeal to critical and sociolegal academics, scholars in the environmental humanities, environmental lawyers, students, and others with interests in the pressing issues of ecology, multispecies justice, climate change, the intersection of AI platforms and the law, and the future of law in the Anthropocene.

Nicole Rogers is Professor of Law at Bond University, Australia.

Michelle Maloney is Co-Founder and National Convenor of the Australian Earth Laws Alliance.

The Anthropocene Judgments Project

Futureproofing the Common Law

Edited by Nicole Rogers and
Michelle Maloney

Routledge
Taylor & Francis Group
a GlassHouse Book

First published 2024
by Routledge
4 Park Square, Milton Park, Abingdon, Oxon OX14 4RN

and by Routledge
605 Third Avenue, New York, NY 10158

Routledge is an imprint of the Taylor & Francis Group, an informa business

A GlassHouse book

British Library Cataloguing-in-Publication Data
A catalogue record for this book is available from the British Library

ISBN: 9781032485409 (hbk)
ISBN: 9781032485423 (pbk)
ISBN: 9781003389569 (ebk)

DOI: 10.4324/9781003389569

Typeset in Sabon
by Deanta Global Publishing Services, Chennai, India

For Freya, Lillika, and Jess

Contents

Contributors

Foluke Ifejola Adebisi is an Associate Professor at the Law School, University of Bristol, United Kingdom. Her scholarship focuses mainly on the relationship between theories of decolonisation and antiracism and how they do and can interact with legal knowledge. Thus, her scholarly work is concerned with what happens at the intersection of legal education, law, society, and a history of changing ideas of what it means to be human. She founded and runs Forever Africa Conference and Events (FACE), a Pan-African interdisciplinary conference. She blogs about her scholarship and pedagogy on her website 'Foluke's African Skies' at <https://FolukeAfrica.com>. Her monograph *Decolonisation and Legal Knowledge: Reflections on Power and Possibility* was published in March 2023.

Chiara Armeni is Professor of Environmental Law at Université Libre de Bruxelles (ULB), Brussels. Her broad research interests focus on the legal aspects of environmental protection at various levels, particularly at the European level, and the governance of energy technologies. Chiara's research concerns in particular democratic participation in environmental decision-making and the role of environmental rights and environmental justice in the socio-ecological transition. She is also interested in legal imagination and legal futures.

Chiara is a member of the European Law Research Centre (CDE) and of the Interfaculty Institute of Socio-Ecological Transformations (iiTSE) of the ULB. As an environmental lawyer, she is a member of the Belgian High Committee for the Just Transition (federal). Chiara holds a PhD in law from University College London (United Kingdom).

Dr Bindi Bennett (she/her) is a Gamilaraay woman, mother, and social worker and is a Professorial Research Fellow at Federation University, Australia. She is a social justice scholar and activist who is committed to improving and growing cultural responsiveness; re-Indigenising Western spaces; understanding and exploring Indigenous Knowledge Systems in research; and exploring the human-animal bond.

Mark Brady is a Law and Technology expert and Lecturer in Law at Charles Darwin University, Darwin, Australia. Mark researches into regulation of autonomous technologies with a focus on establishing transport and mobility solutions within society. Mark's PhD research into regulation for disruptive technology focused on autonomous land vehicles as a case study examining the regulatory impact these technologies have on law. Mark's publications include the 2021 QUT report into Roadside Enforcement Powers and Automated Vehicles for the National Transport Commission (NTC July 2021), along with journal articles and submissions to the National Transport Commission regarding automated vehicles. His research has been cited in NSW Parliamentary *Hansard* (Report 2/56 of 2016) and by the National Transport Commission (Motor Accident Injury Insurance and Automated Vehicles, October 2018, and On-road Enforcement for Automated Vehicles, July 2022).

Greta Bird has had a long interest in law and marginalised peoples. She currently teaches a course Law and Land in an Aboriginal Master's programme at the University of South Australia. Greta has most recently published, with Dr Jo Bird, on a nuclear waste dump in South Australia and on the award of damages for Aboriginal cultural loss in the Timber Creek judgment in the High Court. Greta has had a long career in the university sector and has been a Commissioner on the Australian Law Reform Commission.

Jo Bird is an interdisciplinary thinker with a background in law and the humanities encompassing cinema studies, sociology, critical theory, and history. She has a PhD in Law from the University of Melbourne on the topic of human rights and bioethics. She has taught international law for many years. Jo has also worked as a research associate to First Nations Pro Vice-Chancellor Professor Irene Watson. She has an interest in post-colonial philosophies with many publications including internationally in the *Cardozo Journal of Law and Gender* (NYC).

Susan Bird is a Senior Lecturer in Law at Charles Darwin University, Darwin, Australia. She is an interdisciplinary research scholar. Her interests lie broadly in law and legal philosophy, in particular legal geographies, and making spaces more inclusive. She works to support diversity at university through the scholarship of teaching and learning, and her teaching practice. Susan is also interested in the way law can be employed to address environmental issues.

Elena Cirkovic is a Senior Research Fellow at the Max Planck Institute Luxembourg for International, European, and Regulatory Procedural Law; Helsinki Institute of Sustainability Science (HELSUS) Arctic Avenue Grant-funded researcher, Ecosystems and Environment Research Programme; and Research Affiliate at the Massachusetts Institute of

Technology (MIT) Media Lab. Her project entitled 'Anthropocentrism and Sustainability of the Earth System and Outer Space (ANTARES)' has been hosted at the University of Helsinki, Finland (since 2017) and Max Planck Institute Luxembourg (since 2022).

Robert Cunningham is Professor, Dean, and Head of Curtin Law School, Australia, barrister within Murray Chambers, and a Fellow of the Australian Academy of Law. He is a Director of the Australian Pro Bono Centre Board and the National Law Lead of the Digital Finance Cooperative Research Centre. Throughout his professional career, he has engaged with the law in his capacity as both a legal practitioner and academic. In his barrister practice within Murray Chambers he specialises in Federal Court matters with an emphasis on matters such as corporate law, consumer protection, native title, and intellectual property. In academia, Robert's pursuits have primarily focused on the way law interfaces with international trade, ecological sustainability, and information governance. He has published numerous academic journal articles and books, as both sole-author and co-author, primarily in the fields of corporate law, intellectual property, and international trade law. Robert is the author of *Information Environmentalism: A Governance Framework for Intellectual Property Rights*, published open access. He maintains teaching and research commitments in corporate law, corporate governance, and World Trade Organization law.

Aoife Daly is a Professor of Law at the School of Law, University College Cork, Ireland, where she specialises in human rights law. Aoife's research focuses on human rights-based approaches and children's rights in areas that include environmental rights, climate activism, and access to justice. In 2023 she secured a European Research Council Consolidator Grant to build a team to carry out a large-scale research study on child/youth climate justice – inside and outside the courts – around the world. She is a member of the Global Network of Human Rights and the Environment, and UCC's Environmental Research Institute.

André Dao is a Postdoctoral Research Fellow with the Laureate Program in Global Corporations and International Law at the Melbourne Law School, Australia. His first novel, *Anam*, was published in 2023.

Oscar Davis has Gamilaraay heritage on his mother's side and New Guinean heritage on his father's side. He lives and works on Kombumerri country as an Indigenous Fellow and Assistant Professor in Philosophy and Australian History at Bond University, Australia. He is interested in how Indigenous knowledges inform our conceptions of what it means to be in relation to the world.

Julia Dehm is a Senior Lecturer in the School of Law, La Trobe University, Australia. In 2023 she was a Member of the School of Social Science,

Institute for Advanced Studies, Princeton. Her scholarship addresses urgent issues of international and domestic climate change and environmental law, natural resource governance, and questions of human rights, economic inequality, and social justice. Her first monograph, *Reconsidering REDD+: Authority, Power and Law in the Green Economy* (Cambridge University Press, 2021), was awarded the 2021 ECR Publication Prize from the Law and Society Association of Australia and New Zealand. She has co-edited two books, *Locating Nature: Making and Unmaking International Law* (with Usha Natarajan) and *Power, Participation and Private Regulatory Initiatives: Human Rights under Supply Chain Capitalism* (with Daniel Brinks, Karen Engle and Kate Taylor) as well as three journal symposium or special issues. She has worked as a consultant to the United Nations Special Rapporteur on the right to adequate housing and her work has been widely published, including in the *Leiden Journal of International Law*, the *European Journal of International Law*, the *Griffith Law Review*, the *Australian Journal of Human Rights*, and others. She holds a Bachelor of Arts, a Bachelor of Laws (Honours), and a PhD from the University of Melbourne.

Kate Galloway is an Associate Professor of Law at Griffith Law School, Brisbane, Australia. Her research focuses on land tenure and the implications of property for justice, including environmental justice. Kate's work extends beyond property theory and doctrine to understand how hierarchies of interests in land adversely affect the land itself with implications for the more-than-human world and the self-determination of First Peoples in Australia. As a lawyer and educator, she also researches in legal education specialising in curriculum designed for values-oriented graduates equipped to face complex challenges, including climate change. Kate has served for many years on the boards of community environmental legal centres including the Environmental Defenders Office (Australia).

Brad Jessup is a Senior Lecturer at Melbourne Law School within the University of Melbourne, Australia. Brad is a descendant of European and British settlers who received early grants of the stolen lands of the Boonwurrung People. Brad continues to have habitat in the unceded lands of the Boonwurrung and Wurundjeri people of the Kulin nations, and Brad works among the humans of the Centre for Resources Energy and Environmental Law. Brad's expertise is in Australian land use laws, environmental risk and protection, and legal geography.

Orla Kelleher is an Assistant Professor at Maynooth University, Ireland. Orla specialises in climate change, and environmental and human rights law. Orla was awarded her PhD without corrections from UCD Sutherland School of Law in March 2022. Her PhD examined the impact of climate change on legal reasoning in rights-based systemic climate change litigation in European countries. Her PhD was awarded dual research scholarships: an IRC Government of Ireland Postgraduate Scholarship (2019–2021) and a UCD School of Law Doctoral Scholarship (2017–2019). Orla is also a non-practising barrister, having been called to the Irish Bar in 2020.

Michelle Lim is an Associate Professor of Law at the Yong Pung How School of Law, Singapore Management University, Singapore. She is Deputy Editor of the *Review of European, Comparative and International Environmental Law* and Deputy Chair of the Biodiversity Law Specialist Group of the IUCN World Commission on Environmental Law.

Clara Mancini, PhD, is Professor of Animal–Computer Interaction in the School of Computing and Communications at The Open University, UK, and founding head of the University's Animal–Computer Interaction (ACI) Laboratory. She has led, supervised, and worked on research projects exploring the interaction between animals and technology; the nexus between technology, animal well-being and justice, and human-animal relations. Her research spans the theory, methodology, practice, and ethics of designing technology for and with both free-living and managed animals. Clara has published in prestigious peer-reviewed computing, interaction design, animal–computer interaction, and animal behaviour and welfare journals and conference proceedings. She has examined and refereed on international ACI research; lectured and given keynotes on ACI worldwide; and advised and consulted for international companies and funding bodies. Her research has appeared in national and international media, including print, radio, and television. Clara is interested in exploring ACI's potential to inform equitable and inclusive multispecies ecosystems.

Michelle Maloney, PhD, is the National Convenor of the Australian Earth Laws Alliance; Adjunct Senior Fellow at the Law Futures Centre, Griffith University, Australia; and Director of Future Dreaming, an Indigenous and non-Indigenous partnership organisation working to share cross-cultural ecological knowledge with people around Australia.

Kelly Menzel is a proud Ngadjuri cis-straight woman with ancestral connections to Bundjalung Nation. She resides on beautiful Widjabul Wiabul Country and is currently the Associate Dean Education, Gnibi College of Indigenous Australian Peoples at Southern Cross University, Australia. In her research, she seeks to expose the causative factors of institutional racism, which, despite policies of diversity and inclusion, continue to exacerbate the socioeconomic, educational, and health gaps between Australia's Aboriginal and Torres Strait Islander and non-Indigenous populations and explore the possibilities of reforms that may facilitate Aboriginal and Torres Strait Islander Peoples' social mobility through professional advancement and achievement.

Zoe Nay is a PhD candidate at Melbourne Law School, Australia, and a Research Fellow at Melbourne Climate Futures – an interdisciplinary initiative at the University of Melbourne which aims to accelerate the transition to a positive climate future. Zoe specialises in climate change law and the law of the sea, with a focus on climate change impacts and responses in the Pacific region. Her doctoral research examines the legal issues related to state responsibility for climate change loss and damage in Pacific Island states.

Christine Parker is a Professor of Law at the University of Melbourne, Australia, and Chief Investigator in the Australian Research Council Centre of Excellence for Automated Decision Making and Society. She is of Anglo-Celtic settler descent, and her current habitat is on the unceded country of the Wurundjeri people of the Kulin nations and also the unceded lands of the Djab Wurrung people, and among the humans of the Centre for Resources Energy and Environmental Law at Melbourne Law School. Her expertise is in business regulation, including 'ecological regulation', animal law, food law, and lawyers' ethics.

Ayesha Riaz is a Senior Lecturer at the University of Greenwich, United Kingdom, a PhD candidate at Queen Mary University of London, and a part-time teacher at the London School of Economics and Political Science.

Nicole Rogers is a Professor of Climate Law in the Faculty of Law at Bond University, Australia. She has been involved in the wild law movement in Australia since 2009, when the inaugural Australian wild law conference was held, and was one of the founding members of the Australian Earth Laws Alliance. From 2014 to 2017, she instigated and co-led the Wild Law Judgment project: a key scholarly intervention in reimagining the common law from a wild, or ecocentric, perspective. She is co-editor of *Law as If Earth Really Mattered: The Wild Law Judgment Project* (2017) and has published widely in the area of wild law. Nicole also researches and publishes in the field of interdisciplinary climate studies. Her 2019 Routledge monograph, *Law, Fiction and Activism in a Time of Climate Change*, was shortlisted for the 2020 Hart-SLSA Book Prize and the inaugural 2020 Australian Legal Research Book Award. Her most recent monograph, *Law, Climate Emergency and the Australian Megafires*, was published by Routledge in 2021.

Tania Sourdin is a Professor of Law, Dean of the University of Newcastle Law School, Australia, and the President of Academic Senate, and has researched in the area of innovation in justice (including issues related to access to justice and the rule of law) for more than two decades. She has previously led large-scale research projects focused on AI and the Law. Professor Sourdin is the author of more than 140 publications including books, articles, and papers that are focused on justice reform issues and is the author of *Judges, Technology and Artificial Intelligence* (2021). She is currently leading a worldwide study of judges and their relationship with technology. Professor Sourdin has expertise in negotiation, collaboration, dispute resolution, high-conflict behaviours, and mediation. She has also worked in respect of complaints handling and has written standards in this area and mediation. She has advised governments and other organisations both in Australia and overseas on the topic of justice innovation and is an active member of a number of Boards and Committees focused on current issues.

Natalia Szablewska, PhD, is Professor in Law and Society at The Open University Law School, UK, and holds further affiliate positions at institutions in Australia, New Zealand, and Cambodia. She currently serves on the Modern Slavery Leadership Advisory Group to the New Zealand Government and as an Executive Committee Member of Australian Lawyers for Human Rights (ALHR). She has a diverse and extensive research portfolio focusing on themes at the intersection of law, public policy, and applied ethics. Applying human rights-based approaches and multiple and mixed research methods, her research has contributed to several areas, including forced/involuntary migration, modern slavery, and gender (in)equalities in post-conflict societies. Natalia's interdisciplinary research has been published in leading journals and with publishers across a number of disciplines and has been reported in government publications, scientific outlets, and a number of national and international media. Her current research focus is on the legal, social, and economic linkages between business and human rights, modern slavery, and sustainable development.

Acknowledgements

We are immensely grateful to the many people who believed in this project and participated in diverse ways. Climate scientists Professor Brendan Mackey and Professor Gretta Pecl presented at workshops and generously shared their scientific expertise with participants. Justice François Kunc, First Nations scholar Professor Mary Graham, and novelists Dr Honni van Rijswijk and Clare Moleta provided important insights into, respectively, the crafting of judgments, First Nations relational philosophy, and the art of creating fictitious futures. Ambelin Kwaymullina gave us permission for her poem 'This Thing Called Anthropocene' to be read at the first project workshop; we would like to acknowledge and thank her for this, and also to thank Karinda Burns for reading the poem. Christina Meyers helped with workshop organisation. Holli Edwards provided invaluable research assistance.

We are very grateful to Colin Perrin at Routledge for his support of the project.

Nicole would also like to thank her colleagues at the Faculty of Law at Bond University, and Professor David Schlosberg and Professor Danielle Celermajer from the Sydney Environment Institute where she was an Honorary Visiting Fellow for the first four months of the project.

Chapter 1

Judging the Future and the Future of Judging: The Anthropocene Judgments Project

Nicole Rogers

The Anthropocene Judgments project forms part of an ever-expanding international collection of critical judgment projects,[1] in which the process of judging has been reinvigorated by scholars 'dressing up as judges'.[2] Rather than revisiting the judgments of the past, as has commonly been the practice in other critical judgment projects, the Anthropocene Judgments project anticipates the practice of judging in, and for, the future; or, as we have contextualised it, judging in and for the Anthropocene. For the purposes of the project, we have adopted the concept of the Anthropocene; we have not done so uncritically. The term is shorthand for signalling a radically different Earth, one shaped and disfigured by human technology and human activities. The term is also a testament to human hubris, a quality highlighted and lamented in a number of the judgments in the project. The concept captures much about our current global predicament, but it is also highly problematic and, for that matter, highly problematised.

The Troubling and Troubled Concept of the Anthropocene

What is meant, then, by the Anthropocene? This term signifies our entry into a new epoch, distinguished by the massive human imprint on Earth and Earth systems; as historian Dipesh Chakrabarty pointed out in 2009, '[h]umans now wield a geological force'.[3] Rupture is a word that surfaces frequently in scholarly discussions about the Anthropocene. For instance, Clive Hamilton and Jacque Grinvevald have described the Anthropocene as representing

1 See 'Critical Judgments', *UNSW Sydney: Gilbert & Tobin Centre for Public Law* (Web Page, 6 April 2022) <https://criticaljudgments.com>.

2 Margaret Davies, 'The Law Becomes Us. Rediscovering Judgment' (2012) 20 *Feminist Legal Studies* 167, 172.

3 Dipesh Chakrabarty, 'The Climate of History: Four Theses' (2009) 35(2) *Critical Inquiry* 197, 206.

DOI: 10.4324/9781003389569-1

a 'dangerous shift, and a radical rupture in Earth's history'.[4] Humanity is, Rakhyun Kim argues, 'stumbling towards' a precipice; falling from this precipice 'will mark an irreparable rupture in time'.[5] It is arguable that there has to be a clear rupture for a new epoch to be officially recognised by the International Commission on Stratigraphy: the international body charged with epoch-naming responsibilities.[6] Acknowledgement by this authoritative body is still pending, although in 2019, the Anthropocene Working Group, a component body of the Subcommission on Quaternary Stratigraphy which is, in turn, a constituent body of the International Commission on Stratigraphy, voted in favour of recognising the Anthropocene as a formal chrono-stratigraphic unit.[7]

The Anthropocene follows or replaces the Holocene era, which spanned over 11,000 years and encompassed the global adoption of agriculture, the establishment of large-scale human settlements, and the rise and fall of civilisations. Climatically speaking, the Holocene, which began at the end of the last Ice Age, was what climate activist Bill McKibben has called 'the sweetest of sweet spots'.[8] Confronted as we are today by an ever-expanding litany of climate disasters – unprecedented fires, floods, hurricanes – we may well view that prolonged period of climate stability with nostalgia. We cannot return to it; nor can we return to the comfortable and dangerous illusion that there is something separate to humanity called 'Nature' which is subject to human control, able to be disciplined, tamed, and made productive.

Paul Crutzen, a Nobel Prize-winning atmospheric chemist, proposed the term 'the Anthropocene' in 2000,[9] and it immediately found favour in popular discourse. The onset of the Anthropocene remains a matter of some debate. Some argue that it commenced with the Industrial Revolution; some suggest even earlier, in 1610, following the 'collision of the Old and the New Worlds', when the colonisation of the Americas and resulting genocide led to an observable decline in carbon dioxide in the atmosphere.[10] There is no doubt that humanity's impact on the biophysical systems of Earth dra-

4 Clive Hamilton and Jacques Grinevald, 'Was the Anthropocene Anticipated?' (2015) 2(1) *The Anthropocene Review* 59, 62.

5 Rakhyun E Kim, 'Taming Gaia 2.0: Earth System Law in the Ruptured Anthropocene' (2021) *The Anthropocene Review* 411.

6 Cameron Harrington, 'A Quantum Anthropocene? International Relations Between Rupture and Entanglement' in Joana Castro Pereira, and André Saramago (eds), *Non-Human Nature in World Politics: Theory and Practice* (Springer International Publishing, 2020) 53, 56.

7 Working Group on the Anthropocene, 'Results of Binding Vote by AWG', *Subcommission on Quaternary Stratigraphy* (Web Page, 21 May 2019) <http://quaternary.stratigraphy.org/working-groups/anthropocene>.

8 Bill McKibben, *Eaarth: Making Life on a Tough New Planet* (Black Inc Books, 2010) 3.

9 Paul J Crutzen and Eugene F Stoermer, 'The "Anthropocene"' (2000) 41 (May) *Global Change Newsletter* 17.

10 See Simon L Lewis and Mark A Maslin, 'Defining the Anthropocene' (2015) 519 *Nature* 171, 175.

matically increased during the 'Great Acceleration', which began after World War II. This is a term devised by a group of scientists to denote the massive upsurge in global greenhouse gas emissions and other human impacts in the post-war period. Climate scientist Will Steffen and his colleagues have described this period as the 'leading candidate' for the commencement of the Anthropocene.[11] They assert that this starting point is 'by far the most convincing from an Earth System science perspective', since it is then that fundamental human-caused shifts in Earth system processes became discernible.[12]

The advent of the Anthropocene has unsettled established concepts and belief systems, including the pervasive modernist dichotomy of humanity/nature. The Anthropocene has been distinguished by a renewed philosophical focus upon entanglement,[13] upon the agency of the nonhuman and more-than-human and what Jane Bennett refers to as 'vibrant matter',[14] and upon interspecies collaborations and interactions. As three scholars have recently put it, '[i]n the disorientating moment of the Anthropocene ... the curtain is drawn back, the scenery and props come to life, and the distinction between the human and inhuman is no longer assured'.[15] However, in focusing upon the role of humanity as, seemingly, a cohesive whole in reshaping Earth and its systems, users and proponents of the concept, in the words of one writer, '[risk] erasing histories of violence and dispossession of the wretched of the modern world and [disavow] the multiple worlds inhabited with(in) "Anthropocenes"'.[16]

The term conceals the profound socio-economic inequities in the causal forces behind the transformation of Earth system processes. As Andreas Malm and Alf Hornborg point out, we cannot 'write off divisions between human beings as immaterial to the broader picture, for such divisions have been an integral part of fossil fuel combustion in the first place'.[17] The term also masks the socio-economic inequities in how these radical changes in Earth system processes are experienced and addressed. As Indigenous scholar Zoe Todd has pointed out, 'it is important to query which humans or human systems are driving the environmental change the Anthropocene is meant to

11 Will Steffen et al, 'The Trajectory of the Anthropocene: The Great Acceleration' (2015) 2(1) *The Anthropocene Review* 81, 83.

12 Ibid 93.

13 See, eg, Donna J Haraway, *Staying with the Trouble: Making Kin in the Chthulucene* (Duke University Press, 2016).

14 Jane Bennett, *Vibrant Matter: A Political Ecology of Things* (Duke University Press, 2010).

15 Daniel Matthews, Kathleen Birrell, and Tim Lindgren, 'Sovereignty in the Anthropocene' (2022) 31(3) *Griffith Law Review* 435, 438.

16 Marie-Catherine Petersmann, 'Response-abilities of Care in More-Than-Human Worlds' (2021) 12 *Journal of Human Rights and the Environment* 102, 103.

17 Andreas Malm and Alf Hornborg, 'The Geology of Mankind? A Critique of the Anthropocene Narrative' (2014) 1(1) *The Anthropocene Review* 62, 66.

describe' and 'who is dominating the conversations about how to change the state of things'.[18]

It has been suggested that the plural term '[a] billion black Anthropocenes' is more accurate,[19] and/or that the term should be abandoned altogether. There has been no shortage of alternative suggestions for our current era; these include the Capitalocene, the Plantationocene, the Chthulucene,[20] and, provocatively, the Absurdocene.[21]

Given the controversy that has occurred in the context of the usage and application of the term 'Anthropocene', we as editors have adopted it with some wariness, seeking to acknowledge contemporary and historical inequities in judgments that address the unique challenges of our particular time and our daunting future.

Future Legal Imaginings

The Anthropocene has been described as a no-analogue state,[22] and Clive Hamilton and Jacques Grinevald are not alone in concluding that 'the Holocene can be no guide to the Anthropocene geologically or intellectually'.[23] This extends to Anthropocene law and justice. To quote two other scholars, 'Holocene conceptions of justice are not apt to deal with justice in the Anthropocene'.[24]

Anna Grear has written of the need to develop a legal imaginary *for* the Anthropocene, an imaginary markedly different from the dominant legal imaginary *of* the Anthropocene in which the human is 'the only actor in the world whose agency counts'[25] and law is 'complicit in ongoing forms of eco-violence, economic predation and the unparalleled imposition of precarity on humans and non-humans alike'.[26] She proposes an alternative legal imaginary for the Anthropocene which involves new ways of living with the

18 Zoe Todd, 'Indigenizing the Anthropocene' in Heather Davis and Etienne Turpin (eds), *Art in the Anthropocene. Encounters Among Aesthetics, Politics, Environments and Epistemologies* (Open Humanities Press, 2015) 241, 244.

19 See Kathryn Yusoff, *A Billion Black Anthropocenes or None* (University of Minnesota Press, 2018).

20 See Donna Haraway, 'Anthropocene, Capitalocene, Plantationocene, Chthulucene: Making Kin' (2015) 6(1) *Environmental Humanities* 159.

21 Hugh Sillitoe, 'Absurd Performance and Climate Justice' (Conference Paper, Performances Studies International PSi #22 Conference 'Performance Climates', University of Melbourne, 5–9 July 2016).

22 See Steffen et al (n 11) 94.

23 Hamilton and Grinvevald (n 4) 62.

24 Colin Hickey and Ingrid Robeyns, 'Planetary Justice: What Can We Learn from Ethics and Political Philosophy?' (2020) 6(100045) *Earth System Governance* 6.

25 Anna Grear, 'Legal Imaginaries and the Anthropocene: "Of" and "For"' (2020) *Law and Critique* 351, 363.

26 Ibid 355.

Earth.[27] Michelle Lim, a participant in the project, has elsewhere emphasised the urgent need for 'more hopeful, inclusive, and radical imaginations and imaginaries for environmental law in the Anthropocene'.[28]

We have assumed here that new legal imaginaries and 'conceptual innovation'[29] must and will become part of judging practices, and have sought to incorporate these into the project. Participants were tasked with identifying the wicked problems of the Anthropocene, recognising possible legal issues, and resolving these with, in many instances, radically different and new legal principles and ways of viewing law. In order to do these things, we were required to think imaginatively, into the future, and to think creatively, working out how justice can and will be served in future disputes.

Future imaginings play a key role in all forms of adaptation to the challenges of the Anthropocene, including legal adaptation. As Kathryn Yusoff and Jennifer Gabrys have written:

> In the dislocations of the present, the imagination of the future is a catalyst for thinking about how we might be in the world in other, hitherto unthought ways. Future imaginings can thus be thought of as a process for developing adaptive capacities and emotional resilience within changed environments.[30]

In constructing scenarios for the purpose of judging the future, the interplay of law, science, the humanities, and the creative arts becomes important. Writers of speculative fiction are experts in the area of future imaginings, and there is a plethora of works, particularly in the genre of climate fiction, that have assisted us in envisaging changes in the Earth systems of the future and possible social and political responses to such changes. Climate scientists are also adept at future modelling, charting future scenarios associated with different emissions pathways and what they will look like. We utilised the expertise of both groups of futurists in the project, with workshops featuring presentations from Intergovernmental Panel on Climate Change climate scientists Professor Gretta Pecl[31] and Professor Brendan Mackey[32] and writers

27 Ibid 356.

28 Michelle Lim, 'Fiction as Legal Method – Imagining with the More-Than-Human to Awaken Our Plural Selves' (2021) 33(3) *Journal of Environmental Law* 501, 502.

29 See Hickey and Robeyns (n 24) 6.

30 Kathryn Yusoff and Jennifer Gabrys, 'Climate Change and the Imagination' (2011) 2(4) *Wiley Interdisciplinary Reviews: Climate Change* 516, 522.

31 Gretta Pecl is a professor of climate change ecology and marine ecology at the University of Tasmania and leads the Future Seas project; see 'Future Seas', *Future Seas* (Web Page, 2023) <https://futureseas2030.org>.

32 Brendan Mackey is a climate scientist and forest ecologist. He is the Director of the Griffith Climate Action Beacon at Griffith University.

of speculative fiction Dr Honni van Rijswijk and Clare Moleta.[33] In workshops, we also heard from Indigenous elder and philosopher Professor Mary Graham on the relational philosophy of Australian First Nations people, and from Justice François Kunc of the New South Wales Supreme Court[34] on the art of judgment writing.

Contributors utilised a rich array of interdisciplinary sources in constructing their future judgments. Scientific predictions shaped the visioning process, with many writers referring to future destructive fire seasons, flooding and inundation, and incorporating material from the recent reports of the Intergovernmental Panel on Climate Change in so doing. Natalia Szablewska and Clara Mancini drew upon Clara's expertise in animal-computer interactions to portray a 'smart' dairy farm of the future. Brad Jessup and Christine Parker have incorporated the research of ecologists and conservation scientists into their chapter, in which the habitat of a currently extinct and resurrected species, and a currently endangered species, is reconceived and protected.

A number of writers were influenced by works of speculative fiction. André Dao, for instance, incorporated nonhuman characters from *The Swan Book*,[35] the widely acclaimed novel by Australian First Nations writer Alexis Wright. The Chief Judge of the New South Wales Land and Environment Court, Judge Brian Preston, has emphasised the importance of art and aesthetics in 'chang[ing] how we view change',[36] and Chiara Armeni's contribution to the Anthropocene Judgments project was inspired by an artwork: an image entitled 'Piccadilly Circus Water Lilies', by Robert Graves, in which well-known London landmark Piccadilly Circus has been transformed by rising waters and wind turbines spin above its water lilies. Another judgment writer, Michelle Lim, was influenced by musical works.

Despite the departure from conventional judging practices in this collection and, in some instances, quite radical departures from and a deliberate rejection of conventional judging formats, the line between imagining future judgments and delivering 'real' judgments is not as clear-cut as some may think. Sitting judges are not anchored entirely in the present and past in their decision-making processes; nor are they impervious to the temptation of imagining the future. Admittedly not all judges are prepared to take on

33 Honni van Rijswijk's young adult, speculative fiction novel *Breeder* was published in 2021 by Blackstone Publishing. Clare Moleta is the author of the climate fiction novel *Unsheltered* (Simon & Schuster, 2021).

34 Justice Kunc is also a former President of the Law and Literature Association of Australia.

35 Alexis Wright, *The Swan Book* (Giramondo Publishing, 2013).

36 Brian Preston, 'Changing How We View Change: the Artist's Insight' (Conference Paper, IUCN Academy of Environmental Law 2022 Annual Colloquium, QUT, 15 July 2022) <https://www.lec.nsw.gov.au/publications-and-resources/judicial-speeches-and-papers.html#Chief0>. See, also, Benjamin J Richardson, *The Art of Environmental Law. Governing with Aesthetics* (Hart Publishing, 2019).

the persona of a hypothetical judge in an imaginary case set 20 years in the future, as was Judge Preston in a mock trial organised by the Victorian Environmental Defenders Office in 2012. This judgment, in proceedings brought by Green Sea Turtles against the Commonwealth and Queensland governments in order to obtain a remedy for the acts and omissions that destroyed their habitat, was reproduced with the judge's permission in the Wild Law Judgments edited collection.[37]

Reflecting there upon the process of being a hypothetical judge rather than a real judge, the judge indicated that he found it an enjoyable experience: '[t]here are no disappointed litigants to appeal the judgment, no rebukes from appellate courts and no censure in the media'.[38] Yet even in orthodox judging practices, judges can stray into future imagining.

The following haunting passage from Judge Josephine Staton, in which the judge extrapolates from current omissions and obfuscations of both a political and judicial nature, is illustrative of judicial future imagining in the climate context. In *Juliana v United States*, the constitutional and atmospheric public trust lawsuit brought by young litigants against the United States government for its failure to mitigate, Judge Staton was the dissenting judge in the Ninth Circuit Court of Appeals. While her colleagues, the majority judges, were prepared to concede that the world is heading for an environmental apocalypse,[39] they felt that it was not the court's role to 'step into [the] shoes' of the political branches of government.[40] Judge Staton disagreed and looked ahead in so doing:

> If plaintiffs' fears, backed by the government's *own studies*, prove true, history will not judge us kindly. When the seas envelop our coastal cities, fires and droughts haunt our interiors, and storms ravage everything between, those remaining will ask: Why did so many do so little?[41]

It is not only the stark reality of climate change that has given rise to such articulated judicial forebodings on what lies ahead. For instance, in the *Tasmanian Dams* case, Australian High Court judge Justice Murphy was quite prepared to go beyond the immediate constitutional issues and speculate upon the environmental and legal dimensions of what might have seemed,

37 Brian Preston, 'Green Sea Turtles by Their Representative, Meryl Streef v the State of Queensland and the Commonwealth of Australia' in Nicole Rogers and Michelle Maloney (eds), *Law as if Earth Really Mattered: The Wild Law Judgment Project* (Routledge, 2017) 31.

38 Brian Preston, 'Writing Judgments "Wildly"' in ibid 19, 25.

39 *Juliana v United States* 947 F3d 1159 (9th Cir 2020) 1164 (Hurwitz CJ).

40 Ibid 1175 (Hurwitz CJ).

41 Ibid 1191 (Staton J) (emphasis in original).

in 1983, a somewhat dire futuristic scenario. From today's perspective, his scenario looks uncannily prescient. He hypothesised that:

> suppose that in the next few decades, because of the continuing rapid depletion of the world's forests and its effect on the rest of the biosphere, the survival of all living creatures becomes endangered. This is not a fanciful supposition (see The Global 2000 Report to the President of the United States (1980)). Suppose the United Nations were to request all nations to do whatever they could to preserve the existing forests. Let us assume that no obligation was created (because firewood was essential for the immediate survival of people of some nations). I would have no doubt that the Australian Parliament could, under the external affairs power, comply with that request by legislating to prevent the destruction of any forest, including any State forest.[42]

Justice Murphy was correct in predicting that the voracious pursuit of resources and profit would come to endanger all living creatures, and that deforestation would play a critical role in this. At the time of writing, he surely could not have envisaged that, despite the clearest of warnings from the United Nations as well as from the world's scientists, governments would have continued to permit the unchecked production and use of fossil fuels to the point where climate change now poses a threat to all life forms and their habitats.

Precursors and Antecedents

The Anthropocene Judgments project continues a tradition of critical judgment projects, which began with feminist judgment projects but now includes a range of different critical approaches.[43] Judgment rewriting is an established methodology in critical socio-legal scholarship. There is a growing array of feminist judgment projects,[44] in which the template devised in the first feminist judgment

42 *Commonwealth v Tasmania* (1983) 158 CLR 1, 170–171 (Murphy J).

43 See, eg, Gabrielle Appleby and Rosalind Dixon (eds), *The Critical Judgments Project: Re-Reading Monis v The Queen* (Federation Press, 2016); Eva Brems (ed), *Diversity and European Human Rights: Rewriting Judgments of the ECHR* (Cambridge University Press, 2012); Stephen W Smith et al (eds), *Ethical Judgments: Re-Writing Medical Law* (Hart Publishing, 2017); Helen Stalford, Kathryn Hollingsworth and Stephen Gilmore (eds), *Rewriting Children's Rights Judgments: From Academic Vision to New Practice* (Hart Publishing, 2017); Rogers and Maloney (n 37); Eva Brems and Ellen Desmet (eds), *Integrated Human Rights in Practice: Rewriting Human Rights Decisions* (Edward Elgar Publishing, 2017); and Nicole Watson and Heather Douglas (eds), *Indigenous Legal Judgments: Bringing Indigenous Voices into Judicial Decision Making* (Routledge, 2021).

44 See, eg, Rosemary Hunter, Clare McGlynn and Erika Rackley (eds), *Feminist Judgments: From Theory to Practice* (Hart Publishing, 2010); Heather Douglas et al (eds), *Australian Feminist Judgments: Righting and Rewriting Law* (Hart Publishing, 2014); Kathryn M

projects continues to apply: that is, past judgments are reworked within the framework of existing laws but from a feminist perspective. There has, however, been some subsequent experimentation with form and genre,[45] most noticeably in a 2021 published collection of Australian Indigenous judgments.[46] Furthermore, in the Wild Law Judgment project, participants by agreement embraced the art of writing judgments wildly. Wild law judging means putting Earth and the protection of Earth systems and processes at the centre of all judgments, a challenging task given the profoundly anthropocentric or human-centred nature of legal systems. Not all participants, but quite a few, were prepared to invent new laws and create standing for an expanded group of nonhuman and more-than-human litigants, in the process of judging wildly.

The Anthropocene Judgments project diverges from its predecessors, the other critical judgment projects, in two key respects: firstly, in its future orientation and secondly, in its interdisciplinary methodology. Although the Wild Law Judgment project incorporated some futuristic judgments,[47] the Anthropocene judgments are *all* devised around future scenarios. Participants were required to envisage the future but also to create plausible legal and extra-legal frameworks which might have developed between now and that future point: new laws, new precedents, new policy developments, and perhaps new forms of international relationships.

Given its forward-looking focus, the Anthropocene Judgments project is aligned with the scholarly enterprise in *The Cabinet of Future Laws*, in which a diverse group of contributors envisaged 'decisions to be' and 'experiment[ed] with law'.[48] As Peter Goodrich, one of the editors of that collection, queried in the Prelude, '[o]f what use … is a justice that cannot dance

Stanchi, Linda L Berger and Bridget J Crawford (eds), *Feminist Judgments: Rewritten Opinions of the United States Supreme Court* (Cambridge University Press, 2016); Máiréad Enright, Julie McCandless and Aoife O'Donoghue (eds), *Northern/Irish Feminist Judgments: Judges' Troubles and the Gendered Politics of Identity* (Hart Publishing, 2017); Loveday Hodson and Troy Lavers (eds), *Feminist Judgments in International Law* (Hart Publishing, 2019); Elisabeth McDonald et al (eds), *Feminist Judgments of Aotearoa New Zealand – Te Rino: A Two-Stranded Rope* (Hart Publishing, 2017); and Sharon Cowan, Chloe Kennedy and Vanessa E Munro (eds), *Scottish Feminist Judgments: (Re)Creating Law from the Outside In* (Bloomsbury Publishing, 2019).

45 See, eg, Irene Watson, 'First Nations Stories, Grandmother's Law: Too Many Stories to Tell' in Douglas et al (n 44) 46 and Nicole Watson, 'In the Matter of Djaparri (Re Tuckiar) [2035] FNCA 1' in Douglas et al (n 44) 442.

46 Watson and Douglas (n 43).

47 Preston (n 37); Cormac Cullinan, 'Great Barrier Reef v Australian Federal and State governments and others' in Rogers and Maloney (n 37) 39; Benedict Coyne, 'The Fraught and Fishy Tale of Lungfish v The State of Queensland' in Rogers and Maloney (n 37) 56; Robert Cunningham, 'Information Environmentalism and Biological Data: A Thought Experiment' in Rogers and Maloney (n 37) 355.

48 Peter Goodrich and Thanos Zartaloudis, 'Prelude' in Peter Goodrich and Thanos Zartaloudis (eds), *The Cabinet of Imaginary Laws* (Routledge, 2021) xvii, xxiv.

backwards and forwards from the past and into the future?'[49] Extraordinary legislative innovations and fantastical hypothetical institutions[50] were canvassed therein. In the Anthropocene Judgments project, by way of contrast, participants focused specifically upon future judging rather than legislative and policy innovations.

As a product of the project, this collection consists of 16 judgments and two reflections on law and justice in the Anthropocene and post-Anthropocene periods. It features the work of 26 writers from Australia, the United Kingdom, Europe, and Asia and, in one judgment, the contribution of an Artificial Intelligence system. We have grouped the judgments and reflections under the following themes for ease of reference, although many canvass a number of themes.

Themes

Multispecies Justice

Multispecies justice is a term devised to enable its advocates 'to contest the exclusive classificatory politics of anthropocentric justice'.[51] In the words of a group of interdisciplinary scholars, multispecies justice 'seeks to understand the types of relationships humans ought to cultivate with more-than-human beings so as to produce just outcomes'.[52] In this section, contributors have explored what the term means in a legal context, specifically as interpreted by fictional judges. To adopt a multispecies justice lens means to acknowledge the more-than-legal, and to accept and encourage cross-fertilisation between disciplines and different ways of thinking.

In this part of the collection, Brad Jessup and Christine Parker envisage an expanded reading of the demonstrably deficient *Environment Planning and Biodiversity Protection Act 1999* (Cth) to protect the habitat rights of endangered Tasmanian devils and newly resurrected Tasmanian tigers[53] in the Tarkine wilderness. This iconic wilderness area, currently threatened by mining and logging activities, is, in their futuristic scenario, slated for suburban development to accommodate a wave of climate refugees from mainland Australia.

Natalia Szablewska and Clara Mancini, after setting out the surveillance regime and behavioural controls in an intensive smart dairy farm in 2057,

49 Ibid xxi.

50 These included a Treasurer of Injustice, a Court of the Monuments, a Law in the Round Act, a Proof of Judicial Omniscience Act, and the Statute of Snouts.

51 Danielle Celermajer et al, 'Multispecies Justice: Theories, Challenges, and a Research Agenda for Environmental Politics' (2021) 30(1–2) *Environmental Politics* 119, 120.

52 Ibid.

53 This development is anticipated in a work of climate fiction – James Bradley's novel *Ghost Species* (Hamish Hamilton, 2020), which is about the resurrection of various extinct species, including but not restricted to the Tiger, in an isolated scientific laboratory in Tasmania.

create a fictional court, the World Court for Rights, which adjudicates the application of various rights to the nonhuman inhabitants of the farm. These rights, currently recognised as human rights, are the right to dignity and to be treated with respect at all times, the right to privacy, and the right to freedom from slavery, servitude, and forced labour.

Foluke Adebisi's chapter provides a voice for the ocean itself, its nonhuman inhabitants, and the humans who have drowned therein as a consequence of the actions of other humans. In an underwater court, the Keeper of the Flood pronounces its judgment on humanity by applying the law of the sea. Although the proceedings have a dreamlike, poetic character, the starting point for the judgment is a real historical case,[54] decided in 1783; here an appellate court assessed, for the purpose of an insurance claim, the alleged necessity of a ship captain's decision to discard the 'cargo' of African slaves into the ocean.

Finally, in this section, André Dao has contributed a judgment written about and around Alexis Wright's extraordinary climate fiction work, *The Swan Book*.[55] Key characters in the book, wild swans, are seeking climate refuge in Australia. The question addressed in the judgment, as to whether wild swans should be permitted entry through closed borders, highlights issues of interspecies climate displacement, and the likely limitations of algorithmic decision-making in the context of multispecies justice. The judges also address the relevance of narrative and, in particular, First Nations storytelling in decision-making processes.

Intergenerational Justice

In recent years, it has been impossible to ignore the voices of young people, in courtrooms, in public fora, and on the streets, as they present the case for intergenerational climate justice. They are articulating the outrage and pain that arises from the certain knowledge that they, and generations to come, will bear the brunt of climate impacts. As Justice Bromberg of the Australian Federal Court famously said in 2021, in the *Sharma* case, 'the devastation that the plausible evidence ... forecasts for [Australian children] ... will largely be inflicted by the inaction of this generation of adults, in what might fairly be described as the greatest inter-generational injustice ever inflicted by one generation of humans upon the next'.[56] His decision that the Australian Minister for the Environment has a duty of care to prevent death and injury to Australian children through climate impacts and that this duty applied to the exercise of her statutory decision-making power under the *Environment Protection and Biodiversity Conservation Act* 1999 (Cth)

54 *Gregson v Gilbert* (1783) 3 Doug KB 232.
55 Wright (n 35).
56 *Sharma v Minister for the Environment* (2021) 248 LGERA 330, 411.

in relation to a proposed coal mine extension was overturned on appeal in March 2022;[57] two short weeks after catastrophic, unprecedented flooding in Lismore and the Northern Rivers traumatised an entire community, including local children.

In the first judgment in this section, Kate Galloway revisits this case from a future perspective, integrating an imagined concept of quantum entanglement into her discussion of tortious duties of care and thereby rejecting the reasoning of the appellate court. She draws, inter alia, upon eco-philosopher Timothy Morton's neologism of hyperobjects to highlight the porous, non-linear, complex relationship between climate culprit and climate victim, and the need to move beyond traditional legal understandings of causation and proximity.

A similar case follows, one that has been mounted in Ireland in 2033 by a group of young people, at a time when over one-fifth of the country's landmass is now underwater. This also builds upon an earlier domestic decision, *Friends of the Irish Environment v Government of Ireland*.[58] In this future judgment, however, as written by Aoife Daly and Orla Kelleher, the Court is prepared to hold that the Irish government has an implied constitutional duty to protect the environment, including the climate. Furthermore, the government must take the best interests of children and future generations fully into account in formulating its climate mitigation policies.

Intergenerational climate injustice is compounded by the uneven distribution of climate impacts. As one commentator has put it, '[c]limate change will spare no country, but no country will face hardships like those awaiting SIDS'.[59] According to a report co-written by economist Nicholas Stern and presented at the 2022 COP27, developing countries, with the exception of China, will need 1.75 trillion pounds a year by 2030 to address climate adaptation and the effects of climate breakdown.[60] This amount of climate financing is, at present, not forthcoming from the Global North. At that COP, loss and damage was placed on the agenda for the first time. Nevertheless, John Kerry, the United States' special presidential envoy on climate, explained that the United States does not view this as compensation but rather as part of 'our efforts to try to help countries to adapt'.[61]

57 *Minister for the Environment v Sharma* (2022) 400 ALR 203.

58 *Friends of the Irish Environment v Government of Ireland* [2020] IESC 49.

59 Ori Sharon, 'State Extinction Through Climate Change: Reflection 3: Disappearing States' in Benoit Mayer and Alexander Zahar (eds), *Debating Climate Change* (Cambridge University Press, 2021) 349, 364.

60 Fiona Harvey, 'Developing Countries "Will Need $2tn a Year in Climate Funding by 2030"', *The Guardian* (online, 8 November 2022) <https://www.theguardian.com/environment/2022/nov/08/developing-countries-climate-crisis-funding-2030-report-nicholas-stern>.

61 Quoted in Fiona Harvey and Nikhita Chulani, 'Loss and Damage Must Be at Heart of COP27 Talks, Experts Say', *The Guardian* (online, 6 November 2022) <https://www

In the judgment of Zoe Nay and Julia Dehm, a hypothetical Truth and Reparations Commission of Oceania, sitting in 2035, addresses the question of reparations for the climate injustices experienced by three young plaintiffs from Oceania. Here, the emphasis is upon storytelling and the amplifying of stories, upon distributive justice and self-determination. The tribunal members query the legitimacy of legal processes, legal authority, and the official dispensation of legal remedies; they propose, instead, that a representative Youth Justice Forum on Climate Change should be established to decide on necessary steps for reparative justice.

In my judgment, I explore the legitimacy of acts of property damage to fossil fuel infrastructure, as carried out by a group of young climate activists who call themselves Last Generation Queensland. This strategy was popularised by Swedish theorist Andreas Malm,[62] and in a 2022 film based on his work. I am interested here in unpacking the legal dimensions of necessity, as enshrined in a common law defence and the statutory extraordinary emergency defence, after the megafires of a future Summer of Terror have highlighted the devastating impacts of climate change on the ecosystems and population of the Australian mainland.

Chiara Armeni, inspired by the Piccadilly Circus artwork mentioned above, explores the tension between attachment to so-called 'non-places' and the need for renewable energy. Her judges of the future must assess the value of place experience, when familiar places have been transformed by climate change, and the extent to which this should outweigh the importance of renewable energy projects.

Attachment to place is something that has already come up in climate lawsuits, including in numerous windfarm lawsuits.[63] It has also surfaced in the context of Australian coal mining developments which would irrevocably destroy familiar surroundings; for instance, part of Justice Brian Preston's reasoning in rejecting the Rocky Hill coal mine at Gloucester in 2019 was based on place. He concluded, on the basis of the visual, amenity, and social impacts of the mine, that the idyllic[64] rural setting of Gloucester was the 'wrong place' for the mine.[65]

.theguardian.com/environment/2022/nov/05/loss-and-damage-must-be-at-heart-of-cop27-talks-experts-say>.

62 Andreas Malm, *How to Blow up a Pipeline: Learning to Fight in a World on Fire* (Verso, 2021).

63 See, eg, *Taralga Landscape Guardians Inc v Minister for Planning and Anor* [2007] NSWLEC 59.

64 *Gloucester Resources Limited v Minister for Planning* [2019] NSWLEC 7 [2].

65 Ibid [699].

In his earlier decision relating to the Warkworth mine at Bulga, the judge utilised a word invented by eco-philosopher Glenn Albrecht: solastalgia.[66] Solastalgia is the emotional response to 'the lived experience of the physical desolation of home'.[67] Albrecht was compelled to create a new concept to capture this emotion, after witnessing the distress of people confronting a vast expansion of mining and power generation activities in their region. Robert Macfarlane explains it thus: '[s]olastalgia speaks of a modern uncanny, in which a familiar place is rendered unrecognisable by climate change or corporate action: the home become suddenly unhomely around its inhabitants'.[68]

We are all likely to experience solastalgia, if we have not yet done so.

Postcolonial Justice

In the third section, Australian First Nations scholars Oscar Davies, Bindi Bennett and Kelly Menzel contribute an insightful reflection on Indigenous relational philosophy:[69] the philosophy that informs First Nations law and offers more sustainable ways of being and interrelating with the more-than-human than those embedded in Western legal systems.

Writing at a time when controversy rages over the inclusion of an Indigenous Voice in the Australian Constitution, Robert Cunningham bypasses contemporary hesitations, misunderstandings, and bigotry to catapult us into a future in which the *Uluru Statement from the Heart*[70] has been fully implemented. He envisages a not-so-distant future when Aboriginal sovereignty is formally recognised in the Australian legal system and proceeds to interrogate questions of identity and rights of traditional knowledge for First Nations Peoples in Australia and the Torres Strait Islands.

Greta and Jo Bird present a critique of international law from a postcolonial perspective. In their version of the future, 200 years from now, humanity has evolved through technological means into chimeras: beings which are, genetically, both human and tree. The chimeras do not deliver a judgment,

66 *Bulga Milbrodale Progress Association Inc v Minister for Planning and Infrastructure and Warkworth Mining Limited* [2013] NSWLEC 48 [420]-[430].

67 Glenn Albrecht et al, 'Solastalgia: The Distress Caused by Environmental Change' (2007) 15(1) *Australasian Psychiatry* 95.

68 Robert Macfarlane, 'Generation Anthropocene: How Humans Have Altered the Planet for Ever', *The Guardian* (online, 1 April 2016) <https://www.theguardian.com/books/2016/apr/01/generation-anthropocene-altered-planet-for-ever>.

69 A number of chapters in this collection were influenced by Professor Mary Graham's inspirational presentation on First Nations relational philosophy at the inaugural project workshop.

70 The *Uluru Statement from the Heart* was a historic document ratified at the National First Nations Constitutional Convention at Uluru, Australia in 2017. It contains three central tenets: the Voice, Treaty, and truth telling. See 'The Uluru Statement' (Web page, 2023) <https://ulurustatement.org/the-statement>.

instead engaging in deep listening and 'yarning' with a river and with First Nations people to devise a 'collective response to the climate crisis'.

Inspired by the governance system and relational ethos of Australian First Peoples, Michelle Maloney imagines a future legal system with a focus upon bioregional governance and respect for Country. Her judgment is handed down by the Indigenous and non-Indigenous members of a Bioregional Tribunal. They adopt a consultative approach and apply a relationist ethos and benchmarks of ecological health, in assessing and rejecting a commercial proposal to grow cotton in the Darling Plains bioregion.

Finally, Ayesha Riaz considers international refugee law in a future era of massive climate displacement. Given the unequal geographical distribution of climate impacts, it is likely that vast numbers of people from the Global South will seek to settle in the Global North: a global reversal of the historical movements of colonisation. Riaz recounts the narrative of a Pakistani woman whose journey across the European continent is fraught with danger and hardship. Drawing upon John Lanchester's climate fiction novel *The Wall*,[71] she anticipates continuing obstacles to legal recognition of the plight of climate refugees. In this novel, the still functional Britain of the future is surrounded by a formidable coastal wall, guarded and reinforced to ensure that there is no ingress for the less fortunate who arrive by boat.

After the Anthropocene

In this section, contributors look beyond the Anthropocene, to a time of judgment by the posthuman or more-than-human. Tania Sourdin, enlisting the assistance of ChatGPT, explores how an AI platform might judge the Anthropocene and assign liability for humanity's misdeeds.

Taking Kafka's *After the Law*[72] as a starting point, Elena Cirkovic contemplates what may lie beyond human-made law. In this provocation, which embraces the posthuman and post-Anthropocene, there is no space for formal judgment with its embedded colonialist and patriarchal assumptions. Drawing upon sources as diverse as literature, physics, Earth System Science, and posthumanist philosophy, Cirkovic concludes with ruminations on a Creature: the merger of the human with its natural and artificial surroundings.

Susan Bird and Mark Brady have extrapolated beyond this and envisage an Intergalactic Court of Justice passing judgement (in .012 seconds) on the corporation: described by the authors as 'one of the law's most destructive fictions'. From a legal perspective, corporations are immortal; David Whyte writes that 'the corporation will not die unless it is killed. And it can only be killed by some form of state intervention or legal procedure that liquidates

71 John Lanchester, *The Wall* (Faber and Faber, 2019).
72 Franz Kafka, *Before the Law* (Kurt Wolff, 1919).

it'.[73] In this distant future, when Earth has become uninhabitable for humans, the abstract corporate bodies that played a dominant role in creating this outcome still exist.

Finally, Michelle Lim's chapter takes place in a future Terracene. For Rock, narrator and Judgment Day Returning Officer, the time of the Pale Skin-Beings was but a short interlude in the history of Earth, or Planet 3. On Judgment Day in 2050, following an event of planetary fission, three more-than-human beings deliver Judgments of Radical Possibility through the interface of Rock. These meditative narratives from Matang Mountain, Kilimakyero Lichen, and Pakan Island Bush Warbler, on living harmoniously and sustainably on Planet 3, diverge radically from conventional judgments as delivered by Skin-Beings. This is the future, and past, of stories as law.

Conclusion

Both utopian and dystopian visions of the future, and future law, feature in this collection. To some extent, these judgments act as a stark warning. They highlight the monumental ethical and legal issues that await us in the Anthropocene. Yet the judgments can also be viewed as a form of prefigurative politics, in representing judging practices that would characterise an epoch of justice between and beyond humans and the more-than-human. As Zoe Nay and Julia Dehm write in their chapter, 'we hope to make potential future trajectories that might currently seem unimaginable, appear as real potentialities that could be collectively struggled for and enacted'.

Despite the great diversity in approaches, and the spectrum of futuristic visions, the contributors have all left behind law as it evolved in the epoch of stability and embarked on a journey into new terrain – although as Oscar Davis, Bindi Bennett, and Kelly Menzel point out, much of this terrain is familiar to First Nations people. Foluke Adebisi suggests in her commentary that '[t]o survive at all, we may have to completely abandon our idea of law'. Law and judging will, undoubtedly, be radically transformed in the Anthropocene. This project is part of the daunting task of intellectual and conceptual transformation required to address the challenges of a new epoch.

73 David Whyte, *Ecocide: Kill the Corporation Before It Kills Us* (Manchester University Press, 2020) 23.

Part I

Multispecies Justice

Chapter 2

Takayna/Tarkine and the EPBC Act

From Heritage Frameworks to Habitat Thinking

Brad Jessup and Christine Parker

Commentary

A Place of Value, Historical Controversy, and a Future for Protection

Takayna, also formally (and only recently) known as the Tarkine (the Europeanised spelling of the Aboriginal word *Takayna*), is described as a wilderness and as a place of heritage value. These are descriptions that orient the landscape as a historical relic. They have meant little in the law. It remains a region of diverse flora and fauna and relative scarcity of humans. If the Tarkine is to be categorised as something, it is certainly habitat. Unlike elsewhere in Australia where biodiversity is threatened by the expansion of human population centres, the present threats to Takayna are its vulnerability to climatic change[1] and the richness of its timber and mineral resources.[2]

Over recent decades, Takayna has been a site of human conflict about the environment played out in the courtroom.[3] Over a longer timeframe it has been a place of species decline: a site of human on more-than-human-species conflict. Tasmanian Tigers (*Thylacinus cynocephalii*) roamed the grasslands, woodlands, and rainforests of the area before they were hunted to extinction in the early 1900s.[4] The impotence of the law in protecting the Thylacine is lore. Yet the Tigers' sister species, Tasmanian Devils (*Sarcophilus harrisii*),

1 Brendan Mackey et al, 'Assessing the Risk to the Conservation Status of Temperate Rainforest from Exposure to Mining, Commercial Logging, and Climate Change: A Tasmanian Case Study' (2017) 215 *Biological Conservation* 19; Ben Walter, 'Speak for the Trees: Hope and Hopelessness Mingle in the Singed Tarkine' (2017) 76(1) *Meanjin* 72.

2 Ibid. Sean T Cadman, 'Tasmanian Temperate Rainforests' in Dominick A DellaSalla and Michael I Goldstein (eds), *Imperilled: The Encyclopaedia of Conservation* (Elsevier, 2022) 53.

3 Most recently in *Tarkine National Coalition Incorporated v Director, Environment Protection Authority* [2023] TASSC 3.

4 Jaxon Waterhouse and Chantelle Mitchell, '"Has Anybody Seen a Tasmanian Tiger Lately?": Ethical and Ontological Considerations of Thylacine De-extinction' (2022) 26(1) *Green Letters* 14.

DOI: 10.4324/9781003389569-3

are now imperilled by a facial tumour disease,[5] contributed at least in part by human activity.[6] Existing legal frameworks are just as defective as those that failed to protect the Thylacine. The national *Environment Protection and Biodiversity Conservation Act 1999* (Cth) ('*EPBC Act*') has been used to stymie landscape protection for the devils and has accommodated the approval of mining developments in their habitat, despite decision-makers' understanding that those developments will increase the pace of decline of the species' populations. It seems decision-makers have not learned the lessons of history.

In this commentary, we speculate about a different future for the tigers and devils, now known as *kaparunina* (Thylacine) and *purinina* (*Sarcophilus harrisii*) in *palawa kani*, the language of Tasmanian Aborigines, with thanks to the Tasmanian Aboriginal Centre for permission to use these terms.[7]

In the judgment that follows we demonstrate how a shift in legislation accompanied by evolving conceptualisations of property and place might re-define habitat to realise protection for those species inhabiting Takayna. Throughout our judgment we make critical points about Australia's *EPBC Act*, now acknowledged as unable to arrest species decline after more than two decades in operation.[8]

Uncertainties, Speculation, and a Place of Habitat

As we write this commentary in the summer of 2022/23, the future for purinina (Devils) is uncertain. The species is endemic to lutruwita (Tasmania) where the population numbers less than 20,000. Since 1996, the facial tumour disease afflicting the species has spread from one side of the state, across its territory, and is now present in virtually all devil habitat. There has been a rapid decline of population with a more than 80% reduction in purinina density,

5 CE Hawkins et al, 'Emerging Disease and Population Decline of an Island Endemic, the Tasmanian Devil Sarcophilus Harrisii' (2006) 131(2) *Biological Conservation* 307.

6 Gregory M Woods et al, 'Two Decades of the Impact of Tasmanian Devil Facial Tumor Disease' (2018) 58(6) *Integrative and Comparative Biology* 1043, highlight especially the role of roadkill. See also Statement of Reasons for Approval of Nelson Bay River Magnetite and Hematite Mine, near Nelson Bay River, north-west Tasmania (EPBC 2011/5846), 27 August 2013.

7 See 'palawa kani, the only Aboriginal language in *lutruwita* today', *Tasmanian Aboriginal Centre* (Web Page) <http://tacinc.com.au/programs/palawa-kani> (stipulating that seeking permission to use Aboriginal language is an act of courtesy and respect for Aboriginal culture and protocols). We use the Western and Aboriginal names alternately in this commentary and assume in the judgment that the palawa kani names have become commonplace.

8 Graeme Samuel, *The Independent Review of the Environment Protection and Biodiversity Conservation Act 1999: Final Report* (Commonwealth of Australia, 2020); Commonwealth of Australia, *Australia: State of the Environment* (Report, 2021).

including in Takayna,[9] previously considered a stronghold for the species. A sense of emergency has led to the creation of an 'insurance population' in a place Devils never previously inhabited, leading to unintended consequences for other species.[10] Governments have directed developer project proponents to offset their harm to purinina habitat through donations to ex-situ species preservation.[11] Yet there is emerging ecological research indicating that the species is adapting to disease, with genetic indications that the population may resist the disease.[12] Already classified as endangered under state and federal environmental laws, Tasmanian Devils are now less likely to become extinct.[13] There is a possibility, embraced by us for the purpose of this judgments exercise, that purinina recover and continue to inhabit Takayna, now and into the future. But if that does occur, the species will need a place of habitat.

Within this commentary, we also speculate about the return of the Thylacine (alternatively Tasmanian Tigers). Kaparunina (Thylacine) were deliberately hunted to extinction by colonial settlers concerned about the threat to their flocks from this 'large predator'.[14] The law enabled this process, with private bounties on the Tigers in the 1840s, entrenched by colonial government-sponsored extermination programmes in the 1880s. The last Tasmanian Tiger probably died in captivity in 1936, only weeks after Tasmanian laws supposedly protecting them from extinction were finally

9 CX Cunningham et al, 'Quantifying 25 Years of Disease-caused Declines in Tasmanian Devil Populations: Host Density Drives Spatial Pathogen Spread' (2021) 24(5) *Ecology Letters* 958.

10 Phil Wise et al, 'The Conservation Introduction of Tasmanian Devils to Maria Island National Park: A Response to Devil Facial Tumor Disease (DFTD)' in Pritpal S Soorae (ed), *Global Re-Introduction Perspectives: 2016: Case-Studies from Around the Globe* (IUCN, 2016) 166; Donna Lu, 'Tasmanian Devils Wipe Out Thousands of Penguins on Tiny Australian Island', *The Guardian* (online, 21 June 2021) <https://www.theguardian.com/environment/2021/jun/21/tasmanian-devils-wipe-out-thousands-of-penguins-maria-island-australia>.

11 *Tarkine National Coalition Inc v Minister for Sustainability, Environment, Water, Population and Communities* (2013) 214 FCR 233 ('Tarkine case').

12 S Hendricks et al, 'Conservation Implications of Limited Genetic Diversity and Population Structure in Tasmanian Devils (*Sarcophilus harrisii*)' (2017) 18(4) *Conservation Genetics* 977.

13 Katherine A Farquharson et al, 'Restoring Faith in Conservation Action: Maintaining Wild Genetic Diversity Through the Tasmanian Devil Insurance Program' (2022) 25(7) *iScience* 104474.

14 Alistair Evans, Douglass S Rovinsky and Justin W Adams, 'The Tasmanian Tiger Was Hunted to Extinction as a "Large Predator" – But It Was Only Half as Heavy as We Thought', *The Conversation* (online, 19 August 2020) <http://theconversation.com/the-tasmanian-tiger-was-hunted-to-extinction-as-a-large-predator-but-it-was-only-half-as-heavy-as-we-thought-144599>. Finding that the Tigers' average size, and hence the danger they posed to larger prey like sheep and the impressiveness of killing them, was almost certainly exaggerated in contemporaneous reports.

passed.[15] Kaparunina were made vulnerable by the colonisation of Tasmania and the despoiling of their habitat for cattle and sheep grazing, for mining, for forestry, and the associated roads and infrastructure, which also now impact purinina.

Nevertheless, the romantic image of the Thylacine as Australia's missing apex predator, and a sense of guilt at having deliberately extinguished their future, has promoted a crypto-zoological obsession with the fantasy that a small population remains hidden away in a kind of Antipodean animal Atlantis somewhere.[16] In recent years, the fabled hope of 'resurrecting' the species has taken a very literal turn in the form of plans to 'de-extinct' kaparunina using genetic editing technology.[17] There is a possibility, embraced by us for the purpose of this exercise (only), that the Tasmanian Tiger will be resurrected within the next decade and will (re-)inhabit Takayna.

We also assume that (the resurrected) Tigers and (the recovered) Devils will live in interspecies conviviality, not as human prejudice often imagines, fighting one another for dominance of the food chain. It is unlikely that the Tigers and Devils directly competed for food in the past, with the kaparunina hunting smaller species such as quolls, and purinina scavenging carcasses, including larger ones, and both likely having a role in 'regulating' populations of smaller mammals and introduced species.[18] Where there is a record

15 Aaron Greenville and Glenda Wardle, 'Will We Hunt Dingoes to the Brink Like the Tasmanian Tiger?', *The Conversation* (online, 21 November 2013) <http://theconversation.com/will-we-hunt-dingoes-to-the-brink-like-the-tasmanian-tiger-19982>. For a recent alternative interpretation of the evidence suggesting a much later survival of a remnant subpopulation up until the 1980s to early 2000s, see Barry W Brook et al, 'Resolving When (and Where) the Thylacine Went Extinct' (2023) 877 *Science of The Total Environment* 162878.

16 Waterhouse and Mitchell (n 4).

17 The resurrection would be based on genome mapping of extant [M]useum samples with missing genetic material to be supplied by the Tigers' closest living evolutionary relatives, Numbats. See Parwinder Kaur, 'We've Decoded the Numbat Genome – and It Could Bring the Thylacine's Resurrection a Step Closer', *The Conversation* (online, 8 February 2022) <http://theconversation.com/weve-decoded-the-numbat-genome-and-it-could-bring-the-thylacines-resurrection-a-step-closer-176528>.We have used this as a device for this piece and pass no judgement – good or bad – on the wisdom of the idea. But note in its reliance on complex technology to reverse human impact on the environment it exemplifies the ethos of the Anthropocene with all its ethical and ontological consequences: see Corey JA Bradshaw, 'Can We Resurrect the Thylacine? Maybe, but It Won't Help the Global Extinction Crisis', *The Conversation* (online, 9 March 2022) <http://theconversation.com/can-we-resurrect-the-thylacine-maybe-but-it-wont-help-the-global-extinction-crisis-178425>.

18 Anna C Lewis, Channing Hughes and Tracey L Rogers, 'Effects of Intraspecific Competition and Body Mass on Diet Specialization in a Mammalian Scavenger' (2022) 12 *Evolution and Ecology* e8338. By contrast, both tigers and devils competed with dingoes, likely contributing to the extinction of both on the mainland of Australia: CN Johnson and S Wroe, 'Causes of Extinction of Vertebrates During the Holocene of Mainland Australia: Arrival of the Dingo, or Human Impact?' (2003) 13 *The Holocene* 941.

of species competition, in the case of introduced and invasive species[19] and landscape modification,[20] it is humans that are central in the endangerment of one species by the other. Yet, when humans attend carefully to animal societies, not only is there abundant evidence of communication and cooperation among consociates of the same species but even between species.[21] Perhaps in the future purinina will reach out to their newly returned sister kaparunina and teach them how to survive and flourish in Takayna once more.

Together, we propose the Tigers and Devils decide to bring a civil action to protect this place – their place – through the mediation of artificial intelligence-powered animal-computer interaction, allowing humans and animals to not merely communicate but to richly understand each other's lifeworlds.[22] Having understood that kaparunina and purinina wish to live together in habitat, we investigate in this judgment how the law might accommodate them: through a combination of recognition and habitat rights.

From Recognition Rights to Habitat Rights, and the Legal Shifts Required

Australian law recognises that only humans with a 'special interest'[23] or active or expert[24] in the environment have standing to commence legal proceedings on behalf of protected animals, plants, habitats and places under the *EPBC Act*, and other protective legislation. Others argue that the law should recognise certain living species and ecological communities, such as river and mountain systems, as having 'rights of nature'.[25] The most celebrated example of this approach being put into action is that of the Whanganui River in Aotearoa/New Zealand, which was legislatively made a legal person under

19 Duojie Jiabu and Weide Li, 'Impact of Different Invasion Methods of Invasive Species on Omnivorous Food Webs' (2023) 475 *Ecological Modelling* 110217.

20 Michael Gilpin, 'Community-level Competition: Asymmetrical Dominance' (1994) 91(8) *Proceedings of the National Academy of Sciences of the USA* 3252.

21 Eva Meijer, *Animal Languages* (MIT Press, 2020).

22 See 'Interspecies Internet', *Interspecies Internet* (Web Page) <https://www.interspecies.io>; Clara Mancini, 'Animal-Computer Interaction: A Manifesto' (2011) 18(4) *Interactions* 69. For a more realistic assessment of how the deployment of artificial intelligence is and likely will be directed by existing powerful interests to harm nonhuman animals, see Christine Parker and Simon Coghlan, 'Harm to Non-human Animals from AI', manuscript on file with authors.

23 *Australian Conservation Foundation v Commonwealth of Australia* (1980) 146 CLR 493.

24 *Environment Protection and Biodiversity Conservation Act 1999* (Cth) ('*EPBC Act*') s 475.

25 Christopher Stone's essay, 'Should Trees Have Standing? Toward Legal Rights for Natural Objects' (1972) 45 *Southern California Law Review* 450, became a rallying point for such calls; Alex Traub, 'Christopher Stone, Who Proposed Legal Rights for Trees, Dies at 83', *The New York Times* (online, 28 May 2021) <https://www.nytimes.com/2021/05/28/us/christopher-stone-dead.html>.

the stewardship of the Whanganui Iwi in response to Māori claims to the river system under the Treaty of Waitangi.[26]

These developments provide a backdrop to the developments we assume in our judgment. However, in granting legal recognition to Tigers and Devils to enforce legal obligations to protect and access their habitat, we draw more directly on two slightly different strands of legal thinking.

First, animals (and other nonhumans) already have some legal entitlements under legislation such as the *EPBC Act* (to protection from development of habitat). Recognition rights would give them legal agency to avail themselves of these existing entitlements in whatever fora are empowered to make decisions concerning them, whether a court, a local council planning committee, or a Ministerial decision. We assume that by 2036 the *EPBC Act* will be amended to grant nonhuman animals (and all other protected species and environmental entities) the right not only to standing in court but also to representation and consultation in all other relevant decision-making fora, including before the Minister, to argue for and protect their (existing) legal entitlements.[27]

It has not usually followed that the recognition of standing provides animal species with any greater rights or entitlements than they previously had.[28] In our judgment we (optimistically) show how recognition of standing in the law could lead to further expansion in legal entitlements, as we find, to habitat. Such a transformation would require human decision-makers to understand that rights that they have to property are not exclusively individual

26 Andrew Geddis and Jacinta Ruru, 'Places as Persons: Creating a New Framework for Māori-Crown Relations' in Jason Varuhas (ed), *The Frontiers of Public Law* (Hart Publishing, 2019) 255; Erin O'Donnell and Julia Talbot-Jones, 'Creating Legal Rights for Rivers: Lessons from Australia, New Zealand, and India' (2018) 23(1) *Ecology and Society* 7.

27 A fictional piece of amending legislation is cited in the judgment. The Animal Law Defence Fund has made similar arguments for animals' legal standing to enforce existing legal entitlements in court in the case of Justice the Horse (so far unsuccessful) and in the case of the descendants of Pablo Escobar's Columbian hippopotamuses (successful): see 'Justice the Horse Sues Abuser', *Animal Legal Defense Fund* (Web Page, 7 March 2023) <https://aldf.org/case/justice-the-horse-sues-abuser> and 'Animals Recognized as Legal Persons for the First Time in U.S. Court', *Animal Legal Defense Fund* (Web Page, 20 October 2021) <https://aldf.org/article/animals-recognized-as-legal-persons-for-the-first-time-in-u-s-court>. See also the Islamabad High Court case concerning Kavaan, 'the loneliest elephant', a bear and stray dogs: Nicole Pallotta, 'Islamabad High Court Holds that Animals Have Legal Rights', *Animal Legal Defense Fund* (Web Page, 2 October 2020) <https://aldf.org/article/islamabad-high-court-holds-that-animals-have-legal-rights>.

28 See, eg, the work of animal legal scholars Karen Bradshaw and Angela Fernandez who both argue for such (restricted) conceptions of animal rights to standing to enforce existing rights: Karen Bradshaw, *Wildlife as Property Owners: A New Conception of Animal Rights* (University of Chicago Press, 2020); Angela Fernandez, 'Not Quite Property, Not Quite Persons: A Quasi Approach for Nonhuman Animals' (2019) 5 *Canadian Journal of Comparative and Contemporary Law* 155.

and may be translatable to other species who share similar connections and depend for community welfare and group protection on place.

One of the few court decisions anywhere in the world to grant legal standing to a nonhuman animal was an Argentinian court decision recognising the right of Cecilia the chimpanzee to be released from inadequate confinement in a zoo and moved to a sanctuary. The court accepted that Cecilia was entitled to habeas corpus on the basis of Article 41 of the Argentine Constitution which makes the right to the environment a collective right, noting that the law must evolve to recognise Cecilia's 'right to live in an environment and in conditions adequate to her species'.[29] We see in this approach the glimpses of potential for a constitutionalised form of environmental rights to evolve beyond an anthropocentric and instrumental right (for example, to clean air and water) to an obligation of respect for nature and all its interacting species and communities.

A Future Legal Framework for Species Protection

The processes of (un-)protecting under the *EPBC Act* currently privilege a colonising conception of land and its 'resources' (including fauna) as valuable when propertised and commodified by private interests for economic value. That is a regime in which land is made available to one interest who is empowered to exclude others and to exploit and alienate the land and its resources for exchange value. This is starkly illustrated in recent mining approvals in the Tarkine.[30] A recommendation for listing the Tarkine as a national heritage place was rejected by the federal/national Minister for the Environment in 2013[31] because a listing would have impeded resource extraction activities.[32] There would be no 'balance' between habitat values

29 A full summary is provided in English at 'Acción de hábeas corpus presentada por la Asociación de Funcionarios y Abogados por los Derechos de los Animales (AFADA)', *Animal Legal & Historical Center* (Web Page, 2016) <https://www.animallaw.info/case/afada-habeas-corpus-cecilia>. Note however that the court did caution that this did not mean they were recognising personhood for all fauna and flora, nor that they were granting great apes similar rights to humans.

30 Australian Government, Approval Nelson Bay River Magnetite and Hematite Mine, near Nelson Bay River, north-west Tasmania (EPBC 2011/5846), 29 July 2013; Australian Government, Approval Riley DSO Hematite Mine Project, Tasmania (EPBC 2012/6339), 3 August 2013.

31 Instead, the 'Western Tasmania Aboriginal Cultural Landscape' was included as a National Heritage Place: see 'National Heritage Places – Western Tasmania Aboriginal Cultural Landscape', *Australian Government Department of Climate Change, Energy, the Environment and Water* (Web Page, 10 October 2021) <https://www.dcceew.gov.au/parks-heritage/heritage/places/national/western-tasmania>.

32 Attempts to add the Tarkine to the National Heritage List on an emergency basis were also rebuffed as recently as 2021: see 'Ley Rejects Takayna National Heritage Listing', *Tasmanian*

and resource opportunities, according to the federal government.[33] It is a point worth dwelling upon. It instructs us that the conception of ecologically sustainable development driving the operation of the *EPBC Act* is one less attentive to principles of biodiversity conservation and more to achieving the ongoing exploitation of natural resources via the recognition of private property rights. It is a process of assessment that legitimises and authorises destruction and despoliation by considering and dismissing the potential that a rarely granted exception (of protection) will apply against an overwhelming presumption of 'improvement' and 'development' (exploitation).

A new conceptual framework for protection is needed. For this work, we take inspiration from Davies, Godden, and Graham,[34] who have argued that the concept of real property needs to be realigned 'with the habitat of human and nonhuman life' in order 'to restore its decreasing functionality and re-establish property as both fundamental and relational, essential and dynamically connected to place and things'. As Davies puts it in another place, the law must 'see property as part of a wider socio-ecological network of connections rather than as separate from and imposed on a "natural" world', 'mobile and diverse co-becomings'.[35] We would go further and argue that the law needs to see habitat, and to see it as other than property. The law ought to structure interests over geographies accordingly.

Davies, Godden, and Graham cite Kearney in support of their argument that the ecological meaning of habitat connects life forms to geographical places recognising variability through time.[36] We could say 'habitat' makes space for qualitative understanding of what it means to live or potentially live in a particular place connecting both biotic and subjective features. Davies speaks of 'habitat as the contextual surround and bedrock of the interconnected life forms within which human societies and human beings are positioned'. Habitat must therefore become central to law as it is the basis for life and (ecological) community.[37] Occupation or being, which is the central

Times (online, 13 August 2021) <https://tasmaniantimes.com/2021/08/ley-rejects-takayna-national-heritage-listing>.

33 Andrew Darby, '"No" to Tarkine Environment Listing', *Sydney Morning Herald* (online, 8 February 2013) <https://www.smh.com.au/environment/no-to-tarkine-environment-listing-20130208-2e2bo.html>. What followed was approvals of mining activities, unsuccessfully challenged by an environment group represented by the Tarkine National Coalition, and later the Bob Brown Foundation.

34 Margaret Davies, Lee Godden and Nicole Graham, 'Situating Property within Habitat: Reintegrating Place, People, and Law' (2021) 6 *Journal of Law, Property, and Society* 1.

35 Margaret Davies, 'Habitat and Home' in Margaret Davies, Nicole Graham, and Lee Godden (eds), *The Routledge Handbook of Property, Law and Society* (Routledge, 2022) 203, 211.

36 M Kearney, 'Habitat, Environment and Niche: What Are We Modelling?' (2006) 115 *Oikos* 186 (distinguishing the 'mechanistic' notion of a 'niche' for a species from a descriptive account of 'habitat' as a place, actual or potential, where an organism can thrive).

37 Davies (n 35) 211.

determining factor in legal definitions for real property, therefore is too narrow a framework for understanding habitat.[38]

For the devils and tigers, habitat is a non-fungible entitlement to a place, a home, that allows development of their survival and flourishing, not any place, but this place. To securely have a home, to have habitat in this world means not merely a niche in the environment that is protected, but a place in society and culture, a place within legal institutions.[39] In this judgment we recognise such a place for devils and tigers, a place in Takayna but also a place that is fully protected in our law, places that should have been recognised a century and more ago, and when the last known tiger died.

JUDGMENT

Purinina and Kaparunina v Minister for the Environment [2036] FCFCA 1
Jessup and Parker JJ

Introduction

1. This case is about the meaning of 'habitat' within law. It is a term described as being ubiquitous but lacking consensus in wildlife ecology.[40] Since the turn of the century, there has been no agreement on the definition of the term, and there continues to be no universal view among scientists as to its meaning. In 2001,[41] habitat was defined as 'the part of the biosphere where a particular species can live, either temporarily or permanently'. By 2021, it was apparent that definitions of habitat were too often crafted to suit human agendas, with a tendency to define habitat in a way that marginalised nonhuman species. Habitat was just those

38 The United States Supreme Court in *Weyerhaeuser Co v US Fish and Wildlife Service* 586 US ____ (2018) recognised that under the *Endangered Species Act* 'critical habitat' could include unoccupied areas, but left undecided whether 'habitat' must be areas capable of currently supporting species, not simply areas suitable into the future for species. See further: Chris Wilson, 'Weyerhaeuser v. U.S. Fish and Wildlife Service: Swirling Uncertainty around the Definition of Habitat' (2020) 47(2) *Ecology Law Journal* 761.

39 Davies (n 35), at 207–8, refers to habitat as 'resources and conditions in an area that produce occupancy – including survival and reproduction – by a given organism'. This means not just physical survival (bare physical health) but flourishing. Moreover, where human society and human law exist and govern a space, it must mean that the animals' habitat must include resources and conditions within human society, culture, and law to enable and ensure its survival and flourishing. This is a process, not a mere state, and one which endures or degrades according to Davies. So what processes are available and accessible to the species?

40 Michael S Michael and Mark Hebblewhite, 'Carnivore Habitat Ecology: Integrating Theory and Application' in Luigi Boitani and Roger A Powell (eds), *Carnivore Ecology and Conservation: A Handbook of Techniques* (Oxford 2012) ch 13.

41 Charles J Krebs, *Ecology: The Experimental Analysis of Distribution and Abundance* (Benjamin Cummings, 5th ed, 2001), cited in Bamford and Calver, below (n 43).

places where humans were not developing and where species were permitted to take refuge.[42]

2. The need for greater precision in the meaning of habitat has long been argued for,[43] with the law especially identified as lacking consistency and cogency in its use of the term.[44]
3. In this judgment, we give legal meaning to 'habitat' and confirm a legal right of species to their habitat not simply as a 'natural' space where species might be located but as the land, water, and air within and whereupon species have the greatest opportunity to flourish. Habitat ceases to be a record of where species are. It instead becomes understood as a place where humans constrain their activities so that species can be, to inter-relate, and to evolve.
4. Habitat becomes the notion through which there is a reorientating of interests away from humans and their property towards other species in land, water, and air. We seek guidance from Davies et al whose work in 2021 resituated ideas of property and habitat as a 'socio ecological complex of resources and relationships needed by organisms to sustain life and flourish'.[45] They instruct us that life ought to reorient property.
5. This case was initiated under the *Judiciary Act 1903* (Cth) ('*Judiciary Act*') s 39B and supported by the *Environment Protection and Biodiversity Conservation Act 1999* (Cth) ('*EPBC Act*') under section 475A of that Act, as amended by *Environment Protection and Biodiversity Conservation (Species Standing) Act 2028* (Cth) which clarified that an 'interested person' for the purpose of that Act includes all species potentially affected by any of the environmental harms that the Act seeks to avoid. Relief is sought under the *Judiciary Act* and under the *Federal Court of Australia Act 1976* (Cth) s 21(1).
6. Precipitated by a state government proposal to develop the region for housing, the case is brought by the two charismatic marsupials of lutruwita (Tasmania) that have, through the development of science, beaten extinction events precipitated by human activity: the devil *Sarcophilus harrisii*, now known as purinina in palawa kani, the language of Tasmanian Aboriginals; and the tiger Thylacine, now known as kaparunina. These two species, co-habiting in the Takayna region of lutruwita (Tasmania), seek:

42 Kennedi Fichtel, 'The Key Deer Is Headed for Extinction: How Repealing a Trump-Era Federal Rule Defining "Habitat" Could Allow Assisted Migration to Save Species Threatened by Climate Change' (2022) 34(2) *St Thomas Law Review* 109.

43 Michael Bamford and Michael Calver, 'A Precise Definition of Habitat Is Needed for Effective Conservation and Communication' (2014) 37(2) *Australian Zoologist* 245.

44 Wilson (n 38).

45 Davies, Godden and Graham (n 34).

a) a declaration that succeeding governments have failed to adequately recognise and protect their rights to habitat in Takayna despite knowing since at least 1937 that the Takayna if protected would have been an appropriate place of refuge for first the tigers and then the devils, and that this failure left them vulnerable to extinction; and
b) an order to immediately protect what remains of the Takayna and recognise it as non-exclusive habitat for the tigers and devils to not merely prevent their extinction but allow their flourishing.

7. The court grants the relief sought by purinina and kaparunina for the following reasons.

The Facts

8. The Tarkineer people were one of the Aboriginal communities that occupied the land, rivers, and shores that now are described by their name and in their language: Takayna.
9. In 1937 it was proposed that a sanctuary be created within Takayna for the Tasmanian tiger.[46] It is now accepted that the tiger was likely extinct in the wild by 1936, though speculated sightings continued in Takayna (and elsewhere, but particularly Takayna) for decades afterwards.[47] A proposal for a sanctuary was reconceived in the 2000s for the betterment of the devil, with its local population then unaffected but at risk from facial tumour disease.[48]
10. Takayna at various times in more than a century, as it is now, is a refuge, a place for reconsolidation, respite, and renewal. Not for bushwalkers or cinematographers experiencing and recording a landscape absent human settlements, but for the species, including the two representative species who initiated this claim, that need habitat in a place most suited to their persistence, resilience, and at times for their survival.
11. Humans have never given species that which they have needed from Takayna.
12. Decades ago, Takayna was a site of legal controversy. Then, it was decided that minerals considered essential to human advancement should be extracted from the landscape.[49] Beforehand, in the 1990s, there was

46 Australian Heritage Council, *National Heritage Assessment: Final Report – The Tarkine* (Commonwealth of Australia, 2011) 35
47 Elizabeth Claire Alberts, 'Bob Brown: Australia's first leader of the Greens', *Ecologist* (online, 30 December 2013) <https://theecologist.org/2013/dec/30/bob-brown-australias-first-leader-greens>. See also Brook et al (n 15), arguing (controversially) that a sub-population of Tasmanian Tigers could have survived until the 1980s or later.
48 Noted in Australian Heritage Council (n 46) 23.
49 Tarkine case (n 11).

anger at the construction of a road that would be used to facilitate both mineral and timber exploitation.[50] There were refusals to acknowledge Takayna as a place and battles to have it recognised; only resolved in 2013.

13. A series of environmental conservation groups have continued to advocate for the preservation of Takayna, including to secure World Heritage listing. This now and at last appears as imminent as the listing of the Lemonthyme and Southern Forests was in the 1980s, as the final mines close and because logging activities are no longer exempt from the *EPBC Act* following the abandonment in Tasmania of the Regional Forestry Agreement regime, a regime which this court has repeatedly confirmed threatens species through the removal of habitat.[51] The *EPBC Act* now offers the possibility, purpose, and promise of protection from forestry activities, to the relief of the Greater glider and the Leadbeater's possum, the Swift parrot, and the Tasmanian wedge-tailed eagle among others. It does more than give the false impression of protection as it did while logging activities were beyond its remit.[52]
14. The advocacy for preservation, its resource richness, and its distance from metropolitan centres of Launceston on the kanmaluka and nipaluna (Hobart) have kept human presence in the Takayna to a minimum such that today, despite the scars of human extraction and the increased occurrence of extreme weather events and wildfires, it is capable of much the same description as a quarter of a century ago. In 2011, the Australian Heritage Council described it as 'remote' comprising 'wild windswept beaches through to extensive buttongrass plains with stunning vistas to impressive rainforests'.[53]
15. The description, however, was selective and is now incomplete. For the Tasmanian devil (purinina) roamed Takayna and as a species recovered from facial tumour disease to restore a stable population in the region, and the Tasmanian tiger (kaparunina), while extinct at the time that those words were composed, has returned to Takayna through the efforts of scientists on the mainland.
16. Yet migration from the mainland, which spiked at the turn of the century,[54] means that Takayna is now seen as a place for town building, not species

50 Joel Wilson, 'The Road to Nowhere', *Green Left Weekly*, Issue 265 (online, 5 March 1997) <https://www.greenleft.org.au/content/road-nowhere>.

51 *Brown v Forestry Tasmania (No 4)* (2006) 157 FCR 1; *Friends of Leadbeater's Possum Inc v VicForests (No 4)* [2020] FCA 704.

52 Brad Jessup, 'Justice, Recognition and Environmental Law: The Wielangta Forest Conflict, Tasmania, Australia' (2015) 34(1) *University of Tasmania Law Review* 5.

53 Australian Heritage Council (n 46).

54 Tasmanian Government, *Consultation Paper - Refreshing Tasmania's Population Strategy* (January 2023).

building. Residents of Warrane (Sydney) and Naarm (Melbourne), living with uncomfortable heat, enduring summers of haze-filled skies, and managing with insecure and unpredictable water supplies, are among the future residents planned for the region. This is the next phase of climate adaptation led by individual behaviour[55] because governments failed to first mitigate climate change, and then failed to plan for climate adaptation.

17. To cope with the influx of 'climate changers', the lutruwita (Tasmanian) government has released plans that propose turning Takayna's woodlands into dwellings with pools, its rainforest into schools and cafes, and its heathlands into factories and offices.

The Legal Background

18. A conventional case brought under the *EPBC Act* would fail in the absence of an action – a development proposal to turn sod. The Act and the jurisprudence under the Act have long been clear that new or changed government policy is not justiciable,[56] even where government decisions are the primary avenue to environmental harm.[57] There is no action here. Rather, there is a decision of a government to nominate a landscape for a change of use.[58]
19. This proceeding does not fail as previous ones have, however, because the species do not bring to this court a conventional claim under the *EPBC Act*. They are not seeking to challenge a decision on judicial review grounds, a tenuous basis for legal challenge for environmentalists for over a century. They are not seeking injunctive relief to stop an action proceeding in advance of a Ministerial appraisal because of its likely significant impact on the matters of national environmental significance, often an illusory claim for conservationists, including because of the financial impost associated with undertakings and legal costs: ongoing barriers to access to environmental justice in this country.[59]
20. Instead, the species seek declaratory relief which, in essence, concedes that the *EPBC Act* has failed them. They thereafter seek confirmation

55 David King et al, 'Voluntary Relocation as an Adaptation Strategy to Extreme Weather Events' (2014) 8 *International Journal of Disaster Risk Reduction* 83.

56 *EPBC Act* ss 523–4; *Secretary, Department of Primary Industries, Parks, Water and Environment v Tasmanian Aboriginal Centre Incorporated* (2016) 244 FCR 21.

57 *Tasmanian Aboriginal Centre Incorporated v Secretary, Department of Primary Industries, Parks, Water and Environment (No 2)* [2016] FCA 168.

58 *Save the Ridge Inc v Commonwealth* (2005) 147 FCR 97.

59 Anne Kallies and Lee Godden, 'What Price Democracy? Blue Wedges and the Hurdles to Public Interest Environmental Litigation' (2008) 33(4) *Alternative Law Journal* 194.

that they possess a form of proprietary interest in their habitat such that the *EPBC Act* can no longer facilitate its ongoing despoilation.

21. In circumstances such as the residential development foreshadowed by the lutruwita/Tasmania government, the *EPBC Act* operates as a check, but rarely a delay or denial,[60] on development. It requires assessment and approval of actions likely to have significant impacts on any of the matters of national environmental significance.[61] Yet the Act is typically unavailable to provide oversight as policies, proposals, and programs are developed and designed. Intervention is only available once a developer seeks to put concrete into the plans of policymakers. The Act shifts focus and responsibility from government to the private sphere, representing a privatisation of environmental regulation.
22. We recognise that in acceding to the demands of the species that we are also passing judgement on the *EPBC Act*. This is the judgement day for that Act, a judgement that should have been reached within a decade after its enactment when an inquiry into the Act recommended significant reforms to address multiple deficiencies and limitations,[62] none of which were accepted.[63] The Act, despite its oversight of species decline,[64] has only ever been amended in a piecemeal way.[65]
23. Purinina were listed as an endangered species under the *EPBC Act* from 2009[66] until 2032. Kaparunina were never subject to a Commonwealth conservation law regime. The effect of the *EPBC Act* listing was a requirement that there be a plan for conservation and recovery of the devil, yet the devil recovered and ceased being endangered owing in large part to its own genetic diversity and resilience,[67] supported by the ex-situ

60 Andrew Macintosh, Heather Roberts and Amy Constable, 'An Empirical Evaluation of Environmental Citizen Suits under the Environment Protection and Biodiversity Conservation Act 1999 (Cth)' (2017) 39(1) *Sydney Law Review* 85.

61 *EPBC Act* Parts 3 and 9.

62 Alan Hawke, *The Australian Environment Act: Report of the Independent Review of the Environment Protection and Biodiversity Conservation Act 1999: Final Report* (Commonwealth of Australia, October 2009).

63 Commonwealth Government, *The Australian Government Response to the Independent Review of the Environment Protection and Biodiversity Conservation Act 1999 (EPBC Act)* (24 August 2011).

64 Samuel (n 8); Commonwealth of Australia (n 8).

65 See, eg, the *Environment Protection and Biodiversity Conservation Amendment Act 2013* (Cth).

66 Department of the Environment, Water, Heritage and the Arts, *Approved Conservation Advice for* Sarcophilus harrisii *(Tasmanian Devil)* (Advice, 19 May 2009) <http://www.environment.gov.au/biodiversity/threatened/species/pubs/299-conservation-advice.pdf>.

67 Gael L Glassock et al, 'Reducing the Extinction Risk of Populations Threatened by Infectious Diseases' (2021) 13(2) *Diversity* 63.

insurance population program.[68] Conservation advice for the devil was at times ignored.[69] When considered, developments were still allowed to proceed despite the knowledge that the developments reduced the purinina population.[70] The history of the *EPBC Act* has been that determinations have been reached that projects would have a significant impact on the species, such as a notable decline in species,[71] or the loss of habitat of a species endemic to a locale or where the species population is small.[72] Yet when the Minister for the Environment has been called upon to decide whether to permit development in Takayna, the Minister has permitted projects at the expense of the habitat of purinina and in the face of harm to individual devils.[73]

24. If the *EPBC Act* cannot provide a framework for protection of species, and that is the principal law with the purpose of protecting species, we agree with the tiger and the devil that succeeding governments have failed to protect them while knowing and articulating in policy documents since at least 2009 that loss of breeding habitat is a primary threatening process for species. Succeeding governments have also known that a refuge was necessary to separate kaparunina from their colonial hunters, to provide a refuge from human economic-oriented development of grazing and mining taking priority over all other forms of life.
25. We therefore make the declaration sought by the applicants, and based on the following reasons also grant the order sought.

Habitat, Place, and Rights

26. What the two species required at the time of their endangerment and extinction was habitat. As the tigers were being killed, they needed a place separate from humans to replenish their population. As the devils were dying from facial tumour disease, they needed a place undisturbed – to prevent human caused deaths and facilitated spread of the disease, and to bolster their genetic pool. The species needed, and continue to need, a conception of habitat that did not exist at the time. They needed more than an acknowledgement that the species were present in a place. Perhaps a reservation or a heritage listing may have achieved this, but

68 C E Grueber et al, 'A Tasmanian Devil Breeding Program to Support Wild Recovery' (2019) 31 *Reproduction, Fertility and Development* 1296.

69 Tarkine case (n 11).

70 Statement of Reasons for Approval of Nelson Bay River Magnetite and Hematite Mine, near Nelson Bay River, north-west Tasmania (EPBC 2011/5846), 27 August 2013.

71 *Booth v Bosworth* (2001) 114 FCR 39.

72 *Brown v Forestry Tasmania (No 4)* (2006) 157 FCR 1.

73 Statement of Reasons for Approval of Riley DSO Hematite Mine Project, Tasmania (EPBC 2012/6339), 4 September 2013.

such spatial constructs depend on government action which has not been forthcoming.

27. What then is 'habitat' for the species applicants? And how does habitat transform into protection in a way that the promised sanctuaries and recommendations for heritage listings have not?
28. Article 2 of the *Convention on Biological Diversity*[74] defined habitat as 'the place or type of site where an organism or population naturally occurs'. This definition is a relic of its time when there was a very human-centred conception of there being a distinction between the natural and so-called 'human' world. In 2019, the Australian government recognised that urban, human-modified land may constitute habitat.[75] This judgment is part of the narrative that we hope puts an end to the false binary that more-than-human species occupy 'nature' and that their habitat is 'natural'.
29. The *EPBC Act*, s 528 defines 'habitat' as:

    the biophysical medium or media:

    (a) occupied (continuously, periodically or occasionally) by an organism or group of organisms; or
    (b) once occupied (continuously, periodically or occasionally) by an organism, or group of organisms, and into which organisms of that kind have the potential to be reintroduced.

30. The operative provisions of the *EPBC Act*, however, only attend to habitat when it is considered 'critical' to a threatened species.[76] Habitat is only of interest to this law when a species is in decline or when a species is being reintroduced after a typically human-precipitated decline event.[77] At that time, a discretion lies in the Minister as to whether or not to identify 'habitat ... as being critical to the survival' of a listed species or ecological community. The experience during Trumpian America was that executive discretion in the characterisation and protection of habitat does not augur well for species,[78] even in the United States' much more favourable legislative framework for species.[79] We therefore shift

74 *Convention on Biological Diversity*, opened for signature 18 March 1996, 1760 UTS 79 (entered into force 29 December 1993).
75 Commonwealth of Australia, *Australia's Strategy for Nature 2019–2030* (2019).
76 *EPBC* Act s 207A.
77 Johnson and Wroe (n 18).
78 Fichtel (n 42).
79 Mark Squillace and Brad Jessup, 'A Tale of Two Species' (Conference Paper, Global Dialogue on Biodiversity Law and Governance, Centre for Environmental Law, Macquarie Law School, 7 July 2021).

from a statutory framework to a conceptual one to understand habitat for law.

31. Our starting point is that species' habitats are more than occupation or presence, and species habitat should not be confined to the land of disutility for humans.[80] The habitat of species changes, it may shrink or expand as climate changes or, too often in the history of humanity, as species seek refuge from human development. No longer can it be that a species habitat is designated, constricted, or rendered by decisions or actions about resource or land use of humans.
32. A second principle for us is that habitat is not fungible, and while there was in recent history an attempt to replace habitat or fund habitat restoration as trade-offs for resource extraction, with devils variously priced in the tens of thousands and their breeding grounds offset through the payments of hundreds of thousands, that regime then,[81] as now, is recognised as flawed – scientifically and morally.[82]
33. Habitat is an integral connection with a place.[83] Humans have habitat:[84] in our homes certainly, but also in our workplaces, areas of spiritual, familial, or historical connection, and in the many places we all share.[85] These places serve us in manifold ways. They are more than where we occupy or present ourselves.
34. For nonhuman species, we understand habitat as a 'socio ecological complex of resources and relationships needed by organisms to sustain life and flourish'.[86] An organism's habitat is not merely some mechanistic niche that could be re-created in a zoo or sanctuary, much as the scientists resurrected the tiger. Habitat consists of 'networks of ecological communities and systems'.[87] It is dynamic and emplaced. It can change and adapt just as species can adapt and grow and interact to be in place.[88] It is habitat, the finding of a place of community with devils

80 'Howard Chips Away in Tasmania', *The Sydney Morning Herald* (online, 7 October 2004) <https://www.smh.com.au/politics/federal/howard-chips-away-in-tasmania-20041007-gdjvgc.html>.
81 *Tarkine National Coalition Incorporated v Minister for the Environment* (2014) 202 LGERA 244.
82 Peter Menkhorst, Nick Clemann and Joanna Sumner, 'Fauna-Rescue Programs Highlight Unresolved Scientific, Ethical and Animal Welfare Issues' (2016) 22(4) *Pacific Conservation Biology* 301.
83 Davies, Godden and Graham (n 34).
84 Miquel Bastons and Jaume Armengou, 'Human Habitat, Space and Place' (2016) 29(4) *Journal of Agricultural Environmental Ethics* 559.
85 Ibid.
86 Davies, Godden and Graham (n 34) 1.
87 Ibid.
88 Sonia E Sultan, *Organism and Environment: Ecological Development, Niche Construction, and Adaptation* (Oxford University Press, 2015).

and the other species of the Takayna that has truly brought the tigers back to life, not the scientists with their gene editing technologies. And the law must thus respect this habitat and its life, rather than constraining access through exclusion of non-legal title holders and authorisation of alienation for economic value.

35. From the point of view of the Minister for the Environment and private property holders, this entails an obligation. Land is not merely there to be utilised as an economic resource (subject to narrowly defined exceptions). Rather we have come to understand that it is very much the other way around. Land is there to provide a space, a place, a home for life in all shapes and sizes and in all its interconnections and collisions. That is: land is habitat. Law and all its accoutrements, including private property, must find their place within habitat, not the other way around.
36. The purinina and kaparunina do not seek their habitat to the exclusion of others. It would not be their habitat without the other species that make it their home. They do, however, seek to protect their habitat in a way that it can continue to become a home for them and the others with whom they live in mutuality and connection. It is not our concern here whether this sometimes involves one species eating another as long as the ecological community continues on balance to provide a home for all. It is, however, our concern if the activities of one species or individual harm that balance and the ability of others to make themselves a home, a habitat. In this case (and in almost every case where habitat is threatened) it is human development that is the threat and the habitat of the devils and tigers that must be protected.
37. We therefore decide that human activity can only occur to the extent it can do so consistent with the many overlapping tenancies and easements[89] enjoyed by the many species who live on this land, Takayna. We therefore declare that in making planning and development assessments, the Minister must give effect to all the entitlements to habitat seeing land as a commonwealth, a common home, for ecological communities of intersecting beings.[90] A mining tenement is a thin right in the context of the thick web of entitlements that other members of the ecological community hold over Takayna.
38. The *EPBC Act* is the principal statute that gave species standing in this court[91] and that provides habitat definitions. However, we find an

89 Davies, Godden, and Graham (n 34) 40 (seeing conservation easements as one way of prefiguring a form of land registration that recognises habitat).

90 Ibid 41 (noting the need to see landscape ecology as a commonwealth that has a real impact on planning controls).

91 *Environment Protection and Biodiversity Conservation (Species Standing) Act 2028* (Cth) (a fictional Act).

overarching, fundamental legal habitat interest that is not sourced in colonial laws and that cannot be diminished through statute other than expressly and consistently with other constitutional requirements. We acknowledge that what we achieve in this judgment is an environmental right that resides in all species, and a rethinking of property law that recognises 'the foundational role of land as habitat supporting human and nonhuman life on Earth'.[92]

39. We do not attempt to equate habitat interests with Aboriginal and Torres Strait native title interests, yet this right to habitat is similarly situated beyond and intersecting with the common law. However, the right to habitat we envisage will oftentimes be consistent with the exercise of native title rights, and the right to habitat while likely viewed as disruptive to colonial settler law is more aligned with many of the legal systems of Australian First Nations, grounded as they are in concepts of mutual respect and reciprocity.[93]
40. We have described already species as having a right to their habitat. We appreciate that the exercise of that right is still unable to be achieved by species, despite advances in digital technologies that facilitate human and more-than-human species communication and entanglement.[94] Those species inhabiting places, therefore, obtain their interest in habitat by imposing responsibilities on humanity not to diminish their habitat.

Conclusion

41. The purinina and kaparunina do not seek a kind of property interest defined by settler colonial law. For instance, exclusive possession of Takayna. Rather, they seek recognition of their entitlement to a place for survival and flourishing, in this place, Takayna, where they survived and flourished for thousands of years. They also seek the space to continue to evolve and flourish in community with other species.
42. In this judgment, we express the interest of species to habitat as a non-exclusive legal right to protection of their habitat. It is given effect by a compatriot responsibility to protect. That is a responsibility held by all humans, and none more so than the individual who occupies the position of Minister for the Environment with powers under the *EPBC Act*.
43. The interest may in the future be expanded or described in different ways. However, a non-exclusive legal right to protection of their habitat

92 Davies, Godden and Graham (n 34) 6.

93 Maneesha Deckha, 'Unsettling Anthropocentric Legal Systems: Reconciliation, Indigenous Laws, and Animal Personhood' (2020) 41(1) *Journal of Intercultural Studies* 77.

94 See references (n 22). See also Jonathon Turnbull et al, 'Digital Ecologies: Materialities, Encounters, Governance' (2022) 0(0) *Progress in Environmental Geography* <doi.org/10.1177/27539687221145698>.

is sufficient for the purposes of deciding this judgment. It means that the Minister cannot countenance approving the residential development of Takayna. It means the state government must abandon that proposal for as long as the devil and the tiger have an ecosystemic community connection to that environment, while Takayna is their place. And both levels of government must turn their minds away from means of human extractive use of Takayna to examining ways to enable the devil and the tiger to flourish there, lest governments continue to fail to protect purinina and kaparunina, as they have done for more than 100 years.

Chapter 3

Are Nonhuman Animals Entitled to Dignity, Privacy, and Non-Exploitation?

A Smart Dairy Farm of the Future

Natalia Szablewska and Clara Mancini

Introduction

Drawing on previous similar projects that aimed to rewrite judicial judgments from different perspectives,[1] we set out to find an answer to the title question. However, there was no actual judgment to guide us and so we undertook to reconceive the law and propose what a similar judgment could or should look like if approached from a non-anthropocentric perspective.[2] This gave us much more flexibility but, at the same time, presented certain challenges. It is one thing to reimagine a particular judgment; it is yet another to create an entire legal context as we see fit with laws that might or might not ensue in the future and then to craft our very own 'future' judgment based on those existing and imaginary developments.

There have been also practical challenges, including that actual judgments often tend to be long, and, in international law, it is not uncommon for them to go into hundreds of pages. This, for obvious reasons, could not

We would like to thank the editors, Nikki and Michelle, for their skilful direction and leadership on this project, the Hon David Ashton-Lewis and Andy Gibson for their most helpful comments and feedback on the earlier draft, as well as Kara and Zena (the huskies), Dennis and Zorro (the cats) and Mishka (the poodle) for being a continuous source of inspiration for us.

1 See, eg, Rosemary Hunter, Clare McGlynn and Erika Rackley (eds), *Feminist Judgments: From Theory to Practice* (Hart Publishing, 2010); Heather Douglas et al (eds), *Australian Feminist Judgments: Righting and Rewriting Law* (Hart Publishing, 2014); Elisabeth McDonald, Rhonda Powell and Rosemary Hunter (eds), *Feminist Judgments in Aotearoa New Zealand: Te Rino – A Two-Stranded Rope* (Bloomsbury, 2017); Bridget J Crawford, Kathryn M Stanchi and Linda L Berger (eds of the series), *Feminist Judgment Series: Rewritten Judicial Opinions* (Cambridge University Press) <https://view.updates.cambridge.org/?qs=a9ad2e84ecc846381e57004efcf48ebad96f3dd4fd23a5ba660578a9635ee6ea2911206432f4ca495e43c364e382ca00cfb2dc337509e61aa80e979a69b15d2c727ab44a48bdadbff3595b05d824cff8>; Loveday Hodson and Troy Lavers (eds), *Feminist Judgements in International Law* (Bloomsbury, 2019); Nicole Rogers and Michelle Maloney (eds), *Law as if Earth Really Mattered: The Wild Law Judgment Project* (Routledge, 2017).

2 On the similarities and differences between this and other similar projects, see the introductory chapter in this book: Nicole Rogers, 'Judging the Future and the Future of Judging: Reflections on the Anthropocene Judgments Project'.

DOI: 10.4324/9781003389569-4

be recreated for the purpose of our judgment. Thus, we focused on three key counts – rights to dignity, privacy, and non-exploitation – that we felt were key to shining the light on issues that have already arisen, but which require further and more nuanced engagement, including via judicial interventions. In our judgment, we rely on existing case law and secondary sources at the time of writing which are then referenced in full, with some cases and events being an outcome of our imagination which, nonetheless, was guided by recent scientific and legal developments in this area. Another challenge we encountered was that we have different backgrounds. That diversity of perspectives proved to be complementary for the purpose of this project, yet it was something that we needed to navigate. Engaging in writing collaboratively a legal judgment by a court that does not exist, and relying on a mix of existing and imaginary laws, proved to be a challenging but invaluable experience.

To address some of the above-mentioned challenges, we start by introducing the background to our scenario to provide a wider context within which the judgment proposed by us is set. This is then followed by a brief overview of the scientific and legal contexts, covering both the actual as well as imaginary laws that nevertheless could be developed in the future. The final section is the fictional judgment that focuses on three legal questions we set to tackle in this project:

1) Can the right to dignity, and to be treated with respect at all times, apply to nonhuman animals?
2) Can the right to privacy apply to nonhuman animals?
3) Can the right to freedom from slavery, servitude, and forced labour apply to nonhuman animals?

We did not cover the issue of remedies and what they might entail in the context of our scenario due to the space limitation. It will, however, require further due consideration as the law in the area of animal rights, or nonhuman persons in general, will continue to evolve.

Commentary

We tackle the three legal questions above with reference to a future smart dairy farm, in which technology plays a key role in animal husbandry and dairy production. We based our scenario in the year 2057 not because the date has any particular significance but rather because it gives over 30 years for some of the matters we discuss to mature and, we believe, could make such a claim quite feasible. It is possible that, along with technological developments already experienced today, our understanding of animals' experience and the implications it has for their quality of life will advance further.

Smart Dairy Farm Scenario 2057

Human population and global urbanisation are at their historical highest and, as a result, natural ecosystems are under enormous pressure. Society has long accepted that animal farming has a very significant impact on an increasingly depleted natural environment. Historically, impacts have included, for example, deforestation to create pastures, greenhouse gas emissions from farm animals' digestion, land pollution from manure, and biodiversity loss from the loss of habitats.

Nevertheless, due to the size of the human population, the demand for meat and dairy is still very high, even though the proportion of humans adopting a vegan diet has steadily increased over the past decades. Cultivated meat and dairy products are now widely available in many countries and marketed at a relatively competitive price. However, many consumers still prefer farmed animal products to cultivated ones, and animal farming lobbies are still successful in promoting the industry.

To minimise the environmental impact of animal farming while still allowing the industry to continue to operate and meet consumer demand, extensive outdoor farming systems have been outlawed and entirely replaced by intensive indoor smart farming systems. These are highly technologised environments, designed to maximise production while optimising animal welfare.

In the smart dairy farm in question, cows live all year round in indoor quarters that are completely insulated from the outside world. This prevents the spread of disease, in case of outbreaks, and allows farmers to fully control the farm's environmental conditions, to make the most efficient use of energy, and to address some of the animal welfare concerns. Animals are fed organic feed, and the use of antibiotics is almost obsolete, following the introduction of global restrictions on the use of antibiotics for humans and nonhuman animals.

An air circulation system pumps in clean air and pumps out saturated air, which is mechanically and chemically filtered before being released into the atmosphere; the air filtration system also captures the carbon dioxide produced by the animals and maintains constant thermal conditions. Additionally, artificial illumination also reproduces exact day-night cycles to support the cows' circadian rhythm. Moreover, to compensate for the complete lack of stimuli from the outside world, animated images are projected on many of the farm's walls to provide visual stimulation for the cows.

To maximise the use of the real estate, the space available per cow is very limited. To alleviate this limitation, the farm is divided into different areas (for example, milking parlour, sleeping quarters, eating area, toileting area, insemination area, parturition area, and calves' nursery), which the animals are regularly moved to as required. Each cow is fitted with an identification tag which allows her to access her individual feeding stations and sleeping cubicles. Each feeding station delivers a feed mix exactly tailored for the cow that accesses it. Her identification tag also allows each cow to access one of the robotic milking stations, always the same, at specific times during the day, to avoid crowding. Tightly controlling the cow's access to the farm's facilities and resources is necessary to limit the risk of conflict between cows, posed by the limited space availability and by the cows' hierarchical social dynamics.

In such a tightly regulated system, and with so many animals living locked in at such closed quarters, monitoring welfare is paramount. To this end, sensors are embedded in the environment and in bovine wearables to record data about the animals' physiology and activity. For example, pressure and temperature sensors are incorporated in each cow's bedding structures to monitor how much she sleeps and to assess the quality of her sleep; triaxial accelerometers embedded in each cow's head harness monitor rumination activity; and bracelets on each cow's front legs monitor locomotion and capture any gait abnormalities. Machine learning techniques are used to identify early signs of disease or poor welfare and promptly cull the animal to minimise the risk of outbreaks.

On the smart farm, calves are taken away from their mothers almost immediately after birth, before they have a chance to form strong bonds, so that the cow can begin to be milked. The newborns are first fed colostrum from their mothers to boost their immune system and before being moved on to milk formula. Females and males are reared in nurseries, the former destined to become dairy cows, the latter to be moved to a beef farm. To mitigate the welfare impact on grieving cows and calves, deepfake techniques are used to produce realistic simulacra of the mother cow (for the calf) and of the calf (for the mother).

To limit the risk of disease contagion and the onset of epidemics or even pandemics, all tasks on the farm are executed by robots, including cleaning, animal handling, and medication. Thus, human contact with the animals is no longer needed, as farmers can run the farm remotely by controlling its networked computing systems via a single interface.

The Science

This scenario is not entirely imaginary. Intensive indoor farming of dairy cows and other farm animals is a common practice, with so-called 'zero grazing' systems in place in many countries. This tendency is growing as the perceived risk of epidemics from animal-borne diseases and the perceived impact of extensive outdoor farming on the environment are increasing.

In more advanced 'zero grazing' systems, at least some animal-related processes and environmental conditions are highly controlled by or with the help of *smart* computing systems.[3] What is commonly known as 'precision live-

3 Pavan Sikka et al, 'Wireless Adhoc Sensor and Actuator Networks on the Farm' in S Wicker, P Gibbons, and J Paradiso (eds), *Proceedings of the 5th International Conference on Information Processing in Sensor Networks* (Association for Computing Machinery USA, 2006) 492.

stock farming'[4] already employs a wide range of technologies to automate and control aspects of animal husbandry and production. For example, robotic milking systems have been in use since the early 1990s,[5] allowing cows to access them to be milked but also to receive individually tailored medication or food supplements.[6] Likewise, feeding stations exist that can dispense individually tailored food quantity and quality. Tags embedded in cows' collars are used to identify them and give them access to different resources; and to monitor activities such as locomotion and rumination to assess welfare issues such as lameness or even fertility,[7] leveraging machine learning to recognise data patterns.[8] Ambient sensors already used to monitor human wellbeing could plausibly be used to monitor cows' well-being; for example, pressure sensors fitted to mattresses to monitor sleep quality[9] could be embedded in cows' bedding, provided a robust enough design. Animal farming researchers are also exploring the use of deepfake techniques to generate large datasets for training machine learning models to monitor animal health and emotional states, and even foster positive states by displaying digital conspecifics.[10] Air filtration systems are already used in some farmed animal environments, such as live transport trucks.[11] Similarly, time-controlled lighting and temperature regulation are increasingly used in controlled environment agriculture[12] and intelligent microclimate control is being explored for livestock farming too.[13] Finally, just as some production processes are almost entirely controlled by

4 António Monteiro, Sérgio Santos and Pedro Gonçalves, 'Precision Agriculture for Crop and Livestock Farming – Brief Review' (2021) 11(8) *Animals* 2345.
5 Luiz Marcos Simões Filho et al, 'Robotic Milking of Dairy Cows: A Review' (2020) 41(6) *Semina: Ciências Agrárias* 2833.
6 See 'Lely Astronaut A4: The Natural Way to Milk', *Lely* (Web page, 2023) <https://www.lely.com/solutions/milking/astronaut-a4>.
7 See 'Lely Qwes: 24/7 Information on Your Cows', *Lely* (Web page, 2023) <https://www.lely.com/solutions/milking/qwes>.
8 See 'Farming with Data: Data Driven, Smart Farming, Its Yours by Choice', *Lely* (Web page, 2023) <https://www.lely.com/solutions/farming-with-data>.
9 See 'Sleep Analyzer', *Withings* (Web page, 2023) <https://www.withings.com/uk/en/sleep-analyzer>.
10 Suresh Neethirajan, 'Is Seeing Still Believing? Leveraging Deepfake Technology for Livestock Farming' (2021) 8(7) *Frontiers in Veterinary Science* 40253.
11 Cindy Wenke et al, 'Impact of Different Supply Air and Recirculating Air Filtration Systems on Stable Climate, Animal Health, and Performance of Fattening Pigs in a Commercial Pig Farm' (2018) 13(3) PLOS ONE e0194641.
12 See David Cutress, 'Indoor or In-Field: Climate Impacts of Controlled Environment Growth', *Welsh Government: Farming Connect* (online, 9 February 2022) <https://businesswales.gov.wales/farmingconnect/news-and-events/technical-articles/indoor-or-field-climate-impacts-controlled-environment-growth>.
13 Yuri Ivanov and Nikolai Novikov, 'Digital Intelligent Microclimate Control of Livestock Farms' (2020) 175 *E3S Web of Conferences* 11012.

robots, it is not far-fetched to envisage an indoor farming system entirely run by robots.

However, current research in this area shows that the above scenario poses a number of welfare and ethical challenges. The cows are kept entirely isolated from the natural environment which would enable them to express their natural behaviour, such as grazing, and which would provide them with natural stimulation, such as smells, tastes, and sounds.[14] They are kept under a highly controlled regime limiting their ability to exert their agency by exploring complex surroundings, making choices, and gathering novel information, which have been shown to be important for animal welfare.[15] The cows are kept in crowded conditions, which prevent these highly social animals from managing their hierarchical social interactions and from managing their privacy boundaries by implementing distance-setting mechanisms, which have also been shown to be important for welfare.[16] Enormous amounts of data about the animals are collected to extract information about them[17] without their consent and used to control every aspect of their lives, and for example decide when to impregnate or kill them, which is ethically problematic.[18] In a highly automated environment, the loss of contact with the cows and their remote management is likely to reduce the empathy with which farmers might otherwise manage the animals, and instead risks turning the animals into mere cogs in the farming machine.[19] These fundamental welfare and ethical challenges, which are clearly brought to the fore by our scenario, already affect modern animal farming and motivate the three legal questions on which we focus.

14 J Michael Wilkinson et al, 'Some Challenges and Opportunities for Grazing Dairy Cows on Temperate Pastures' (2020) 75(1) *Grass and Forage Science* 1.

15 Sarah Webber, Mia L Cobb and Jon Coe, 'Welfare Through Competence: A Framework for Animal-Centric Technology' (2022) 9 *Frontiers in Veterinary Science* 885973.

16 Patrizia Paci, Clara Mancini and Bashar Nuseibeh, 'The Case for Animal Privacy in the Design of Technologically Supported Environments' (2022) 8 *Frontiers in Veterinary Science* 784794.

17 Dirk van der Linden, 'Interspecies Information Systems' (2021) 26 *Requirements Engineering* 535.

18 Clara Mancini, Orit Hirsch-Matsioulas and Daniel Metcalfe, 'Politicising Animal-Computer Interaction: An Approach to Political Engagement with Animal-Centred Design' (2022) *Proceedings of the Ninth International Conference on Animal-Computer Interaction*, ACI2022, ACM DL, Article 1.

19 Juliette Schillings, Richard Bennett and David Christian Rose, 'Exploring the Potential of Precision Livestock Farming Technologies to Help Address Farm Animal Welfare' (2021) 2 *Frontiers in Animal Science* 639678.

The Law

We could have used some of the existing legal institutions, but we felt that they are not fully fit for purpose in that they are rooted in a human-centred view of the world, often dominated by Western legal systems, and thus we decided to invent a new institution for a new world. An institution that is progressive in the application of the law, and how it operates, without being confined to a particular location and any particular worldview or cultural identity. It might be far-fetched that such an institution will ever exist, or that it can be established within the next 30 years, but we felt that it is important that we attempt to push the normative as well as the institutional boundaries in our project. Consequently, the following institution emerged.

The World Court for Rights is a first-of-its-kind forum with virtual hearings taking place across the globe, where anyone with a valid claim based on human, animal, or other emerging rights law has standing to bring an action and can draw on any of the international and regional rights instruments as well as relevant domestic legislation and case law. Judges are selected based on their expertise and cultural competency by the Global Panel of Experts and a global popular vote, and no states or other public or private entities are involved in the decision-making as to the appointment of the judges or composition of the bench in particular cases. The structure of the Court consists of the Chamber that makes an initial judgment in relation to the alleged violations, which then can be appealed to the Grand Chamber that re-examines the case. The Court's judgments are enforceable via the existing international, regional, and domestic mechanisms.

Due to the space limitations, we do not delve into the fundamental issue of the application of human rights to non-state actors[20] but accept that such developments are feasible within the foreseeable future, including due to the current advances in the realm of business and human rights.[21] Thus, we contend, it would not be too unrealistic to imagine a shift from an expectation of respecting human rights to holding human rights obligations by actors beyond states. Similarly, the recognition of nonhuman animals and nature/ecosystem at large as a 'legal person' has been growing, including in judicial rulings.[22] Consequently, accepting that animals have standing in legal actions

20 See, eg, Andrew Clapham, *Human Rights Obligations of Non-State Actors* (Oxford University Press, 2006).

21 See, eg, United Nations, Office of the High Commissioner for Human Rights, 'Guiding Principles on Business and Human Rights: Implementing the United Nations Protect, Respect and Remedy Framework' (United Nations, 2011); Natalia Szablewska et al, 'An Opportunity for Impact: Recommendations for Regulating Modern Slavery in Supply Chains in Aotearoa New Zealand' (Report, May 2022) <https://modernslaveryrecommendations.nz/Recommendation%20paper%20-%20An%20Opportunity%20for%20Impact.pdf>.

22 For example, *Animal Welfare Board of India v A Nagaraja & Ors* [2014] (6) SCALE 468 by extending the right to life (under Article 21 of the Constitution) to animals; see also *Narayan*

is no longer considered to be particularly novel. One can assume that within the next 30 years, there will be more such cases, including those involving computational and non-biological persons, thus this is not treated as a legal question for this Court's review.[23]

The application of human rights to nonhuman animals posed a bigger challenge. With the recognition of animal/nature legal personhood,[24] it should not be, however, seen as a bar to reimagining human rights to be applicable to nonhuman persons. Rather than relying on animal law[25] or the rights of nature,[26] we applied the human rights discourse and corresponding jurisprudence as if it were applicable to nonhuman persons. One of the key criticisms of the current system is its anthropocentric focus,[27] which in many ways the

---

Dutt Bhatt v Union of India [2018] PIL No 43 of 2014 (India); *Karnail Singh & Ors v State of Haryana* [2019] *SCC ONLINE P&H 704* (India); *Mohd Salim v State of Uttarakhand* (2017) PIL No 126 of 2014 (India); *Salim v State of Uttarakhand HC* (2013) PIL No 126 of 2014 (India); *Te Awa Tupua (Whanganui Claims Settlement) Act 2017* (New Zealand); *Yarra River Protection (Wilip-gin Birrarung murron) Act 2017* (Vic) s 5(b) (Australia); Tercer Juzgado de Garantías de Mendoza 3 November 2016, EXPTE NRO P-72.254/15 (Argentina); for more resources see 'The Non-Human Rights Project', *Nonhuman Rights Project* (Web Page, 2023) <https://www.nonhumanrights.org>; see also Animal Protection Index for corresponding legal claims 'Animal Protection Index', *World Animal Protection* (Web Page) <https://api.worldanimalprotection.org/#>.

23 It is beyond the scope of this chapter to engage in any meaningful way in the debate about the basis for animal rights, including the correlativity and reciprocity of rights and duties as a prerequisite for any meaningful legal rights, but see Saskia Stucki, 'Towards a Theory of Legal Animal Rights: Simple and Fundamental Rights' (2020) 40(3) *Oxford Journal of Legal Studies* 533; Visa AJ Kurki, 'Why Things Can Hold Rights: Reconceptualizing the Legal Person' in Visa AJ Kurki and Tomasz Pietrzykowski (eds), *Legal Personhood: Animals, Artificial Intelligence and the Unborn* (Springer 2017) 79.

24 See the landmark case in Argentina in 2015 where a Sumatran orangutan from the Buenos Aires Zoo was recognised as a 'nonhuman person', with the court declaring she had the right to life, liberty, and freedom from harm, Federal Chamber of Criminal Cassation (18 December 2014) SAIJ id: NV9953; see also Loreley Gaffoglio, 'Conceden un Hábeas Corpus a Una Orangutana' (trans 'They grant a habeas corpus to an orangutan'), *La Nation/The Nation* (online, 21 December 2014) <https://www.lanacion.com.ar/sociedad/conceden-un-habeas-corpus-a-una-orangutana-del-zoologico-porteno-nid1754353>; as well as Fred Motson, 'Should Animals Have Legal Standing?' *The Open University* (25 October 2021) <https://www.open.edu/openlearn/society-politics-law/law/should-animals-have-legal-standing#:~:text=One%20group%20of%20living%20beings,in%20court%20by%20their%20owners>.

25 See, eg, Anne Peters, 'Global Animal Law: What It Is and Why We Need It' (2016) 5(1) *Transnational Environmental Law* 9; Stucki (n 23).

26 Cormac Cullinan, 'A History of Wild Law' in Peter Burdon (ed), *Exploring Wild Law: The Philosophy of Earth Jurisprudence* (Wakefield Press, 2011) 12; Vandana Shiva, *Earth Democracy: Justice, Sustainability, and Peace* (Bloomsbury Publishing, 2016).

27 See, eg, Conor Gearty, 'Do Human Rights Help or Hinder Environmental Protection?' (2010) 1(1) *Journal of Human Rights and the Environment* 7; John H Knox, 'Climate Ethics and Human Rights' (2014) 5 (Special Issue) *Journal of Human Rights and the Environment* 22; Stephen A Plass, 'Exploring Animal Rights as an Imperative for Human Welfare' (2010)

rhetoric of human rights exemplifies as, after all, it holds the 'human' at its centre. However, human rights consist of not only enforceable rights and associated underpinning principles but also create an opportunity for a 'discursive elaboration' allowing it to settle conflicts based on diverse values and translating them into effective practices.[28] Also, as argued by Connor Gearty, the intrinsic drive of the human rights agenda towards empathy for others[29] creates further opportunities for its transformative nature beyond that applied to 'human' beings. As such, we see added value in using the established, albeit contested, discourse of human rights in relation to nonhuman animals.[30] We accept, nevertheless, that the exercise is not an easy one as traditionally legal regimes, be it human rights law or international humanitarian law, have been designed to see value in protecting the natural environment for the sake of the human population and so 'animal welfare is instrumentalised: it is understood not as a good in itself, but is instead valued for its implications for human welfare and rights'.[31]

It is a good starting point but insufficient for transitioning from animal welfare that is based on satisfying 'public interest'[32] to animal rights centred around their inherent autonomy and entitlement to equality and fairness. Recognition of the basic yet fundamental values has been the underpinning driving force behind the human rights agenda, and, as such, there is no reason why it cannot be used successfully for securing the interests of nonhuman animals. Thus, we hope we gave justice, albeit only in an introductory manner, to what is undeniably a complex area of granting nonhuman animals the rights to dignity, privacy, and non-exploitation, and how these entitlements could potentially be argued when blending the existing legal orders with what can be achieved if we were to (re-)imagine the world where humans become decentralised. Such an exercise is justified to account for the anthropocentric impact that has led to the catastrophic climatic and environmental develop-

112(2) *West Virginia Law Review* 403; Tibor R Machan, 'Do Animals Have Rights?' (1991) 5(2) *Public Affairs Quarterly* 163; Phil S Elder, 'Legal Rights for Nature: The Wrong Answer to the Right(s) Question' (1984) 22(2) *Osgoode Hall Law Journal* 285; Stephen Keim and Jordan Sosnowski, 'Human Rights v Animal Rights: Mutually Exclusive or Complementary Causes' (2012) 8 *Australian Animal Protection Law Journal* 78.

28 Natalia Szablewska and Krzysztof Kubacki, 'A Human Rights-Based Approach to the Social Good in Social Marketing' (2019) 155(3) *Journal of Business Ethics* 871.

29 Gearty (n 27); see also Chris A Robinson, 'Biological Foundations of Human Rights' in Dinah Shelton (ed), *The Oxford Handbook of International Human Rights Law* (Oxford University Press, 2013) 54.

30 See also Tom Sparks, 'Protection of Animals Through Human Rights: The Case-Law of the European Court of Human Rights' in Anne Peters (ed) *Studies in Global Animal Law/Beiträge zum ausländischen öffentlichen Recht und Völkerrecht* (Springer, 2020) 153.

31 Ibid 153.

32 See, eg, *Steel and Morris v UK* (2005) 4 Eur Court HR, Application No 68416/01, [88]; *PETA Deutschland v Germany* 5 Eur Court HR (2012), Application No 43481/09, [47].

ments that are threatening the very survival of our and other species.[33] If nothing else, such an action is needed for the sake of humanity.

Judgment

Rights for Animals Watch (on behalf of farm animals) v Positive Farms Animals Ltd.
(Application no 343883/57)

The World Court for Rights, sitting as a Chamber consisting of:
Natalia Szablewska and Clara Mancini

Judgment
This version was certified on 1 August 2057 under Rule 62 of the Rules of Court.

Online

1 August 2057

The Facts

1. The applicant is an international non-governmental organisation whose aims include defending the rights of nonhuman animals ('animals') worldwide.
2. The applicant engaged in strategic litigation on behalf of farm animals kept at a smart dairy farm operated by the defendant.
3. The parties engaged in a non-compulsory mediation facilitated by the Human-Animal World Forum prior to this Court's proceedings, but which yielded no tangible results.
4. On 15 March 2055, the applicant started proceedings at this Court claiming that the animals at Positive Farms Animals Ltd suffered abuse due to a number of welfare and ethical challenges that constitute violations of their rights to dignity, privacy, and freedom from slavery.
5. The facts relating to the treatment of the smart dairy farm animals are not disputed by the parties and were accepted as such at the preliminary hearing of 13 March 2057.
6. The application has been accepted as meeting the statutory eligibility criteria for admission to a full hearing by this Court.

33 See, eg, Valérie Masson-Delmotte et al, *Climate Change 2021. The Physical Science Basis* (Intergovernmental Panel on Climate Change, Report, 2021).

Alleged Violation of the Right to Dignity

The Parties' Submission

7. The applicant submitted that a violation of the principle of dignity, which is a right on its own and essential for the realisation of other fundamental rights, recognised in the *Universal Declaration of Human Rights 1948*[34] Article 1; *International Covenant on Civil and Political Rights 1966*[35] Article 10; *International Covenant on Economic, Social and Cultural Rights 1966*[36] Preamble; and *Convention on the Rights of the Child 1989*[37] Preamble among others, including many regional instruments, for example, the *European Union Charter of Fundamental Rights 2000*[38] Article 1, and it was also part of the *2030 Agenda for Sustainable Development*.[39] As such, it has been now recognised under international law, being part of customary law, and, arguably, is a *jus cogent* norm[40] from which no modification, derogation, or abrogation is allowed. It has also been part of the constitutional law of many countries and has been incorporated into many constitutional documents, including in Canada, Germany, Poland, Colombia, and Peru among others.[41]
8. As it is further argued, the right to dignity implies worthiness and equal treatment and respect for all, which must also encompass nonhuman animals in an equal manner to the human person.

The Court's Assessment

(a) *Recapitulation of the Principles*

9. In the recent case of *Orangutans Habitat Protection v Mount Venture Corporation* (No 48609/55, 18 June 2055) before this Court, with similar earlier decisions across world jurisdictions, including *Animal Welfare*

34 *Universal Declaration of Human Rights*, GA Res 217A (III), UN GAOR, UN Doc A/810 (10 December 1948) ('Universal Declaration').

35 *International Covenant on Civil and Political Rights*, opened for signature 19 December 1966, 999 UNTS 171 (entered into force 23 March 1976) ('ICCPR').

36 *International Covenant on Economic, Social and Cultural Rights*, opened for signature 16 December 1966, 993 UNTS 3 (entered into force 3 January 1976).

37 *Convention on the Rights of the Child*, opened for signature 20 November 1989, 1577 UNTS 3 (entered into force 2 September 1990) ('CRC').

38 *Charter of Fundamental Rights of the European Union*, OJ C 202/389 (entered into force 1 December 2009) ('EU Charter').

39 *2030 Agenda for Sustainable Development*, GA Res 70/1, UN Doc A/RES/70/1 (adopted 25 September 2015).

40 See Mark W Janis, *An Introduction to International Law* (Aspen Publishers Inc, 1999) 53.

41 See also Aharon Barak, *Human Dignity: The Constitutional Value and the Constitutional Right* (Cambridge University Press, 2015).

Board of India v A Nagaraja and Ors,[42] it has been clearly established that nonhuman animals have the inherent dignity and a right to live peacefully, including being protected from unnecessary pain and suffering. From 2039 onwards, the number of Constitutions containing a specific reference to providing the natural environment with protection and living creatures with compassion has increased considerably,[43] attesting to the recognition of the application of the right to dignity to nonhuman animals and not being confined to human beings only.

10. Recognition that nonhuman animals are entitled to honour and dignity is neither a novel nor a particularly controversial proposition anymore. As nonhuman animals are now endowed with legal personality, those in charge of their well-being and safety have both direct and indirect duties towards them.
11. Even if the application of this principle to nonhuman persons can be persuasively argued, this Court accepts that the concept of 'human dignity' has not been applied consistently throughout jurisdictions and its very notion as a 'right' has been challenged.[44] As the exact scope of the right in relation to human beings has not been fully established, having its consequences for its application to nonhuman animals, it needs to be assessed contextually on the merit of each case.

(b) *Application to the Present Case*

12. The applicant alleges that the fulfilment of the right to dignity prevents the defendant from keeping the animals away from their natural habitat and expressing their natural behaviours. The use of modern technology is not sufficient to ensure that the physiological and psychological needs of the animals are fully met. Similarly, as with the successful challenge to using animals for medical experimentation as causing them unnecessary pain and suffering (for example, the Swiss case of granting fundamental rights to primates[45]), high-intensity farming should also be considered as

42 [2014] (6) SCALE 468 (India).

43 See, eg, *Constitution of India 2020* (India) art 51A(g).

44 See, eg, Christopher McCrudden, 'Human Dignity and Judicial Interpretation of Human Rights' (2008)19(4) *European Journal of International Law* 655; Conor O'Mahony, 'There is No Such Thing as a Right to Dignity' (2012) 10(2) *International Journal of Constitutional Law* 551; Hanna-Maria Niemi, 'The Use of Human Dignity in Legal Argumentation: An Analysis of the Case Law of the Supreme Courts of Finland' (2021) 39(3) *Nordic Journal of Human Rights* 280; Biruta Lewaszkiewicz-Petrykowska, 'The Principle of Respect for Human Dignity – Proceedings of the UniDem Seminar organised in Montpellier, 2–6 June 1998' (Council of Europe Publishing, 1999); cf Paolo G Carozza, 'Human Dignity and Judicial Interpretation of Human Rights: A Reply' (2008)19(5) *European Journal of International Law* 931.

45 See *Vischer & Ors v Ness & Ors* (2020) *Bundesgericht* [Constitutional Court of Basel-Stadt], 15 January 2019, VG2018.1 and Swiss Federal Supreme Court, Judgment, 16 September

illegally depriving animals of their rights, including to a life of intrinsic worth, dignity, and honour.

13. The Court accepts the applicant's proposition that nonhuman animals have an equal right to dignity as humans, as well as that animal testing is not only cruel to nonhuman animals but also constitutes unnecessary suffering due to the significant limitations presented by medication developed through animal testing as well as the existence of viable alternative testing methods.[46]
14. The Court, however, cannot agree that both situations, that is of using animals in scientific research and for food production, are fully analogous. In both instances, the animals can be exposed to pain and suffering, but in the Court's view, the extensive technological adjustments taken by the defendant, including to control the air purity, feeding, and milking practices, along with the use of machine learning technology to identify early signs of poor health and to offer the animals a variety of visual stimuli, minimise the number of the welfare issues that had been raised in relation to animal testing. This is not to say that animal farming might not be considered an outdated and unnecessary practice in light of the technological developments, including the highly advanced production of cultivated meat and like by-products, but this is not the question put before this Court.
15. The issue at hand is also not whether the humans' rights triumph over the animals' rights, as they no longer do, but rather about balancing competing interests and meeting the test for breaching or otherwise the right to dignity. On a balance, the applicant has not made a convincing case that the use of technology by the defendant does not alleviate many of the negative impacts on animals of historically practised intensive farming and that, in the current form, does not allow the animals to exercise their right of personality and status to a degree allowing for meeting the legal threshold for the said right. In this Court's view, the life of the farm animals at the defendant's farm does not amount to, in the language of *Geneva Conventions 1949* Common Article 3,[47] 'humiliating and degrading treatment'. To make the position of this Court

2020, 1C_105/2019; see also Charlotte Blattner and Raffael Fasel, 'The Swiss Primate Case: A Judicial Unicorn or Justice in Action?' (2021) 11(1) *Transnational Environmental Law* 1.

46 See, eg, Aysha Akhtar, 'The Flaws and Human Harms of Animal Experimentation' (2015) 24(4) *Cambridge Quarterly of Healthcare Ethics* 407; Pablo Perel et al, 'Comparison of Treatment Effects Between Animal Experiments and Clinical Trials: Systematic Review' (2007) 334(7586) *British Medical Journal* 197; Hugh LaFollette, 'Animal Experimentation in Biomedical Research' in Tom L Beauchamp and Raymond G Frey (eds), *The Oxford Handbook of Animal Ethics* (Oxford University Press, 2011) 812.

47 *Geneva Conventions I-IV*, 12 August 1949, 75 UNTS 31; 75 UNTS 85; 75 UNTS 135; 75 UNTS 287 (respectively).

clear, it is not that the alleged infringement is justified on the grounds of some greater competing interests but that the infringement of the right to dignity in the circumstances has not been made out.

16. The Court concludes that nonhuman animals at the defendant's farm need to be treated with honour and dignity at all times, but the current facts do not support the applicant's claim that the conditions amount to a violation of their right to dignity.

Alleged Violation of the Right to Privacy

The Parties' Submission

17. The applicant also claimed a violation of the right to privacy, which provides as follows:
    1 The right to privacy is foundational for ensuring one's dignity and autonomy by being able to live a fulfilling life free from unwarranted physical and psychological interference;[48]
    2 Personal data should be subject to processing restrictions, including in relation to storing and sharing private and confidential information;[49] and
    3 Any limitations on the right to privacy need to be necessary, proportionate, and based on an appropriate legal basis.[50]

The right to privacy is incorporated in most international and regional human rights instruments.[51] States across the world have it protected under their Constitutions, and there are further laws protecting this right,[52] as well

48 See, eg, ICCPR (n 35) art 17.

49 See, eg, Special Rapporteur on the Right to Privacy, 'Principles Underpinning Privacy and the Protection of Personal Data' UN Doc A/77/196 (20 July 2022).

50 See, eg, Human Rights Committee, *Toonen v Australia*, UN Doc CCPR/C/50/D/488/1992 [8.3].

51 Based on the Universal Declaration (n 34) art 12; the ICCPR (n 35) art 17; the CRC (n 37) art 16; *European Convention for the Protection of Human Rights and Fundamental Freedoms*, opened for signature 4 November 1950, 213 UNTS 221 (entered into force 3 September 1953) art 8 ('ECHR'); *American Convention of Human Rights 'Pact of San José, Costa Rica'*, opened for signature 22 November 1969, 1144 UNTS 123 (entered into force 18 July 1978) art 8 ('ACHR'); *Arab Charter on Human Rights* 2004, arts 16 and 21; *International Convention on the Protection of the Rights of All Migrant Workers and Members of Their Families*, opened for signature 18 December 1990, 2220 UNTS 3 (entered into force 1 July 2003) art 14; *African Charter on the Rights and Welfare of the Child* (entered into force 29 November 1999) art 10; the *ASEAN Human Rights Declaration* 2012 (entered into force 19 November 2012) art 21.

52 At the time of writing, over 130 states have included specific constitutional provisions on the right to privacy and over 100 have introduced specific laws on some form of privacy and data protection.

as a number of specific regional and institutional instruments pertaining to the protection of privacy.[53]

18. Along with technological advancements and an increase in the development of interactive technologies targeting nonhuman animals, calls have been made to extend privacy considerations to nonhuman animals.[54]

The Court's Assessment

(a) *Recapitulation of the Principles*

19. The Court observes that privacy is a concept that is now well-established in relation to humans, and the growing understanding of what it might entail in relation to animals has been on the rise since the 1960s. There is bourgeoning research indicating that there is much merit in asserting some form of privacy right, necessitating protection, to nonhuman animals accounting for their psychological well-being, at a minimum. In the human context, privacy is considered foundational for building and maintaining relationships with one another as well as fulfilling one's agency allowing for adapting to and transforming the immediate environment; the same biological triggers apply to nonhuman animals striving to protect their private spheres when their territorial space becomes invaded by the same or different species.[55]

20. The Court also notes that the general perception of privacy has changed over time, including the public understanding of the importance of the right to privacy, not the least for the fulfilment of other rights, in relation to humans, as has it in relation to nonhuman animals. The right to privacy does not protect only from an unwanted inference, such as the use of personal data for inferential analytics in profiling or automated

53 For example, *Council of Europe Convention 108 for the Protection of Individuals with Regard to the Automatic Processing of Personal Data*, opened for signature 28 January 1981, ETS 108 (entered into force 1 October 1985); *Asia-Pacific Economic Cooperation (APEC) Privacy Framework* 2005; *Updated Principles of the Inter-American Juridical Committee on Privacy and Personal Data Protection with annotations* (2021), adopted by the General Assembly of the Organisation of American States; Organisation for Economic Co-operation and Development, *Recommendations of the Council concerning the Guidelines on the Protection of Privacy and Transborder Flows of Personal Data of the Organisation for Economic Co-operation and Development* (adopted on 23 September 1980, amended on 11 July 2013).

54 See, eg, Paci et al (n 16); Angie Pepper, 'Glass Panels and Peepholes: Nonhuman Animals and the Right to Privacy' (2020) 101(4) *Pacific Philosophical Quarterly* 628; Ewa Haratym, 'Animals' Rights to Privacy' (2017) 85 *World Scientific News* 73; Brett Mills, 'Television Wildlife Documentaries and Animals' Right to Privacy' (2010) 24(2) *Continuum: Journal of Media and Cultural Studies* 193; Alan F Westin, *Privacy and Freedom* (Atheneum, 1967).

55 See, eg, Westin (n 54).

decision-making[56] but also from the consequences of such an inference, including the harm done due to the data associated with the (non)human person falling into the hands of those who are not authorised or otherwise approved to be in possession of that information. It is not difficult to imagine that, for example, data collected by scientists about the migratory paths of some protected wild animals could be misused if it were to fall into the wrong hands. The right to privacy is more than merely a human-oriented attempt to secure 'free development and expression of an individual's personality, identity and beliefs',[57] it has very practical implications.

21. The right to protect animals' privacy must be balanced against the unintended consequences of its application, including when protecting animals' other rights. In 2015 a journalist filed a public record information request to enquire about the necropsy of a bottlenose dolphin being found in a river,[58] leading to his euthanasia based on humanitarian grounds, which led to a public debate on whether the release of the medical report would be against the dolphin's right to medical privacy.[59] As animals rely on humans to protect their rights, a balance must be struck between granting such rights which might have unintended consequences for preventing their other rights from being realised. However, resolving conflicts between rights, or rights and other values,[60] is neither a novel problem nor prohibitive for recognising new rights.

(b) *Application to the Present Case*

22. Based on the facts provided, the animals at the defendant's farm have been subject to extensive surveillance, some of which is justified on the basis of ensuring that their health and well-being needs are met and are proportionate to the achievement of the legitimate aim. Some instances,

56 For the impact in the context of the use of big data analytics and artificial intelligence (AI), see Sandra Wachter and Brent Mittelstadt, 'A Right to Reasonable Inferences: Re-Thinking Data Protection Law in the Age of Big Data and AI' (2019) 2 *Columbia Business Law Review* 494.

57 'Special Rapporteur on the Right to Privacy', *United Nations Human Rights Office of the High Commissioner* (Web page, 2023) <https://www.ohchr.org/en/special-procedures/sr-privacy>.

58 See <https://www.muckrock.com/foi/new-jersey-229/nj-dolphin-necropsy-20430>.

59 Jason Koebler, 'New Jersey Says Releasing Dolphin's Autopsy Would Infringe Its Privacy', *Vice* (online, 12 January 2016) <https://www.vice.com/en/article/vv7egy/new-jersey-says-releasing-dead-dolphins-autopsy-would-infringe-its-privacy>.

60 See, eg, Federica Giovanelle, *Copyright and Information Privacy: Conflicting Rights in Balance* (Edward Elgar, 2017) in particular ch 1; Olivier De Schutter and Françoise Tulkens, 'The European Court of Human Rights as a Pragmatic Institution' in Eva Brems (ed), *Conflicts Between Fundamental Rights* (Intersentia, 2008)169; Daniel Shapiro, 'Conflicts and Rights' (1989) 55(3) *Philosophical Studies: An International Journal for Philosophy in the Analytic Tradition* 263.

however, including the monitoring of activities such as locomotion and rumination for the purpose of inferring when the cow is in oestrous so that she can be successfully inseminated and thus produce more milk,[61] or to infer when the cows' production potential is declining to inform the decision-making when they are no longer worth keeping, cannot be justified on such grounds. As acknowledged now in relation to humans, limitations on involuntary neural surveillance need to be considered in this case.[62]

23. The defendant has put forward no justification for not affording the farm animals in their care privacy in relation to their farm life. Asserting simply that the right 'does not apply to nonhuman persons' is not sufficient and unreasonable in light of the scientific discoveries in this domain. This Court accepts that the scope of the right to privacy applicable to nonhuman animals might not be the same as for human persons, including as certain aspects of the right to privacy might simply not apply to nonhuman animals, such as respect for one's correspondence; it nevertheless does not negate that nonhuman animals can and do experience consequences due to their privacy being interfered with (as per para 22 above).
24. Consequently, this Court finds in favour of the applicant that the animals' right to privacy at the defendant's farm has been subject to a number of unjustified interferences and, as such, it requires remedying.

Alleged Violation of the Right to be Free from Slavery

THE PARTIES' SUBMISSION

25. The applicant's final submission alleged that the animals at the defendant's farm were deprived of their right to freedom from slavery, servitude,

61 Kathrin Wagner et al, 'Effects of Mother Versus Artificial Rearing During the First 12 Weeks of Life on Challenge Responses of Dairy Cows' (2015) 164 *Applied Animal Behaviour Science* 1.

62 This is not an entirely imaginary argument as the right to cognitive liberty, complementing the rights to privacy, freedom of thought, and self-determination, has been claimed to be a new emerging human right requiring recognition, see Zoë Corbyn 'Interview: Prof Nita Farahany: "We need a new human right to cognitive liberty"', *The Observer* (online, 4 March 2023) <https://www.theguardian.com/science/2023/mar/04/prof-nita-farahany-we-need-a-new-human-right-to-cognitive-liberty>.

and forced labour[63] pursuant to numerous international[64] and domestic instruments.[65]

26. The defendant challenged the claim as having no legal basis, including that this Court lacked jurisdiction to hear the claim, considering the lawsuit filed in the United States District Court for the Southern District of California in San Diego by PETA (People for the Ethical Treatment of Animals) on behalf of five wild-born killer whales against Sea World for 'involuntary servitude' or 'slavery' in violation of the 13th Amendment to the United States Constitution. The case was dismissed as it lacked subject matter jurisdiction as whales were not 'persons' and, since the 13th Amendment applied to persons only, they could not be offered constitutional protection.[66]

ADMISSIBILITY

27. The Court notes that it had already determined at the preliminary hearing of 13 March 2057 that this complaint was not manifestly ill-founded and, as it fell within the scope of this Court's jurisdiction under Article 5 of the Rules of Court, it must therefore be declared admissible.

THE COURT'S ASSESSMENT

(a) *Recapitulation of the Principles*

28. As this Court stated on many occasions, freedom from slavery is a well-established right in international law, from which no exceptions are permitted, with more than 300 international agreements to that effect and it is criminalised across the world jurisdictions. Under the *League of Nations Slavery, Servitude, Forced Labour and Similar Institutions and*

63 For the purpose of this chapter, the three categories are specifically mentioned, but with an understanding that these are not exclusive forms of what is referred to at the time of writing as 'modern slavery' and they also have a different threshold for application. It is beyond the scope of this chapter to go into the details, but see Paul M Taylor, *A Commentary on the International Covenant on Civil and Political Rights: The UN Human Rights Committee's Monitoring of ICCPR Rights* (Cambridge University Press, 2020) 218-239.

64 The Universal Declaration of Human Rights (n 34) art 4, the ICCPR (n 35) art 8, the ECHR (n 51) art 4, the ACHR (n 51) art 6, *African Charter on Human and Peoples Rights*, opened for signature 27 June 1981, 1520 UNTS 217 (entered into force 21 October 1986) art 5.

65 At the time of writing, some half of world jurisdictions have made slavery illegal under their criminal laws, see Jean Allain and Katarina Schwarz, 'Antislavery in Domestic Legislation', *Antislavery in Domestic Legislation, University of Nottingham Rights Lab, and Monash University Castan Centre for Human Rights Law* (Web page, 2023) <https://antislaverylaw.ac.uk>.

66 *Tilikum v Sea World Parks & Entertainment*, 842 F Supp 2d 1259 (SD Cal 2012).

Practices Convention 1926, slavery is defined as 'the status or condition of a person over whom any or all of the powers attaching to the right of ownership are exercised' (Art 1(1)), which is reproduced in its substance in the *Supplementary Convention on the Abolition of Slavery, the Slave Trade, and Institutions and Practices Similar to Slavery 1956* and the *Statute of the International Criminal Court 1998*. There are a number of other international instruments that are relevant to the claim, including the International Labour Organisation (ILO) *Forced Labour Convention 1930* (No 29), in addition to how these offences have been codified in many domestic jurisdictions.

29. Even though relevant to this case, the condition of 'ownership', an element of chattel slavery, is no longer required, and it is sufficient that there must be some form of coercive 'control' in place or abuse of a position of a person's vulnerability. The uniting element for the different forms of slavery is 'exploitation' as their purpose, but this has not been defined in international law, and thus its specific parameters continue to be debated.[67] As prior domestic jurisprudence suggests, the meaning of 'exploitation' is context-specific and depends largely on the circumstances of each case.[68] Also, there needs to be an element of temporal indeterminacy in the control exercised over the enslaved person.[69]
30. The Court notes that, despite the right to be free from slavery being widely adjudicated, this is the first case before this Court where the right is invoked in relation to nonhuman persons. It is, however, not a new topic, and it has been debated vigorously, including by national parliaments in the past decade. Historically, comparing animal to human 'slavery' was considered ill-founded, if not offensive;[70] an objection that is no longer considered to be persuasive or, in fact, necessary, considering legal developments in granting nonhuman animals personhood in many other areas. This Court wishes to reiterate that it is the legal similarities and the relevant economic motivation behind the practice that are to be considered analogous in these circumstances.

67 See United Nations Office on Drugs and Crime, Issue Paper: 'The Concept of "Exploitation" in the Trafficking In Persons Protocol' (Issues Paper, 2015); Marija Jovanovic, 'The Essence of Slavery: Exploitation in Human Rights Law' (2020) 20(4) *Human Rights Law Review* 674.

68 See, eg, Supreme Court of the Netherlands (2009), LJNBI7079, 08/03895 (the Netherlands).

69 Monti Narayan Datta and Kevin Bales, 'Slavery in Europe: Part 1, Estimating the Dark Figure' (2013) 35(4) *Human Rights Quarterly* 817; see also the *Bellagio-Harvard Guidelines on the Legal Parameters of Slavery*, Guideline 3 <https://www.monash.edu/__data/assets/pdf_file/0004/2263693/Bellagio-Harvard-Guidelines-English.pdf> [4].

70 See, eg, Seren Sensei, 'Factory Farming, Though Vile, Does Not Equate To Chattel Slavery', *Riot Material* (online, 13 June 2018) < https://www.riotmaterial.com/factory-farming-not-chattel-slavery/>.

(b) *Application to the Present Case*

31. The applicant put forward that the animals in the defendant's farm are subjected to exploitation in the form of female cows being forcibly impregnated and then separated from their newborns for their milk production and for as long as they can produce milk following which, along with male newborns, they are being disposed of. If we accept that animals have agency and are entitled to dignity, as argued under the applicant's first claim, then the plight of livestock animals in factory farms should not be ignored. What the animals are expected to do and how they have been treated in the process becomes part and parcel of granting legal rights to nonhuman animals on an equal grounding to a human person. The animals also experience the alleged exploitation on a continuous and indefinite basis, as the very reason why they are kept as farm animals is for the purpose of their exploitation.
32. The Court notes that, even if the conditions in which the animals are kept at Positive Farms Animals Ltd do not meet the threshold for humiliating and degrading treatment to be in violation of the animals' right to dignity, the purpose of them being kept is for the exclusive gain of the defendant. Being farm animals, they do not benefit from being kept in whatever good conditions, thus there is no comparative threshold for mutually beneficial interactions.[71] On that account, ceasing the enslavement of farm animals amounts to the annihilation of farm animals in the future.
33. Having regard to all of the foregoing factors, this Court finds that the treatment of animals at the defendant's farm, despite the technological investment and improvements in their care and treatment, is still exploitative. It abuses the position of vulnerability of the animals, and the control exercised over them is of indeterminate nature. In this matter, we find that animals were deprived of their right to freedom from slavery, servitude, and forced labour.

71 See Robert Mayer, 'What's Wrong with Exploitation?' (2007) 2(2) *Journal of Applied Philosophy* 137.

Chapter 4

The Sea Casts Its Net of Justice Wide

A Speculative Judgment for What Has Been Left to the Waters of Despair

Foluke I Adebisi

Secured ownership of land and legal control over underwater resources was and continues to be essential to enacting colonial conditions that persist into the present and are exemplified by a lack of harmonious respect for all life, nature, and Earth. This contrasts with restorative and transformative knowledge systems. For example, in Andean Indigenous thought, 'pacha' describes global space-time and Pachamama is Earth (land and water) as well as a goddess from whom all life emerges. These types of philosophies do not place the human above Earth in importance. Conversely, anthropocentrism in Euro-modern legal knowledge has meant that the law has thus far proven insufficient to protect Earth from the result of harmful human activity – especially activity connected to the accumulation of capital. This is because Euro-modern legal knowledge often prioritises and is fashioned to protect and regulate the accumulation of capital over and above respect for all life and Earth. Consequently, knowledges beyond Euro-modern law that do not suffer from these predilections may provide us with a better lens to rethink environmental protection for the future.

Using (among others) the frameworks of Black/African Science Fiction, Indigenous Knowledge, Earth Laws, and anticolonial logics, this chapter imagines the sea and its inhabitants addressing a petition against humanity for what it has thrown into the waters. Therefore, this judgment, relying on ideas from the law and literature movement, will be written in the form of a fictional 'court case', proceeding under the sea. Taking the 1783 case of *Gregson v Gilbert*[1] as its entry point, the petition that forms the focus of this judgment will include, *inter alia*, prayers for remedies for harm from debris thrown overboard during armed conflict on the high seas, repair for oil spills in conflict and peacetime, justice for bodies left to the waters across space-time (including enslaved persons thrown overboard and refugees trying to get to Europe across the Mediterranean), as well as the cleaning up of plastic refuse on the ocean floor and in underwater life. By imagining the ocean (as

1 *Gregson v Gilbert* (1783) 3 Doug KB 232.

DOI: 10.4324/9781003389569-5

well as its inhabitants) as sentient petitioners, this judgment seeks to explore a wide range of questions such as: what is the purpose of the law – profit or environmental protection? Are the processes, concepts, and structures of Euro-modern legality adequate to respond to the past, present, and future of environmental protection and repair? Who can/should speak for Earth? What would be adequate remedies for environmental disasters? What should the purpose of remedies be in relation to the environment – punishment, restitution, or transformation?

Commentary

In *Staying with the Trouble*, Donna Haraway points humanity to desirable responses to the dangers of the Anthropocene – imagination, community, future flourishing, survival, and sanctuary.[2] The conception of the 'Anthropocene' recognises that human activity has had a significant detrimental effect on the environment. This epoch was inaugurated through humanity's unequal overemphasis on the importance and exceptionalism of the human in the world. This exceptionalism proceeds unequally by casting many out of the category 'human' through varied uses of classification – race, gender, class, and ability, for instance. Over-exploitation in human relationships with each other and with nature is managed by what can be described as 'colonial-capitalism'. The colonial structure brutally acquired the key means of production through dispossession – the kidnapping and enslavement of Africans and the seizure of Indigenous land and natural resources. Therefore, this evolving historical conjunction ('colonial-capitalism') is characterised by evolving use of unfree/coerced labour, destruction of nature, over-consumption, and false scarcities, which are all then implicated in environmental disasters and climate emergencies.[3] Essentially, the redefinition of the human and '*the negation of earth*' (my emphasis).[4]

By significantly disrupting and endangering planetary relations, the Anthropocene brings with it new challenges, and it is recognised that new thinking within the law is necessary to handle those challenges. This need for new thinking is significant, considering the legal disruptiveness of climate change as well as law's complicity in heralding the advent of the Anthropocene. The law has always regulated what is considered legal – the

2 Donna J Haraway, *Staying with the Trouble: Making Kin in the Chthulucene* (Duke University Press, 2016) 100.

3 Farhana Sultana, 'The Unbearable Heaviness of Climate Coloniality' (2022) 99 *Political Geography* 102638.

4 Rolando Vázquez, 'Precedence, Earth and the Anthropocene: Decolonizing Design' (2017) 15(1) *Design Philosophy Papers* 77, 78; see also Mark Jackson, 'On Decolonizing the Anthropocene: Disobedience via Plural Constitutions' (2020) 111(3) *Annals of the American Association of Geographers* 698–708.

trade in enslaved humans, the dispossession of Indigenous peoples of their lands and resources, as well as the legitimation of the methods of accelerated profit-making from the foregoing in the aftermath of the foregoing.[5] Thus, it is important to note law's duality, both as structural underpinning of the human activity that shepherds in the Anthropocene and also as hope for deliverance from its effects. This paradox indicates that due to the entanglement of Euro-modern legal knowledge with reproducing the conditions for the Anthropocene, our legal imaginary may be unable to adequately respond to the exact nature of environmental harms that will eventuate in the near and far future and who should bear responsibility for them. Specifically, law's relation to time is such that it divorces itself from causes that are too remote in the past, and harms that are too far in the future, while also placing limits on who or what it considers to be harmed or even what amounts to harm. This all has implications for how justice within the Anthropocene can be conceived through the language and framework of Euro-modern law.

My entry point into this judgment is the case of *Gregson v Gilbert*.[6] This case, decided on appeal in 1783, was concerned with whether insurance could be claimed on a 'cargo' of more than 130 kidnapped Africans who had been thrown overboard due to a presumed 'necessity', occasioned by dwindling food and water supplies onboard the slave ship *Zong*. At the trial of first instance, the claim for insurance was accepted by the court. On appeal, Justice Mansfield ordered a new trial due to the presentation of new conflicting evidence (that there had been some water supplies during the killings) and to determine the precise cause of the deaths – whether they resulted from some external fortuity. It should be noted that despite this, and the often-celebratory reading of the case as a win for abolition, Justice Mansfield did not discount the possibility that a sufficient necessity could eventuate in such a circumstance. The judgment was founded on the fact that it was not clear that a sufficient necessity had eventuated in this case. The failure to engage with the killing of the Africans underscores how Mansfield foresaw exculpatory conditions for mass murder within Euro-modern legal knowledge to protect the legal doctrine of necessity as well as the wider processes of insurance, trade, and capital.[7]

Therefore, *Gregson v Gilbert* raises many questions about the nature of law and its evolving functions and outcomes – especially its role in the (non) protection of life and nature. As Robert Cover tells us, '[l]egal interpretation

5 JM Moore, '"Law", "Order", "Justice", "Crime": Disrupting Key Concepts in Criminology through the Study of Colonial History' (2020) 54(4) *The Law Teacher* 489, 492–502.
6 (1783) 3 Doug KB 232.
7 Jeremy Krikler, 'The Zong and the Lord Chief Justice' (2007) 64(1) *History Workshop Journal* 29, 37.

takes place on a field of pain and death'.[8] In this case, the self-confessed killing of 130 people was not the main concern of the law and its agents. This illustrates how the primacy of property-making, as well as the protection and maximisation of profit within the law, operate in tandem to perpetually put life and nature in jeopardy. This happens by breaking the bonds which bind planetary life forms in the community. The law's nature cannot thus be divorced from the violence it eventuates, it 'is force, law is absolute capitalisation, the hyperbolic appropriation of violence'.[9] So, I reiterate: can we really hope for human and planetary justice from the corpus of law that we have now, which has come to us by way of a history of producing racialised enslavement, exploitative colonisation, and neo-colonial racial capitalism?

Consequently, this judgment seeks to question the centrality of the current structure of Euro-modern law, as the sole source of succour for the harms of the Anthropocene. Therefore, it has as the focal point of its imaginary, the bottom of the waters into which the *Zong*'s souls were cast, what they met there, and what has since then joined them there. I lean into the literary devices I use (described below) and also to provide a comparative lens, by framing the judgment as a first-person account of a law student taken in their dreams to an underwater court to hear the petition. This changes our lens on the human disposition to abandon human bodies and other things/beings considered waste to the waters, and points to a global schema of disposability (underwritten by law) that relies on structures of human hierarchy in relation to each other and to the planet. The switched perspective dislodges this centring by asking us to question everything we consider settled about the law and the futures we can posit from this dislodging. What if nature is centred, and existence under the waters is placed in the driving seat of adjudication? What if the souls from the *Zong* found new life and new worlds under the sea – worlds that are being destroyed because humanity continues to do harm to the waters?

Continuing colonial plunder and pollution of the seas have led to the drastic collapse of rich and dynamic ocean life and the communities that exist in communion with them.[10] Responding to the coloniality of this plunder, this judgment relies broadly on theories of decolonisation, postcolonialism, and racial capitalism for its foundations. Decolonisation as a theory and praxis should be considered forward-looking, a means to think through the future from a present in which humanity and Earth are in jeopardy. This is because an anticolonial/decolonial position responds to the destruction occasioned

8 Robert M Cover, 'Violence and the Word' (1986) 95(8) *The Yale Law Journal* 1601, 1601.

9 Jacques Derrida, *The Death Penalty, Volume II*, tr Elizabeth Rottenberg (University of Chicago Press, 2017) 72.

10 Tero Mustonen et al, 'Who Is the Ocean? Preface to the Future Seas 2030 Special Issue' (2022) 32(1) *Reviews in Fish Biology and Fisheries* 9, 11.

by colonial-capitalism and seeks to imagine ways of being, doing, and living in the world that break from this harm and inaugurate new ways of being in this world that prioritise flourishing life. Therefore, decolonisation can be described as a set of context-dependent strategies designed to unseat colonial modalities of life, however and wherever they may be found.[11] These strategies, placed as they are in opposition to Euro-modernity and its structures, often uncover the inherent conflict between the law's requirement for formality, stability, and predictability in comparison to Earth's, humanity's, and nature's unpredictability.

Therefore, in this judgment I want to explore the use of other ways of thinking to imagine forms of jurisprudence better suited to this volatility. Science fiction may provide such a space.[12] Within this, I specifically rely on 'Black/African science fiction', an umbrella term that describes a literary genre that has 'directly confronted questions of structural racial injustice and the reproduction of race as a global organising principle in space and time'.[13] I also rely on Indigenous jurisprudences that embody alternative ways of thinking about what nature is and humanity's relationship to it. These jurisprudences often adduce a personality to nature. For example, the Indigenous peoples of Colombia often humanise Earth as a 'mother' character.[14] Further, in 2017, New Zealand accepted an Indigenous Māori proposal and recognised the Whanganui River as a legal entity, with rights and responsibilities similar to a human person.[15] This move takes a step forward to disrupt the anthropocentricity of law.

However, just like the adoption of Indigenous principles into constitutions in South America, granting nature personality rights within law comes with inherent limitations. Chief among these is the fact that many Indigenous jurisprudences are fundamentally diametrically distinct from Euro-modern law. Therefore, one could argue that granting Whanganui River legal personality grants the river rights to be a Euro-modern person, rather than fully actualising the Indigenous idea of what a person is. I want to trouble this tendency in this judgment, by exploring how far we can, practically and imaginatively, dislodge Euro-modern personality as the standard around which

11 Foluke Adebisi, *Decolonisation and Legal Knowledge: Reflections on Power and Possibility* (Bristol University Press, 2023) 21–22.

12 Alex Green, Mitchell Travis, and Kieran Tranter, 'Jurisprudence of the Future' (2022) 4(2) *Law, Technology and Humans* 1, 2.

13 Foluke Adebisi, 'Black/African Science Fiction and the Quest for Racial Justice Through Legal Knowledge: How Can We Unsettle Euro-Modern Time and Temporality in Our Teaching?' (2022) 4(2) *Law, Technology and Humans* 24, 28.

14 Bosco Lliso et al, 'Payments for Ecosystem Services or Collective Stewardship of Mother Earth? Applying Deliberative Valuation in an Indigenous Community in Colombia' (2019) 169 *Ecological Economics* 106499, 106508.

15 Mustonen et al (n 10) 9.

legal knowledge revolves. Therefore, this judgment relies on the aforementioned speculative fiction approach to push the boundaries of the limits of the form and content of law to re-examine the question: 'what does it mean to be a person?' This is because, to question the presumptions upon which the world-making of Euro-modern legal knowledge is based, the tools adopted must have the capacity to bend the rules of the law's assumed 'naturals' with impunity. We must do more than bend nature to fit with the demands of law. To survive at all, we may have to completely abandon our idea of law. Speculative fiction provides us with flexible tools to do this, giving us the 'potential to ... imagine alternative worlds' and 'imagine radical alternatives'.[16] Therefore, the scope it provides for creative re-envisioning can be 'a stage on which we imagine humanity's fate, and ... develop the terminology to describe our future'.[17] The devices and vocabularies of speculative fiction allow us to depart from pre-ordained registers to the extent of our imagination.

'To the extent of our imagination' is an important caveat here. In the building of these worlds, I am restricted by the fact that I am writing a 'judgment' with a specified word count. I am also restricted by my own training in the law, the language in which I write and the structure of the story I am telling. Nevertheless, despite these limitations, I still think this a useful exercise to begin to reimagine the nature of the law. A way in which we may begin to unsettle our acceptance of all the structures that underpin our realities through its gaze.

Based on these foundations and through these conceptual and narrative devices, the judgment seeks to push several boundaries – in layers of thought. Firstly, by making the sea the entity that hears the petition I hope to dislodge the dominance of the privileged human as the figure for whom the law is made and who has the right to determine what amounts to harm and who/what can claim to be harmed. As with the *Zong* case, where the court was more concerned with the results and boundaries of recovery of economic loss suffered by the traders in human beings, most environmental protection laws focus on the resulting relative loss to the quality of human life, rather than to nature. The practice of throwing living and dead Africans overboard slave ships was so commonplace during the time of the *Zong*, that the practice altered the behaviour and ecosystem of life under the oceans. Sharks changed their usual migratory patterns to follow these ships, in expectation

16 Carl Death, 'Climate Fiction, Climate Theory: Decolonising Imaginations of Global Futures' (2022) 50(2) *Millennium: Journal of International Studies* 430, 444.

17 Patricia Melzer, *Alien Constructions: Science Fiction and Feminist Thought* (University of Texas Press, 2010) 12.

of a feast of human flesh.[18] In other words, irrespective of what the effect is on humanity, we also need to be asking what the waters, as such, lose by being the receptacle into which humanity disposes its unwanted trash, oil, and even people. This consideration should lead us to question which bodies of laws should govern such a petition. Doing this throws many areas of law into sharp relief and uncovers how Euro-modern law, rather than protecting or allowing for flourishing, often governs, restricts, and dominates that for which it legislates. For instance, the Law of the Sea in practice operates less as law of the sea (protecting and nurturing the interests of the sea) but more as a means by which the man on land dominates and exploits the sea.[19] To examine the layers inherent in responding to these questions, I have framed the judgment as a conversation between representatives of various groups who consider themselves the right person to speak in this case. Finally, in considering what the verdict and sentence can or should be, I am attempting to dislodge our understandings of harm and punishment as viewed through the lens of anthropocentric Euro-modern law. What lies beyond 'guilty' and 'not guilty'? Can a sentence move way beyond punishment and deterrence? What does punishment mean from the perspective of the sea? What role can vengeance play here? Should justice always seek restoration, reparation, and transformation of relationships? What can this judgment tell (or not tell) law thinkers about how to get to our future? Is there a future to be had from where we are now?

Judgment

My name is Ìléríayọ̀ Ejiroghene Máabayọ̀jé[20] and something odd happened to me one night. I had been studying for my environmental law exams and I found myself in a very frustrated state of mind. I had just got off a video call with my parents, and the news they shared with me depressed me a little. My parents had left Nigeria just before I was born. One of their reasons for leaving was because of how difficult it was becoming to find affordable land and housing in Lagos. On my father's side, I am descended from fisher people who made their living in the creeks and lagoons of Lagos. However, industrial development and rising ocean waters are drowning out these traditional professions.

We could have moved out of Lagos to live in the Niger Delta region of Nigeria where my mother is from. But all the creeks are overflowing with

18 Marcus Rediker, 'History from Below the Water Line: Sharks and the Atlantic Slave Trade' (2008) 5(2) *Atlantic Studies* 285, 286-287.

19 I want to thank and acknowledge Michelle Lim for our discussions of this idea.

20 This name in Yoruba Urhobo and Yoruba means A-promise-of-joy Praise-God Let-us-not-spoil-our-joy.

waste from crude oil exploration and the air is filled with black toxic fumes. My grandmother (my mother's mother) lives in the creeks that flow into the Atlantic Ocean. She has always refused to leave her land. According to her, whether by pollution or by the passage of time, she wants to go to sleep from the troubles of this world, in the same place where all her ancestors are buried. She does not want to be separated from them on the other side of existence. I had asked my parents to implore my aging grandmother to come and live with us in the United Kingdom – hoping that her failing health may change her mind. Her lungs are damaged from the fumes. When I speak to her, I can hear the difficulty with which she takes each painful breath. She is being slowly suffocated by oil exploration. But she had once again refused to leave. This was the news that so depressed me that night.

After this video call, I had to turn my mind to my environmental law text. To be honest, I had been finding the reading a bit frustrating. I had chosen to study law, precisely to find a way to help respond to the challenges that climate change and environmental damage posed to the fisher people on the shores of Lagos as well as the people living in creeks in the Niger Delta. The effect of crude oil exploration has been devastating. The mangrove trees are dying. The fish are dead. The water is undrinkable. Children are born with debilitating health conditions. Cancer is rife. In addition to this, it is rumoured that on these shores, from Lagos to the Niger Delta, after the 1807 abolition of the trade in enslaved Africans, slave traders used to throw their captives overboard if they caught sight of the ships of the West Africa abolition Squadron. They preferred wholesale murder to risking prosecution. It is said that souls lost to the seas thus still cry out for a justice that can never come. If the cacophonic drilling of oil and kaleidoscopic flaring of gas ever stopped, you may hear their mournful wails drift up from the deep.

Through all this, the coastline of Nigeria has been bombarded by violent death, waves of pollution, over-exploitation of the ocean's resources, and brutal disruptions to ways of living. For us, environmental emergency is not merely imminent. It is not something we think may happen in the future. It is happening now. To my grandmother, to my family, and through them to me. I had hoped that the answers to these challenges would be found in my legal studies, but the reading I was doing promised me nothing. There was no hope of saving the creeks, the coastlines … or my grandmother. These were the dark thoughts milling around my brain when I fell asleep at my books. And then I had a very strange and vivid dream.

I felt like I was taken below Nigeria's shoreline. As I passed beneath it, I could smell its air, hear its sounds, almost taste its warmth. I passed below the waters for what felt like a million minutes or none. Suddenly a whole different vista opened up to me. There was a great gathering on the sea floor of all types of life – more varieties than I had ever seen in any textbook, all looking more sentient than nature documentaries ever conveyed. When I asked the sea urchin that was close by, I was told that the translucent forms that I could not identify were the souls of humans lost to the waters. As I sat and watched, in the middle of the gathering, a large flame got brighter and brighter. The sea urchin informed me that this was the physical form of Àwọ̀n Olókun Àgbáyé – the keeper of the flood. I listened in awe as a voice rang out from the centre of the flame. This is what it said:

> For 500 years I have been silent and half asleep, listening to the Great Debate between the beings of the deep. They talked about how the humans who live above us, have thrown a great many things into our world, including the souls of their own, altering our diets and customs. Additionally, their varied debris has torn the limbs of many in this world. The humans also pierce the surface of Earth, and the sludge released suffocates and poisons those of us who dwell below. There is enough to eat, but they take too many of us for food, reducing our community and forcing many to seek deeper depths in which to live. The humans send their noisy machines to alarm and offend. Under this onslaught, many lineages of seafolk have completely disappeared. And still, they throw more things into our world – bodies, plastics, sand, oil. In response, the Great Debate has raged, as the inhabitants of the deep try to decide what to do in the face of this Great Desolation. I have listened, and now I will talk.
>
> There have been two main sides in this debate. In addition to this, I will also address the sporadic interjections of Zuan Chabotto, a voyager lost at sea,[21] who has now been hidden in the inner recesses of the deep to spare us his divisiveness. On the one side of the Great Debate, are the souls of the humans who have found their ways to us. They are represented by No Name (she/her), a woman who was cast into this world with her newborn strapped to her chest. On behalf of the human souls, she asks that we respond to and judge the human ways of living that have caused some of the harms to this plane, many of which I have already mentioned. They ask for recompense for the violence that cast them out of their world and is making this one unliveable for their souls. For them, the fact that they were cast into the sea is an injustice that arises from human need for profit, that continues to seek things to be turned into private property for that profit. These human souls have come to us through many means, through being thrown into the sea at a time when the humans traded each other like tools, through death in violent conflict, at times when the humans sought to conquer new lands,

21 Zuan Chabotto (also known as Giovanni Caboto or John Cabot) was a 15th century Italian explorer. He received funding from the British Crown to embark on exploratory voyages that departed from Bristol to the Americas – most notably to North America. He is believed to have been lost at sea. In this judgment, Zuan Chabotto is an avatar of both the man himself, and the practice of discovery and the attendant logics of dispossession that follow.

and also by perishing unsaved and unvalued when trying to cross the plane above without the authority for movement that their restrictive law requires. (Imagine needing my authority to roam the seas!) No Name and her representees have asked that their petition should be heard by a judge, who will give a verdict and pronounce judgement – a punitive sentence if humanity is found guilty. It is out of respect for No Name's grief that I have agreed to set out my response to the Great Debate in this way, as if it were a human judgment.

On the other side of the debate, representing the lifeforms of the sea, is one the oldest beings in this world, Eternity 56 (they/them), an immortal jellyfish. To a certain extent, Eternity 56 agrees with No Name. They accept that the hierarchical profit-seeking of humanity is a planetary nemesis and, if uninterrupted, will eventually bring ruin to the whole planet. While Eternity 56 agrees with No Name about the cause of the harm that she and the other souls have suffered, they argue that the very form of the remedies sought by No Name will not actually heal the rift, but merely extend the time in which humanity will continue to do the harm it does. Eternity 56 argues that there is no law, or structure, that comes from the world above that is sufficient to stay the harm or remedy it. In fact, that there is no human law at present that prescribes that the current harm should be stopped ... merely mitigated. The dominant forms of justice concern themselves with the present, ignoring the contingency of the past and the future. Eternity 56 is of the opinion that the nature of the recognised law in the world above is unable to respond to our needs here on this plane.

I want to mention, at this juncture, Zuan Chabotto's interjections to the Great Debate which included such phrases, as 'you all have no standing before the law', 'it is all in the past, move on, get over it', 'the rule of law is paramount', 'mankind was tasked to be fruitful and have dominion over all Earth'. As these interjections added nothing to the debate but rancour and dismissal of genuine trauma, despite attempts to engage with him in good faith, he was locked away. However, shorn of their divisiveness, Zuan Chabotto's arguments can be summarised thus: the law was and is created by humans to ensure the smooth running of flows of human enterprise. Nevertheless, what Chabotto and human law do not address adequately is how to deal with the harmful effects of maintaining these flows, on humans, on life and on the planet. Therefore, it is imperative that I take a wholistic approach in responding to the questions raised by the debaters.

The first question that the debaters were concerned with was the format in which a response to their petition would be heard. We have chosen to agree with No Name and structure our response in a manner as similar to an Earth court as possible. No Name also asked us to construct our court and judgment along the principles of learning, repair, and growth that the people she comes from follow. Eternity 56 stated that these principles are already embedded into our ways of living and, therefore, there is no need to make those requests as an addition. I agree with Eternity 56, but see no detriment in allowing No Name's requests both on the format of the court and overtly including these principles. When the survival of our shared habitat is in question, we must agree that shared understanding is vital for us to chart paths forward together, as confrontation and inquisition cannot build community.

The next question for concern was which laws have been broken, where they are sourced from and what gives them authority as law. In other words: 'what is law for these waters?' No Name cited her ancestors' traditions as well as contemporary practices as the source of her understanding of law. These traditions rely on principles of mutual respect, communal support, collective development, and care for nature. Chabotto argued that these principles were subjective and nebulous and therefore not useful for certain justice. Eternity 56 claimed that the laws that they believe in originate where they themselves come from, from the foundations of Earth. This source of law could not be verified, argued Chabotto. As Àwọ́n Olókun Àgbáyé, I exist beyond the beginnings of even Eternity 56. All life on Earth has sought to make laws to guide their co-existence on Earth. This is natural and expected. One of the dangers that I have witnessed from this, however, is a tendency for each group that makes law to deify themselves within those laws. Consequently, for example, when the law comes from those humans who have privileged access to make the law, the law they make often cannot properly protect the life and planet that has no access to law-making. This inadequacy goes beyond the content of the laws, but also impacts how the law works, what things mean in law and the structures from which law receives authority. As such, my response to this question is to look at the harm and who has been harmed and put them at the centre of our definition of law. The man on Earth is inviting perdition, but we here in the waters are already living it. No law on the surface of Earth puts us at its centre. So, we must ask ourselves, if this 'law' does not recognise us as equals, does not protect us equally, is it law for us? Do we recognise it? No, we do not. The law we recognise is the law of the sea. Our law, that gives us the authority to protect ourselves – by any means possible. For us harm is a communal concern. As No Name suggests, harm is anything that breaks the bond of community – this includes physical injuries but also social injuries. A harm to one is also a harm to all. Eternity 56 thinks we should consider in our definition of harm how different lifeforms experience time differently. I agree with both of them. Harm is shared, across time and space, when our bonds are broken, before they are broken, after they are broken. We have been drafted into the present continuous of colonial-capitalism time. Not just our bodies have been harmed, but our temporalities. So, our law is defined by our community, our needs, desires and how we have been harmed, across space-time.

So, to my judgment. The next chapter in this Great Debate. Humanity has visited violence upon the deep and its inhabitants. That is a question of both fact and law – their fact and our law. However, despite No Name's request that this finding result in a 'guilty verdict,' I am unable to phrase my conclusions in that way. Humanity is not just guilty under the law; it is ***responsible*** for the harm. As such, there is, to my mind, significant difference between imposing blame and external statements seeking remorse on the one hand and achieving accountability for cessation and repair on the other. For us, justice means an end to the harm and repair of the harm already wrought, not just an acknowledgement of guilt for it. A guilty verdict is not justice if it is directed only to those that stand accused. Justice is produced as a direct result of both the actions we undertake and those we are subject to. What does humanity deserve? What does No Name deserve? What does Eternity 56? Or even Zuan Chabotto? No Name believes that punishing mankind for her plight will bring her solace. But what does she ***deserve***? Can we quantify her grief in monetary sums or years of confinement? I am afraid that the truth is, the law, human or ours, cannot give us what we truly deserve. No Name and her representees deserve to not spend the rest of their existence with us. Eternity 56 and the lifeforms of the sea deserve not to be plunged into the desolation of pollution and extinction. Even Zuan Chabotto deserves a chance to repair his ways and establish community where he has found himself. In this judgment our conception of justice needs to consider the possibility of establishing, repairing and transforming broken relations of community. Consequently, we need to go beyond an outcome that restores parties to the positions they were before the harm. Relations of community did not exist, hence the damage to our world here, and the casting out of No Name into it. The original conditions were already unjust. Justice, for us, means transforming relations on the planet such that the harm is no longer possible. A 'guilty' verdict in response to harm just creates more trauma and harm. But how do we invite humanity into community with us, when humanity only considers us to be commodities … that is, if they even consider us at all?

And so, I feel that for our law, and for this harm, there can be no difference between what No Name calls the verdict and the sentence. In this petition, we find that humanity has significantly harmed our world. For us, restoration may be our only chance of justice … may be our vengeance. If vengeance was our only goal, as Àwọn Olókun Àgbáyé, I would ask that the waters rise up and overwhelm Earth completely. That would bring an end to the harm and punish those who have caused it. However, this also punishes those who have done little to contribute to the harm, those whose lungs have been blackened by human activity.

> This is a response to harm that produces more harm. Such judgment also does not prevent another species of humanity from being established and doing the same all over again. So, for justice, to invite humanity into community, we appeal to a higher power, Àwọ́n Olórí Ọ̀run – the keeper of the galaxy.
>
> Àwọ́n Olórí Ọ̀run, we ask that you make humanity commune with us, make humanity partake of the feast they have set before us. We ask that each night, as the man on Earth goes to sleep, for his soul be taken into the deep. In his dreams, let him be cast into the watery abyss. Let him taste the sludge. Let the plastic clog his lungs. Let him experience abandonment to the waters. Let their waves close over his head. Each time he lays his head to sleep, let the screams of the souls jettisoned ring in his ears. Let him feel the terror they felt as they were bound and cast to the waters. Let him taste the waters of despair. If this yesterday was the price at which his today was bought, take him back to yesterday. Let him never leave it, as we never do. Let him know no rest until he stops this destruction, changes his laws, changes his ways of living, changes his world, and stops destroying the waters. Àwọ́n Olórí Ọ̀run, I beseech you. Àṣẹ.[22]

A great clamour arose from the gathered host and then I was awake to the sound of my alarm. Exam time.

22 Let it be.

Chapter 5

Swan by Her Litigation Representative Bella Donna of the Champions v Administrative Algorithmic Transformer and Minister for Immigration and Border Protection

André Dao

Commentary

Swan by her litigation representative Bella Donna of the Champions v Administrative Algorithmic Transformer and Minister for Immigration and Border Protection ('*Swan*') is set in a future world that takes Alexis Wright's 2013 novel, *The Swan Book*,[1] as its starting point. One hundred years in the future, 'the Mother Catastrophe of flood, fire, drought and blizzard' has changed the world – '[t]owns closed, cities were boarded up, communities abandoned, their governments collapsed'.[2] The ensuing climate change wars have forced climate refugees from the northern hemisphere to take to the oceans in search of liveable land. Some, such as Bella Donna of the Champions, arrive in northern Australia which is still in the grips of a never-ending Intervention.[3]

The Swan Book focuses on a teenage Aboriginal girl, Oblivion Ethylene, or Oblivia, who lives in a kind of detention centre for the local Aboriginal people – little more than a swamp around which the government has erected barbed wire. A rape survivor, Oblivia's mind has been invaded by a 'virus'. On one level, the novel follows Oblivia's attempts to regain sovereignty over her mind – to eradicate the invading virus.

On another level, Wright reckons with a society – indeed, with a planet – that has failed to take seriously humanity's relationship with land. As one reviewer notes, *The Swan Book* is:

> concerned with the entire Earth. It is about the planetary phenomenon of climate change and how it relates to northern hemisphere stories, to

1 Alexis Wright, *The Swan Book* (Giramondo, 2013).
2 Ibid 6.
3 The Intervention, implemented through Commonwealth legislation, encompassed controls and restrictions on First Nations people in the Northern Territory of Australia.

DOI: 10.4324/9781003389569-6

> Western cosmology, whose Christian scientific capitalist schema places 'man' above 'nature' as its namer and master. It considers how the heritage of wandering encoded in the stories of the West, from Homer and the *Bible* onward, might relate to the trespass of Europeans on Aboriginal land and therefore to Indigenous Australian sovereignty. *The Swan Book* writes an Indigenous cosmology – the deep and ancient knowledge of 'nature', of Country, held by the traditional custodians of the swamp – into this global story.[4]

In reckoning with this encounter between rival cosmologies, *The Swan Book* both is and is not an apocalyptic science-fiction novel. That is, Wright works with many of the familiar tropes of such novels: the future setting, spectacular natural catastrophes, and an authoritarian government. And yet, the logic of *The Swan Book* is distinctly non-ruptural. As Honni van Rijswijk puts it, the harms depicted in the novel are 'continuous with the everyday: continuous with racism, assimilation, gender violence, and state law. We are not located in a world of eviscerated institutions and frameworks'.[5] For instance, the authoritarian government which locks up Oblivia's 'swamp people' is not a flat caricature of a totalitarian regime but an Australian Republic. Indeed, this is a future in which an Australian Republic has been forced by a World Court to sign a treaty with its First Nations peoples. A future in which an Aboriginal man, Warren Finch – Oblivia's husband-to-be – is first Deputy President, and then President, of Australia, and 'special old-law rapporteur to the world's highest authority of elders for ancient laws, ancient scriptures, and modern Indigenous law-making'.[6] In other words, this is a future in which the apocalypse has arrived not in the absence of such political and legal institutions, but in spite of – perhaps even *because of* – their continued existence.

Building on Wright's world, I imagine in this judgment that the future Australian Republic has adopted big data techniques to replace voting – effectively using machine learning algorithms to determine the 'will of the people'. Because such algorithms can be constantly deployed, rather than periodic elections leading to broad 'mandates' for political action, every decision – both macro policy decisions and micro administrative ones – is made algorithmically, on the basis that the algorithm can determine what the 'Australian people' would decide in each case.

In *Swan*, Wright's character challenges an algorithmically determined decision to refuse a protection visa to a swan from her climate-ravaged home

4 Jane Gleeson-White, 'Going Viral: The Swan Book by Alexis Wright', *Sydney Review of Books* (Book Review, 23 August 2013) <https://sydneyreviewofbooks.com/review/going-viral>.

5 Honni van Rijswijk, 'Encountering Law's Harm through Literary Critique: An Anti-Elegy of Land and Sovereignty' (2015) 27(2) *Law & Literature* 237, 242.

6 Wright (n 1) 194.

country. The judgment is a judgment of the Full Bench of the Federal Court of the Australian Republic. In the majority decision, Bella Donna's case is treated as an administrative law challenge to the algorithmic decision. The majority affirms the validity of the algorithm's decision and dismisses Bella Donna's case, acknowledging in their judgment the superiority of data-driven decision-making over the archaic, inaccurate, and biased forms of human decision-making that prevailed in the early 21st century. Bella Donna's stories are dismissed as 'mere mythology'. However, the majority does affirm that the applicant swan can be understood to be a 'person' for the purposes of receiving a protection visa under the *Migration Act 2050* (Cth). The discussion on personhood in the judgment responds to calls to recuperate the juristic technique of the law of persons as an art or craft liberated from metaphysical images of the 'natural' person.[7]

The third judgment, a dissent, takes a different approach – there, the judge tries to take seriously Bella Donna's wild stories about swans. This dissent thus becomes an attempt to show the possibilities of law taking seriously Wright's claims to be writing a 'self-governing literature'[8] – that is, what it might mean for law to recognise the deep connection between land and stories, and how such a recognition might be vital to crafting law for or in the Anthropocene. The dissenting judgment draws on three texts by Wright. The first is part of a story told by Bella Donna of the Champions about swans in *The Swan Book* and is reproduced in the judgment as an example of the narratives told by Bella Donna as the applicant's litigation representative before the Court. The second and third texts are essays which are reproduced in the dissenting judgment as if they had themselves been judgments. In doing so, I imagine a world in which such writings have become law, that is, 'coated with a veneer which changes the very nature of that which it covers up'.[9]

I leave the final word to Wright herself. Here, the First Nations people among whom Bella Donna of the Champions lives respond to her stories of climate-induced displacement:

> *Then, without country, imagine that!* Imagining? Can't imagine. For country never leaves its people. This was what the swamp people claimed while seeing some sort of country in her, and dragging it out of her by

7 See, eg, Edward Mussawir and Connal Parsley, 'The Law of Persons Today: At the Margins of Jurisprudence' (2017) 11(1) *Law and Humanities* 44.

8 See Alexis Wright, 'A Self-Governing Literature', *Meanjin* (Winter 2020) <https://meanjin.com.au/essays/a-self-governing-literature>.

9 Robert M Cover, 'The Folktales of Justice: Tales of Jurisdiction' (1984–1985) 14 *Capital University Law Review* 179, 181. For Cover, law is 'a bridge in normative space connecting … "the world-that-is" … with our projections of alternative "world-that-might-be"' (at 181). In the present case, Wright's writing can be understood as jurisprudential insofar as they represent one way in which a group of people might get from one world to another.

listening, like scavenging rubbish out of a bin, rubbish lying everywhere – hard to imagine where it all came from.[10]

Judgment

Court: Full Bench of the Federal Court of the Australian Republic
Date: 30 August 2118
Judges: Nguyen J, Eliot J, and Donaldson J

THE COURT ORDERS THAT:

1. The application is dismissed.
2. The applicant pay the respondents' costs, as agreed or assessed.

Nguyen J

Introduction

1. Before the Court is an application for review of a decision of the Administrative Algorithmic Transformer (AAT), made on 15 February 2117, refusing to grant a protection visa to A Swan (the applicant). The application is brought by Swan's litigation representative, Bella Donna of the Champions. The visa was refused under s 501(56B) of the *Migration Act 2050* (Cth) (the Act). Pursuant to s 501(56B)(4) of the Act, decisions made by the AAT under s 501(56B) are deemed to have been made personally by the Minister for Immigration and Border Protection.
2. The proceedings are within the Court's original jurisdiction.

Statutory Provisions

3. The relevant statutory provision, s 501(56B), was in the following terms:

Section 501(56B): Algorithmic Decision-Making in Visa Refusals

(1) The Administrative Algorithmic Transformer may, in accordance with the criteria established by the Minister, refuse to grant a visa to any non-citizen.
(2) The criteria established by the Minister for the use of the Administrative Algorithmic Transformer in visa refusal decisions shall include, but not be limited to, the considerations set out in section 501(56B)(3) of this Act.
(3) The Administrative Algorithmic Transformer shall consider:

10 Wright (n 1) 26.

(i) the likelihood that the non-citizen will engage in conduct harmful to the security, defence or international relations of Australia;
(ii) the likelihood that the non-citizen will become involved in criminal conduct or other conduct of concern to the Australian community;
(iii) whether the granting of a visa to the non-citizen would, for any other reason, contravene the will of the people of the Republic of Australia as modelled by the Administrative Algorithmic Transformer under the procedure set out in the *Administrative Algorithmic Transformer Act 2097* (Cth).

(4) In making a decision to refuse a visa under this section, the Administrative Algorithmic Transformer shall have regard to:
(i) the information provided in the visa application, and any past visa applications; and
(ii) all relevant information and data regarding the non-citizen's character, conduct, and associations including any relevant intelligence or security assessments.

(5) The decision of the Administrative Algorithmic Transformer under this section is deemed to be a decision of the Minister for the purposes of this Act and shall have the same effect as if it were a decision made by the Minister in accordance with section 501(45A) of this Act.

(6) The Minister may, at any time, review and amend or reverse a decision made by the Administrative Algorithmic Transformer under this section.

The Decision of the AAT

4. The AAT set out the background of the application at [1]. The applicant is a *Cygnus olor* of the Anatidae family, commonly known as a swan or mute swan. On 12 May 2116 at 10:15am, the applicant lodged, through its litigation representative, an application for a protection visa. On 12 May 2116 at 10:16am, the AAT refused the application. The applicant requested reasons for the refusal on 12 May 2116 at 10:25am, and the AAT delivered its reasons on 12 May 2116 at 10:26am.
5. The AAT stated, at [2], that as per *Smith-Jones v Minister for Immigration and Border Protection* [2088] HCAR 112, there is a presumption that any person from Zone GB is a climate-displaced person for the purposes of s 5HA of the Act. The AAT accepted the evidence provided by the applicant that it was born in Zone GB and that it had resided there since

its birth. The AAT also accepted, at [3], that the applicant was a person within the meaning of the Act.

6. The AAT then considered the criteria set out in s 501(56B)(3) of the Act for refusing the applicant's visa. At [6], the AAT found that there was a low likelihood that the applicant would engage in conduct harmful to the security, defence, or international relations of Australia. The AAT also found, at [7], that there was a low likelihood that the applicant would engage in criminal conduct or other conduct of concern to the Australian community.
7. At [8], the AAT found that granting the applicant a protection visa would contravene the will of the people of the Republic of Australia. The AAT set out the basis for this finding at [9], where it explained that according to its modelling of citizens' preferences, 68 per cent of Australian citizens would oppose granting the applicant a visa. The AAT stated that this finding was the basis for its decision to refuse the applicant's visa.

The Parties' Submissions

8. The applicant was not legally represented. At all times before the AAT and in this Court the applicant has been represented by Ms Bella Donna of the Champions. The applicant, through Ms Champions, did not formulate the legal basis of the present application for review in technical legal terms.
9. At times during the proceedings, the applicant appeared to be seeking a review of the merits of the AAT's decision. On judicial review, it is not for the Court to enter into the merits of the AAT's decision: the Court's role is limited to considering the lawfulness of that decision.
10. Ms Champions' lengthy oral submissions before this Court largely concerned narratives about *Cygni olor* throughout history. Many of the same narratives appeared in the original application for a protection visa. Insofar as the applicant could be understood to be making an application for judicial review of the AAT's decision, the applicant appeared to argue that in failing to take account of these narratives, the AAT made an error of jurisdiction by failing to take account of a relevant consideration.
11. Firstly, the respondent Minister submitted that the narratives told by Ms Champions could not be understood to be relevant to the decision to refuse the applicant a protection visa. The Minister submitted that the narratives were vague and general and shed no further light on the applicant's circumstances. Furthermore, submitted by the Minister, the narratives had no bearing on the criteria set out in s 501(56B)(3) of the Act. In particular, the Minister submitted that the narratives had no bearing on whether the grant of a protection visa to the applicant

would contravene the will of the people of the Republic of Australia per s 501(56B)(3)(iii), being the specific ground for the AAT's refusal.

12. Secondly, the Minister submitted that the applicant is not a person for the purposes of s 5HA of the Act. The Minister submitted that *Ponthus v Minister for Immigration and Border Protection* [2101] HCAR 48, in which a *Bos taurus*, or cattle, was held to be a climate-displaced person, should be distinguished, on the basis that the High Court did not intend to establish a general rule about the personhood of nonhuman life. The Minister submitted that the distinguishing features of *Ponthus* included evidence accepted by the High Court about the relative consciousness of a *Bos taurus*. No such evidence was presented in the present case.

Analysis

13. It is convenient to first deal with the respondent Minister's argument that the applicant is not a person for the purposes of s 5HA of the Act. The Minister's argument is founded on a misreading of *Ponthus*. True enough, the High Court did not, in that case, establish a general rule in the positive sense. However, the High Court did establish that the juridical person is not to be equated with the biological or metaphysical human being.[11] As such, personhood is not to be determined by reference to perceived human attributes, such as rationality or consciousness. Instead, the High Court held that the juridical person is an artifice, akin to a mask, 'a jurisprudential tool to be used to solve specific jurisprudential problems'.[12] As such, personhood must be determined by reference to the specific context of the legislation or case law in which the particular category of person concerned appears.
14. In the present case, the 'climate-displaced person', as it appears in the Act, is an artifice to be used to save life from the ravages of the collapsed climate. In *Ponthus*, recognition of the *Bos taurus* as a climate-displaced person saved the once common cattle, after the complete loss of the cattle population on this continent. The *Bos taurus* was not recognised as a person because of any of its attributes, human-like or otherwise. Rather, recognition of personhood solved the difficult problem of how to grant a sufficient level of protection to life from beyond the jurisdiction of the Republic. There is likewise no reason to presume that the applicant could not be recognised as a climate-displaced person to achieve an analogous goal.
15. I turn now to the applicant's (presumed) argument that the AAT improperly exercised its power to refuse to grant a protection visa by failing to

11 *Ponthus v Minister for Immigration and Border Protection* [2101] HCAR 48 [567].
12 Ibid [641].

have regard for a relevant consideration, being certain narratives about *Cygnus olor*. Regrettably, though the narratives told by Ms Champions were compelling, they cannot be understood as being relevant to the AAT's exercise of power. The Act is clear that the decision to refuse a visa is to be concerned with the future conduct of an applicant, and the will of the people of the Republic of Australia in relation to granting the applicant a visa. Mere mythology about swans has no bearing on either the future conduct of this specific applicant, or on the people's attitudes towards this specific applicant.

16. Furthermore, the great innovation of the AAT, as far as legal science is concerned, is that such determinations about future conduct and public opinion are no longer subject to the caprice of human decision-making. As the High Court held in *Khan v Administrative Algorithmic Transformer and Minister for Immigration and Border Protection* [2111] HCAR 328, algorithmic determinations of the probability of an event, or the distribution of an opinion, have long been of sufficient precision and accuracy as to be able to form the basis of rational administrative decision-making. Further, the nature of such algorithmic decision-making militates against fine-grained analysis by human reviewers, judicial or otherwise.[13] As has long been established by data scientists, the greater accuracy of algorithmic determinations compared to human determinations arises from the fact that an algorithmic model will find patterns in vast stores of data that are illegible to a human observer. Thus, as the High Court held in *Khan*, it is not for the courts to 'pierce the veil of the algorithmic determinations' by inquiring, for instance, whether the algorithm had regard to what might otherwise have been an irrelevant consideration, such as an applicant's shoe size, in determining the probability of the applicant's future conduct.[14] Likewise, it is not for this court to pierce the algorithmic veil by inquiring into whether new information, including mythologies, would alter the AAT's determination.

17. According to the AAT's determination, a majority of the people of the Republic of Australia disapprove of granting the applicant a protection visa. In *Ponthus*, the AAT made the opposite determination, finding that a majority of the people of the Republic of Australia approved of granting the applicant, a *Bos taurus*, a protection visa. Subsequent academic commentary has speculated about the cause of that approval, with some notable scholarship arguing that economic factors, namely, the capacity for the *Bos taurus* to donate the tissue required for producing

13 *Khan v Administrative Algorithmic Transformer and Minister for Immigration and Border Protection* [2111] HCAR 328 [54].

14 Ibid [112].

processed beef, were paramount.[15] Future scholars may wish to make similar inquiries about the causes of the AAT's determination in respect of the applicant in the present case. But such speculation has no bearing on the lawfulness of the AAT's decision to refuse to grant the applicant a protection visa. That decision was made on the basis of the fact that the people of the Republic of Australia would disapprove of granting the applicant a protection visa. That fact is sufficient ground for the decision, which was therefore lawfully made.

Orders

18. The orders I proposed are that the application for judicial review be dismissed, with costs.

Eliot J

19. I agree with the orders proposed by Justice Nguyen and with the reasons giving rise to those orders.

Donaldson J

20. The facts and circumstances of this appeal are set out in the reasons for judgment of Nguyen J.
21. I agree with Nguyen J that the applicant is a person for the purposes of s 5HA of the Act.
22. The only remaining substantive issue is what this Court is to make of Ms Champions' stories about swans. Ms Champions' stories were voluminous. It is not possible to set them out in full here. Nor, given the vast diversity of the stories, is it possible to set out even representative samples. Nevertheless, I reproduce one such story below for illustrative purposes:

> I have seen swans all my life. I have watched them in many different countries myself. Some of them have big wings like the Trumpeter Swan of North America, and when the dust smudges the fresh breath of these guardian angels, they navigate through the never-ending dust storms by correcting their bearings and flying higher in the sky, from where they guide like Whistling Swans whistling softly to each other, then beating their wings harder they fly away. I know because I am the storyteller of the swans.

15 See Alexander Crabb, 'Let the Right (b)o(vi)ne In: The Political Economy of the *Ponthus* Decision' (2113) 34(1) *Algorithmic Administrative Law Review*.

> Where I came from, whole herds of deer were left standing like statues of yellow ice while blizzards stormed. Mute Swans sheltered in ice-covered reeds. The rich people were flying off in armadas of planes like packs of migratory birds. The poverty people like myself had to walk herdlike, cursed from one border to another through foreign lands and seas.
>
> You know girl? I owe the fact that I am alive today to a swan. But anyhow, my story of luck is only a part of the concinnity of dead stories tossed by the sides of roads and fathering dust. In time, the mutterings of millions will be heard in the dirt … I am only telling you my story about swans.[16]

23. The respondent submitted that Ms Champions' stories could have no bearing on the AAT's decision to refuse the applicant a protection visa. That argument is founded, on my view, on an erroneous understanding of the function of storytelling in law. The relationship between narrative and law was set out by Wright J in *Phantom v City of Desperance* [2068] HCAR 1:

> I have been trying to build a self-governing literary landscape from what I have learnt from our ancient library contained in the land, from understanding how we are being continuously shaped by story, and from our knowledge that we have always governed ourselves through the ancestral stories that form our own constitution of sacred laws for this country.[17]

24. As is well known, Wright J's dissenting judgment in *Phantom* has, in subsequent decades, become a classic reference in First Nations jurisprudence. It is now considered axiomatic that there is no bright line between First Nations storytelling and First Nations law.[18]
25. In a less well-known judgment, Wright J addressed the connection between storytelling and the collapsing climate.[19] In a felicitous coincidence, that judgment also concerned swans, which were at the time still widely found on this continent:

> At the time of writing … there seemed to be marked changes in our arid zone environment. Many people will tell you stories when you ask, and I heard stories about how swans were now seen more regularly in the drier northern inland areas of Australia. Swans were often seen in sheets of shallow and fast-drying waters lying over country

16 Wright (n 1) 17.
17 Wright (n 8).
18 *Shell Energy Australia v Commonwealth of Australia* [2091] HCAR 256, [45]-[52].
19 *Minister for the Environment v Rio Tinto* [2061] FCAR 87, [22]-[23].

after unseasonal heavy rains from the changing weather patterns, and the climate had flipped – wetter in the centre, dryer in the south.

The questions in my mind were, what happens when there is no story for swans in the places where they were now migrating? What happens when a swan enters the big stories of brolga country? What did it mean to have no cultural story that fixed you to a place either as a person, a swan or any other animal species? How do you take the ancient spiritual law stories that belonged to you when you move to another place? Then what happens if you mix the stories of the Dreaming? The easy answer is that you create new stories, new song, new Dreaming. Perhaps, and perhaps there are other consequences.[20]

26. Today, one can find no *Cygni olor* on this continent. Was that fact taken into consideration by the AAT when it determined that the Australian people would not wish to grant the applicant a protection visa? If so, how? The AAT is, of course, a black box system. It is not possible for this Court, or any human analyst, to retrospectively ask the AAT how it made its determination. Nevertheless, as Wright J's jurisprudence has shown, the land, and stories of and from that 'ancient library', form part of the corpus of the law of this nation. The process of climate change identified by Wright J in *Rio Tinto* is now well-advanced. Ms Champions' stories have doubtless entered the ancient library. They inform the nation's understanding of the extinction of *Cygni olor* on this continent and therefore are relevant considerations for the AAT's determination of public opinion towards the applicant.
27. For the above reasons, the application should be allowed. The matter should be returned to the AAT, and Ms Champions' stories should be added to the AAT's training dataset.

20 Alexis Wright, 'A Journey in Writing Place', *Meanjin* (Winter 2019) <https://meanjin.com.au/essays/a-journey-in-writing-place>.

Part II

Intergenerational Justice

Chapter 6

The Doctrine of Quantum Entanglement

Kate Galloway

Commentary

At the dawn of 2020 while the bushfire crisis gripped Australia, there was an overwhelming sense that the public had been let down by government.[1] Intuitively it seemed that governments, State and Commonwealth, had been negligent in allowing the land to get to the state where it would burn so widely and destructively, ignoring former fire chiefs, slashing land management budgets, promoting fossil fuels, and failing on climate change mitigation. This sentiment continued unabated during the 2021 flooding in eastern Australia and again when flooding revisited many of those same communities in 2022. As with the bushfires, the effect of long-term failures to take flood mitigation measures was apparently intensified by stagnation in climate policy in Australia.

The compounding effects of climate change were, by mid-2022, being keenly felt by humans left in the wake of climate extremes buttressed one against the next, and by the more-than-human world through coastal erosion, swathes of severely burnt country, habitat destruction, and species devastation.

Globally, litigation had already become a means by which to hold governments accountable for climate change harms.[2] Australia embarked on this approach a little later, evidenced by the 2021 decision in *Sharma*.[3] The case succeeded at first instance in establishing a novel duty of care by the Commonwealth Minister for the Environment towards Australian children

1 Adapted from Kate Galloway, 'Analysis: Holding Government to Account', *Environmental Defenders Office* (Blog Post, 24 January 2020) <https://www.edo.org.au/2020/01/24/holding-govt-to-account>.

2 See, eg, *Pabai Pabai v Commonwealth of Australia* FCA VID622/2021; *Neubauer v Germany*, Bundesverfassungsgericht [German Constitutional Court], BVerfG, Order of the First Senate of 24 March 2021 – 1 BvR 2656/18; *Milieudefensie v Royal Dutch Shell plc* [2019] File 90046903; *Urgenda v Netherlands* [2015] HAZA C/09/00456689; *American Electric Power Co v Connecticut* 564 US 410 (2011).

3 *Sharma v Minister for the Environment* (2021) 248 LGERA 330 ('*Sharma*').

DOI: 10.4324/9781003389569-8

– a duty to protect them from physical and mental harm attendant on climate change through greenhouse gas emissions. It was overturned on appeal.[4]

This judgment, delivered in 2030, revisits the issue in the *Sharma Appeal* of the liability of the Commonwealth government and its Ministers for harms caused by extreme weather events, the intensity and frequency of which are attributable to climate change. The source of liability lies in decisions made by individual Ministers, as well as collective government decision-making while knowing of the likely consequences of those decisions.

The challenge for the law, as discussed in the *Sharma Appeal*, is that policy is 'inconsistent with a co-existent duty of care'.[5] Courts have traditionally been reluctant to determine matters affecting day-to-day affairs of government, in what are classified by the court itself as a matter of 'policy'. Following the *Sharma Appeal*, a novel duty of care could exist only beyond the stipulated considerations of the *Environmental Protection and Biodiversity Conservation Act 1999* (Cth)[6] ('*EPBC Act*') while also being more contained in its scope than that argued in *Sharma*.[7]

The threshold for the imposition of a novel duty of care is whether the Commonwealth owes a duty that is recognised in civil law terms. In examining this duty, questions of proximity arise: the spatial and temporal nexus between government and ministerial decisions, and harm. Justifying an extension of the application of tort law by way of a 'novel duty', contrary to the decision in the *Sharma Appeal*, would normally require a shift in context and possibly a change in community standards or expectations.[8]

In an example of the law's capacity to adapt to shifting contexts, *Donoghue v Stevenson*[9] established, through the tort of negligence, liability beyond the privity constraints of contract law. The case provided for changed economic and social conditions: the rise of manufacturing that removed the direct relationship between producer and consumer. Through the neighbour principle, a manufacturer would be liable for damage caused by a breach owed to 'their neighbour' where that damage was reasonably foreseeable.[10]

Despite freeing liability from the bonds of privity that directly link a promisor and promisee, determining who is one's neighbour, and causation, have become limited in circumstances where damage is globalised, and the causes of damage are distributed.[11] Where damage must arise from identified particularised conduct, the requisite apportionment of causation to that damage

4 *Minister for the Environment v Sharma* [2022] FCAFC 35 ('*Sharma Appeal*').
5 Ibid [15], [172] (Allsop CJ).
6 Ibid [92] (Allsop CJ).
7 The 'indeterminacy' argument as described in ibid [341]–[343] (Allsop CJ).
8 *Sharma* (n 3) 451-2.
9 [1932] AC 562.
10 Ibid 580.
11 *Sharma Appeal* (n 4) [299] (Allsop CJ).

is not always possible, resulting in no liability. Concomitantly, a claimant may not be sufficiently proximate to such conduct to establish a duty of care. Relevantly to the Anthropocene, and as confirmed in the *Sharma Appeal*, causation as a legal prerequisite for damages is not up to the task of linking government decisions, carbon emissions from coal, and the effect of global warming on youth. Consequently, those responsible for carbon emissions both directly and through state sanction do not apparently owe a duty of care to those likely to be harmed.

Two key (imagined) developments during the eight years intervening since the successful appeal in *Sharma* colour this judgment. The first is the relentless recurrence of extreme weather events in Australia. In accordance with the science and modelling over an extended period,[12] the ferocity and frequency of these events are attributable to planetary warming arising from anthropogenic activity. While there are, by 2030, multiple interlocking reasons for the extended harms experienced by the human and more-than-human world, the science accepts that the foundational contributor to climate change and thus to extreme weather is the generation of carbon dioxide and methane emissions.[13] As predicted, Earth's systems reached a tipping point resulting in a cascading effect on ecological systems and the communities that depend on them. The premise of this decision that addresses questions of policy and incoherence raised in the *Sharma Appeal* is that by the date of the judgment (2030) the world, the environment, climate, and society have irrevocably changed. The fundamental – and anticipated – shifts wrought by climate change generate a new context for the application and development of legal principle.

The second development is imagined to have occurred in quantum mechanics through scientific consensus on quantum entanglement evidenced by its application from computing to space exploration.[14] Quantum entanglement represents a revolution in human comprehension of the world and, in upending constructs of time, space, and relational proximity, it affects how we understand causation in legal contexts.

In the same way that classical legal theory followed industrial development and the scientific revolution,[15] so too has the rapid application of the newly understood quantum entanglement generated a new jurisprudential turn. In the face of escalating planetary disaster, the law has rapidly adapted

12 As detailed in David Archer and Stefan Rahmstorf, *The Climate Crisis: An Introductory Guide to Climate Change* (Cambridge University Press, 2010).

13 Hans-Otto Pörtner et al (eds), *Climate Change 2022: Impacts, Adaptation and Vulnerability* (Intergovernmental Panel on Climate Change, Report, 2022).

14 Francisco Javier Duarte and Travis S Taylor, *Quantum Entanglement Engineering and Applications* (IOP Publishing, 2021).

15 See, eg, Jay M Feinman, 'Un-Making Law: The Classical Revival in the Common Law' (2004) 28(1) *Seattle University Law Review* 1, 7.

quantum science to meet the challenges of the Anthropocene with the development of tort to embrace the imagined doctrine of quantum entanglement.

During the time between the *Sharma Appeal* and this judgment, against the backdrop of foreseeable extreme events, ongoing and increasingly distributed harms, and the enhanced comprehension of quantum entanglement, the Commonwealth government through both action and omission contributed to the harms experienced by the plaintiffs. Firstly, there was an absence of policy or enactment of policy that would have enacted the recommended scientific strategies to mitigate the effects of climate change.

Secondly, government action continued to actively support the approval and maintenance of coal and gas extraction and export, obstruct the adoption of renewable technologies that might have mitigated the use of and reliance on carbon, and failed to implement any of a raft of complementary policies that might reduce the effect of carbon emissions on global climate.

This application is framed materially differently from *Sharma*. Firstly, it overcomes the indeterminacy of the claim as the plaintiffs have each suffered demonstrable harm from known events and seek damages rather than injunction. Secondly, the plaintiffs are patently vulnerable, establishing a class more likely to justify a duty. Thirdly, it establishes legal foreseeability through situating the harmful events in the past and linking those events with the decades of expert warnings about the consequences of decisions and action.

The imagined doctrine of quantum entanglement is a doctrinal adaptation of the scientific concept. It draws on the elements of Timothy Morton's 'hyperobjects': things that are 'massively distributed in time and space relative to humans',[16] which correlate to the quantum domain. Morton started their career as a scholar of romantic poetry and has evolved into a professor of philosophy adopting object-oriented ontology.[17] Concerned with eco-philosophy, Morton's ontology of objects challenges the subject/object 'correlationist dyad' and 'two modes of essentialism ... Nature and Matter'.[18] Their work offers the conceptual framework to move beyond Nature and Non-Nature, so that 'objects encounter each other as operationally closed systems that can only (mis)translate one another' whereby '[c]ausation is ... vicarious in some sense, never direct'.[19]

Morton's work is helpful in articulating a legal doctrine that will bridge the limitations of tortious causation to sheet home liability for damage

16 Timothy Morton, *The Ecological Thought* (Harvard University Press, 2012) ('*Ecological Thought*').

17 See Timothy Morton, 'Here Comes Everything: The Promise of Object-Oriented Ontology' (2011) 19(2) *Qui Parle* 163 ('Object-Oriented Ontology').

18 Ibid 167.

19 Ibid 165, citing Graham Harman, 'On Vicarious Causation' in Robin Mackay (ed) *Collapse Volume II* (Penguin, 2007) 171. See also Donna Haraway, *Staying with the Trouble: Making Kin in the Chthulucene* (Duke University Press, 2016).

arising from an activity's contribution to a system. For example, they consider global warming as a hyperobject,[20] helping to resolve challenges for the law in attributing its causation to an identifiable individual activity.

The starting point is establishing the duty that will result in liability if breached. The doctrine then attributes liability to an actor where their conduct observably contributes to a hyperobject and that hyperobject observably generates harm to the plaintiff or to the more-than-human world represented by that plaintiff. The duty is based on the interobjectivity[21] or the connectedness of the actor (defendant) to the harmed (plaintiff) through the conduct. In parallel, the basis for the attribution of liability is that the conduct's effects cannot be contained in space or time; that they are 'viscous', subject to 'temporal undulation', and 'meshed'.[22]

This jurisprudential approach is expansive, recognising the interrelatedness of actors and the harmed, and rejecting the limitations of time, space, and linearity in finding causation. The approach is also constrained, however, in adopting an evolutionary rather than revolutionary analogy from science.

Judgment

EB by her litigation representative and Ors v *The Commonwealth of Australia and Anor* [2030] FCA 39
Date of Judgment: 14 December 2030

A The Plaintiffs and Their Damage

This matter was brought in the Federal Court of Australia by a collection of plaintiffs, all of whom suffered damage arising from a series of extreme natural disasters occurring between 2019 and 2022. The nature of the damage they suffered varies, as does the immediate cause. However, all damage arises because of the changing climate. According to the plaintiffs' argument, the changing climate and its consequences has arisen and continues to arise from the failure of the Commonwealth Government, including the Minister for the Environment, to take reasonable care in exercising their respective functions.

The plaintiffs suffered damage as follows:

1. EB was born prematurely during the Canberra bushfires in 2020. She was a low birthweight baby who has experienced respiratory issues since

20 Timothy Morton, *Hyperobjects: Philosophy and Ecology After the End of the World* (University of Minnesota Press, 2013) 3, 7–9 ('*Hyperobjects*').
21 Ibid 81.
22 Ibid 1.

birth. Her premature birth and ongoing respiratory distress were caused by bushfire smoke.[23]

2. Jane Doe was pregnant during the 2020 New South Wales bushfires. Her placenta upon the delivery of her baby was 'grey, grainy and coming apart'.[24] Expert evidence reveals that her placenta was the equivalent of a 'pack a day smoker'.[25] It was so compromised that she required surgery to remove it. She continues to experience complications arising from the complex delivery. All complications are attributable to the smoke from the bushfires.
3. The Australian Koala Foundation[26] is joined on behalf of the surviving koalas of New South Wales. Koala colonies were decimated in the 2020 bushfires, leaving only remnant koala populations that are unlikely to return.[27] The remaining populations sue for the harm to their habitat and the compromise to their capability to survive as a species.
4. Our Watch[28] joins on behalf of women survivors of the radically increased incidence of family violence occurring in the aftermath of the natural disasters of 2020–22.[29] Following the floods in Lismore, survivors were sufficiently fearful of their safety, and that of their children, that they lived in a tent community for some years following the floods.[30] They sue for damages arising from their homeless state and the physical, psychological, and financial harms experienced.
5. Margaret Doe lives with a disability. She was eventually evacuated from her home during the Gympie floods in 2022, only to find herself without

23 Nicole MacKee, 'Harms from Bushfire Smoke: "Yesterday Was the Time to Talk About It"' Issue 37 (21 September 2020) *InSight* <https://insightplus.mja.com.au/2020/37/harms-from-bushfire-smoke-yesterday-was-the-time-to-talk-about-it>; Abdo M et al, 'Impact of Wildfire Smoke on Adverse Pregnancy Outcomes in Colorado, 2007–2015' (2019) 16(19) *International Journal of Environmental Research and Public Health* 3720.

24 Kate Walton, 'Bushfire Danger: Smoke Puts Lives of Mothers and Babies at Risk', *Al Jazeera* (online, 1 October 2020) <https://www.aljazeera.com/news/2020/10/1/bushfire-danger-smoke-puts-lives-of-mothers-and-babies-at-risk>.

25 Ibid.

26 'Save the Koala', *Australian Koala Foundation* (Web Page, 2020) <https://www.savethekoala.com>.

27 Lily M van Eeden et al, 'Impacts of the Unprecedented 2019–2020 Bushfires on Australian Animals (Report, WWF Australia, November 2020) <https://www.wwf.nl/globalassets/pdf/wwf-impacts-of-the-unprecedented-2019-2020-bushfires-on-australian-animals.pdf>.

28 'What We Do', *Our Watch* (Web Page, 2023) <https://www.ourwatch.org.au/about-us>.

29 See, eg, Sonia Maria Martins Caridade, Diogo Guides Vidal and Maria Alzira Pimenta Dinis, 'Climate Change and Gender-Based Violence: Outcomes, Challenges and Future Perspectives' in W Leal Filho et al (eds), *Sustainable Policies and Practices in Energy, Environment and Health Research* (Springer, 2022) 167.

30 Stephanie Boltje, 'Domestic Violence Spike Expected in Flood-Ravaged Northern NSW', *ABC The Drum* (online, 23 March 2022) <https://www.abc.net.au/news/2022-03-23/nsw-floods-domestic-violence-spike-predicted/100932142>.

> accommodation.[31] She claims damages to compensate her for her home and for the damage to her health arising from her vulnerable state.

The parties agree that the plaintiffs' claims can be heard together as they deal with the same alleged duty, and they agree that the Australian Koala Foundation has standing to represent a nonhuman animal collective litigant.

The Claim

Each plaintiff claims damages for breach of the same attributed duty namely, a duty on the government of the Commonwealth of Australia, and its Ministers personally, to take reasonable care in making decisions and in developing policy to avoid causing personal injury or death to humans and nonhuman animals arising from known consequences of increases in CO_2 emissions.

Integral to this duty is the inherent responsibility of government and Ministers for the safety of humans and nonhuman animals as a distinct implied mandatory consideration that colours each government decision. As observed by Allsop CJ in the *Sharma Appeal*:

> The subjection of governments and public authorities, including Ministers of the Crown, to the rule of law encompasses not only the requirement of legality under s75(v) of the *Constitution*, but also liability for tortious wrongs ascertained in accordance with the application of principle under the common law.[32]

The questions here, however, are:

a) whether such a duty exists; and
b) if so,
   i) whether the appellants have breached it; and
   ii) that breach is the cause of the damage suffered by the plaintiffs.

Duty of Care

In *Sharma*, the alleged tort was described as a 'novel duty of care'[33] arising in 'novel factual circumstances'.[34] Unfortunately the factual circumstances

31 Eden Gillespie, 'Advocates Say People with Disability Are Increasingly "Forgotten" in Emergency Planning', *The Guardian* (online, 11 April 2022) <https://www.theguardian.com/australia-news/2022/apr/11/advocates-say-people-with-disability-are-increasingly-forgotten-in-emergency-planning>.
32 *Sharma Appeal* (n 4) [4] (Allsop CJ).
33 Ibid [99] (Allsop CJ).
34 Ibid [182] (Allsop CJ).

in this matter are by now far from novel. The period in respect of which the plaintiffs suffered harm is now considered the point at which Australia entered a new climatic epoch. Pyne refers to the Pyrocene,[35] signifying the impact not only of humans' introduction of fire to the planet but also the radical effects of the burning of fossil fuels on the Earth. However, it is not only fire but previously unparalleled rain events, droughts, and cyclones that are harming humans, nonhuman animals, and the environment we share. These events are now commonplace.

For some years governments have received scientific modelling pointing to the likelihood of more prevalent severe weather events caused by the burning of fossil fuels.[36] Governments have been provided with evidence of tipping points in various climatic factors and their relationship with extreme weather events.[37] Governments have evidence of the impact of such events on the health and safety of humans, of nonhuman animals, and of the natural and the built environment.[38]

I consider that defendants have for many years been on notice of the likely and foreseeable effects of their decisions upon human and nonhuman animals. The weather events involved in the plaintiffs' claims are not novel circumstances. Now, and for some time, these weather events have been emblematic of an entirely new habitat. Government and its Ministers have contributed to this new habitat and have failed to take action to mitigate its effects on humans and nonhuman animals.

In the context of the radical changes in contemporary society and its institutions the law must keep pace. Trends have been apparent for some time of legal adaptation to supplement or compensate for the acts and omissions of government and corporations in responding to profound changes in the society and environment.[39]

35 Stephen J Pyne, *The Pyrocene How We Created an Age of Fire, and What Happens Next* (University of California Press, 2022).

36 See, eg, William J Ripple et al, 'World Scientists' Warning to Humanity: A Second Notice' (2017) 67 *BioScience* 1026; William J Ripple et al, 'World Scientists' Warning of a Climate Emergency' (2019) 70(1) *BioScience* 8; Clive Hamilton and Jacques Grinevald, 'Was the Anthropocene Anticipated?' (2015) 2(1) *Anthropocene Review* 59.

37 David Armstrong McKay et al, 'Exceeding 1.5°C Global Warming Could Trigger Multiple Climate Tipping Points' (2022) 377(6611) *Science*.

38 See, eg, Australian Academy of Science, *The Risks to Australia of a 3°C Warmer World* (Report, March 2021) <https://www.science.org.au/files/userfiles/support/reports-and-plans/2021/risks-australia-three-deg-warmer-world-report.pdf>;

Australian Institute of Health and Welfare, *Australian Bushfires 2019–20: Exploring the Short-Term Health Impacts* (Report, 2020) <https://www.aihw.gov.au/getmedia/a14c3205-784c-4d81-ab49-a33ed4d3d813/aihw-phe-276.pdf>; Marina Romanello et al, 'The 2021 Report of the Lancet Countdown on Health and Climate Change: Code Red for a Healthy Future' (2021) 398 (10311) *The Lancet* 1619.

39 See (n 2).

Given that government is implicated as an actor both within the changing environment and in generating that change, it is no longer appropriate for the Court to quarantine decisions concerning climate change from the remit of duty. In this case, that duty is to be found in a civil standard attributing to government and its Ministers personally a duty of care in the exercise of their professional roles, to protect vulnerable humans and nonhuman animals from the harmful effects of climate change. The ubiquitous nature of climate change and the locus of governance over its consequences within every sphere of government go to the very heart of the relationship between the State and humans and nonhuman animals who are vulnerable to its effects.

Allsop CJ in the *Sharma Appeal* found that the alleged duty would require the Court to 'pass judgment on the reasonableness of conduct that may involve action on political grounds',[40] conduct which is 'ordinarily decided through the political process'.[41] To the contrary, the decisions complained of are presented as executive decisions. On this basis, they may take place in a political context, but they are decisions made according to the constraints of law. In finding an overarching duty, all government and ministerial decisions are bound by that duty. The lawful process of decision-making will continue to apply under a statute, subject to the higher level constraint of the duty.

There is no incoherence in finding an overarching duty. With respect, the Court itself generated the alleged incoherence through its own framing of the issue. If the duty applies to government business, then there need be no disturbance in the process of decision-making according to the text and purpose of a statute. In joining the Commonwealth as defendant, the plaintiffs have recognised the interrelationships within the processes of government so that any one decision cannot be isolated from decisions taken elsewhere.

Given the seriousness of the circumstances, it is not open to government to make decisions that mitigate climate change on the one hand but promote it on the other. The duty of care recognises the whole-of-government as an institution, and ministerial roles as integral to a coherent institutional approach.

Core to this argument is the sheer scale of climate change. Its effect on every aspect of life – health, education, commerce, housing, industry, agriculture, environment, transport, and so on – generates a duty spanning government operations. No one area of government can be quarantined from an overarching response to this existential issue.

Drawing on the pronouncement of Lord Atkin as to a 'novel' duty of care in *Donoghue v Stevenson*, I find that a government and its Ministers:

40 *Sharma Appeal* (n 4) [16] (Allsop CJ), citing Gleeson CJ in *Graham Barclay Oysters v Ryan* (2002) 211 CLR 540, 553–4.

41 Ibid.

a) in making decisions implicated in the generation of greenhouse gases where there is no possibility that any person or group of people might themselves mitigate the effects of that government action; and
b) where government knew that the absence of reasonable care in the making of its decisions will result in injury to humans or nonhuman animals,

owe a duty to humans and nonhuman animals to take that reasonable care.[42]

While I have gone some way to justify the finding of a duty to take reasonable care in making decisions affecting climate change, I rely in addition on an underpinning and novel doctrinal foundation that supports the existence of the duty and explains how it responds to some of the perceived shortcomings of the claim in *Sharma*. I describe this as the doctrine of quantum entanglement.

The Science of Quantum Entanglement

Jurists and commentators have observed a relationship between law and science. '[T]he image of science has shaped the very structure and organisation of ... common law'.[43] Although legal reasoning as science has been challenged, the trajectory of development of the common law is considered to reflect the development of scientific thought.[44] Despite inevitable epistemic differences, ignoring science in the development of law would represent a concerning divergence in forms of thinking considered valid in our society.

To this end, the law's insistence on the linear treatment of causation and damage, constrained by proximity in various guises, responds by analogy to a classical scientific – or Newtonian – understanding of time and space, as well as of thinking itself. Like Newtonian science, legal reasoning adopts a classical theory of concepts, whereby classification of a concept or its definition is an account of a thing that signifies its essence.[45] Yet, in maintaining this approach, the law might be said to fail to accommodate or respond to the implications of the now widely adopted quantum theory and the consequences of multiple quantum systems in relation to each other, or quantum entanglement.

42 *Donoghue v Stevenson* [1932] AC 562, 599.

43 John Veilleux, 'The Scientific Model in Law' (1987) 75(6) *Georgetown Law Journal* 1967, 1968; R George Wright, 'Should the Law Reflect the World?: Lessons for Legal Theory from Quantum Mechanics' (1991) 18 *Florida State University Law Review* 855, 855–6.

44 Wright (n 43).

45 Aristotle, *Topics*, Book 1, Part 5. See generally, Don Howard, 'What Makes a Classical Concept Classical? Toward a Reconstruction of Niels Bohr's Philosophy of Physics' in Jan Faye and Henry J Folse (eds), *Niels Bohr and Contemporary Philosophy* (Springer, 1994) 201.

Let me explain.

Classical physics assumes particles to be independent, so that knowledge of the state of one particle affords no insights about the state of another.

By contrast, a quantum state envisions particles as wave functions where each can exist in many states at once. Each state has a probability. It is only once the attribute of a particle is measured that its state is determined. A quantum system is a 'catalogue of information ... that contains the probabilities for all possible outcomes of the measurements we can make on the system'.[46]

Observation or measurement of any one attribute of an object obscures any other attributes. Any measurement is therefore partial. Consequently, we are constrained in describing an object's physical properties. Further, the measurement or observation of quanta themselves becomes part of the 'universe of objects'[47] given that the measurement is what produces our understanding of that object.

The strangeness of this theory, however, is that unlike classical understanding of separateness, measurement of objects that are separated by great distances reveals the same results as each other, despite the infinite number of possibilities available. In other words, ostensibly differentiated objects affect one another regardless of their location. This has been confirmed by repeat experimentation.[48]

The implication of this is that knowing the state of one object reveals correlating facts about another object, regardless of the distance apart.

Adapting Knowledge for New Contexts

The law and its institutions face an enormous challenge in the emergence of supercomplex systems, when our model of governance has evolved to deal with human-scale contexts. I understand supercomplexity, a 'proliferation of knowledges' that frequently compete,[49] to fundamentally 'challenge to our frameworks of knowing, being and acting'.[50] This context includes a challenge to a classical approach to legal reasoning.

Morton explains the challenge in comprehending contemporary phenomena in terms of classical approaches, describing 'hyperobjects': 'things that

46 Vlatko Vedral, 'Quantum Entanglement' (2014) 10(4) *Nature Physics* 256, 256.

47 Morton, 'Object-Oriented Ontology' (n 17) 180.

48 Ben Brubaker, 'How Bell's Theorem Proved "Spooky Action at a Distance" Is Real', *Quanta Magazine* (online, 20 July 2021) <https://www.quantamagazine.org/how-bells-theorem-proved-spooky-action-at-a-distance-is-real-20210720>.

49 Ronald Barnett, 'University Knowledge in an Age of Supercomplexity' (2000) 40 *Higher Education* 409, 417.

50 Ronald Barnett, 'Supercomplexity and the University' (1998) 12(1) *Social Epistemology* 43, 47.

are massively distributed in time and space relative to humans'.[51] They suggest that such objects are literally too big to understand within our frames of knowing. Hyperobjects may be directly created by humans (for example, the sum of all plastic bags) or not (for example, climate). Importantly for this analysis, Morton describes three properties of the hyperobject that assist in understanding the limits of classical legal reasoning.

1. Hyperobjects are non-local. While one might observe the manifestation of the hyperobject, that local manifestation is not the hyperobject itself. Thus, one might experience a weather event in a single place. Although that weather might arise because of climate change, the event is not climate change per se.
2. Hyperobjects are 'time-stretched', exhibiting 'temporal undulation'. Their sheer scale renders them visible in a way that signifies temporality, yet they are impossible to hold in one's mind[52] given their massive distribution including over time.[53]
3. The hyperobject is interconnected.[54] 'Nothing is experienced directly, but only as mediated through other entities in some shared sensual space'[55] or systems of related subjects.[56] The hyperobject comprises 'links, and … gaps between links'.[57] In contrast to a linear approach to causation, on this understanding, gaps enable causality. Systems of related objects are interobjective.

Morton's hyperobjects evince the dimensions of quantum entanglement more accessibly than pure physics. Quantum mechanics and its associated theory of entanglement represent a paradigm shift in science into a post-Newtonian era. Posited over a century ago by eminent scientists,[58] the theory has been taken up and developed in various quarters until recent scientific efforts have generated general acceptance. While there are features of quantum theory that can be 'mimicked within classical physics',[59] it is entanglement that differentiates quantum physics from classical.

This is a legal judgment and not a scientific treatise. I therefore rely on Morton's hyperobjects to identify key features of quantum entanglement that

51 Morton, *Ecological Thought* (n 16) 130–5.
52 Morton, *Hyperobjects* (n 20) 56.
53 Ibid 67.
54 Ibid 83.
55 Ibid 86.
56 Ibid 84.
57 Ibid 83.
58 See Francisco Javier Duarte *Fundamentals of Quantum Entanglement* (IOP Publishing, 2019) ch 1; Vedral (n 46).
59 Vedral (n 46) 256.

can by analogy inform legal doctrine. This approach represents a paradigm shift in the law's comprehension of supercomplex phenomena that exceed our (and the law's) capacity to understand and analyse.

The Doctrine of Quantum Entanglement

Advances in understanding the world wrought by quantum theory do not mean that the law attaches liability to random actors by evoking the possibility that they are somehow entangled with another.

Instead, quantum entanglement offers an analogy for resolving questions of causation in a tortious claim concerning a hyperobject. The sheer scale and complexity of the hyperobject renders classical legal approaches inadequate both to analyse and to locate liability in a coherent way.

I therefore state the doctrine of quantum entanglement under three limbs.

1 Where a phenomenon is:
   a) evident all around us so that wherever we go it is apparent (viscous);
   b) distributed over time and space so that only one facet or aspect of the phenomenon is visible at any one time (non-local, time-stretched); and
   c) interconnected with multiple objects and phenomena,

   that phenomenon is known as a hyperobject.
2 Where a legal subject has undertaken activity that contributes to or exacerbates the existence of the hyperobject or its manifestation through any facet or aspect, the law recognises the inherent interconnectedness of the activity to the hyperobject.
3 Where a legal subject suffers harm as a result of an event and that event is part of a hyperobject, the law will recognise the inherent interconnectedness of the event to the hyperobject resulting from its non-locality and temporal undulation.

Consequently, it is a bar to a defence to negligence that an activity contributing to a phenomenon that causes harm is too remote to satisfy causation by way of the activity's distribution over either time or space.

Given the framing of this action, the doctrine of quantum entanglement is relevant only to the tort of negligence, otherwise analysed according to the principles already established at law. The starting point then is whether the claim satisfies the neighbour principle.[60]

60 *Donoghue v Stevenson* [1932] AC 562.

Neighbour Principle

To establish a duty of care requires identification of a relationship of the

> requisite closeness, control and vulnerability for that relationship to warrant the imposition of a duty of care by reference to the legal conception of neighbourhood and whether the relationship was suitable for the imposition of a duty capable of founding liability judged by reference to judicial or curial determination.[61]

There is an existing relationship between each of the plaintiffs and the defendants. The human plaintiffs are citizens, the corporate plaintiffs are reliant on government for their legal personhood, and the animal plaintiffs are protected under various statutes. As categorised, none of these relationships brings a duty of care per se. That is to be established independently of the relationship itself. To establish this threshold requires the State to constitute a neighbour as envisaged by Lord Atkin.

Allsop CJ found that in the broader context of the incremental contributions to climate change, of our system of government, and the statutory context, a relationship of neighbour could not be said to exist.[62]

I find instead that the question of incrementalism – individual actions taken over an extended period – cannot interfere with the neighbour principle. A neighbour consistently taking incremental action in the knowledge of an impending radical danger is a neighbour nonetheless.

The *EPBC Act* provided a foil for liability in *Sharma*, given the nature of the claim and remedy sought.[63] In this case, however, the claim rests on a much broader foundation than the application of statutory discretion. The size and scope of government action over time while cognisant of the effects of its actions has generated a knowable and known radical danger. Government cannot resile from its obligation through citing statutory discretion.

The government's power and resources put it in a position of knowledge about climate change, its causes, effects, and potential mitigation. It was in a special position of knowledge that could foresee the events that caused harm to the plaintiffs.

The plaintiffs were powerless to mitigate the effects of climate change. Consequently, they were vulnerable to the actions of government.

I find a duty in the government and its Ministers to 'take reasonable care to avoid acts or omissions which it has reasonably foreseen would be likely to injure' individuals and nonhuman animals in its jurisdiction.

I turn now to analyse this claim under the three limbs.

61 *Sharma Appeal* (n 4) [211] (Allsop CJ), citing *Caltex Refineries (Qld) Pty Ltd v Stavar* (2009) 75 NSWLR 649.
62 Ibid [346] (Allsop CJ).
63 Ibid.

The Three Limbs

Climate Change as a Hyperobject

As Morton observes, climate change[64] is a hyperobject. It cannot be perceived as a whole, but instead can be modelled[65] and aspects or facets of it can be felt or experienced discretely.[66] It is supercomplex: a system of systems.[67] These systems exist globally and manifest differentially, but actions over time of humans, business and governments have worked in concert to generate feedback loops within these systems that have altered the climate globally and, thereby, its constituent systems.

Although no single action can be identified as the tipping point, the trajectory of classes of action has been clearly modelled for many years.

Climate change satisfies the definition of a hyperobject. The doctrine works then to recognise the interrelationships of actions and actors to generate harm despite classical notions of proximity in time and space.

Government Activity's Contribution

Respondents in matters concerning carbon emissions have a track record of admitting the contribution of their activities to emissions generally.[68] Their defences, instead, tend to focus on the absence of proximity between the emissions as a diffuse, globalised phenomenon, and the occurrence of harm.

So too is this defence proffered here: if it were not Australian government decisions, then it would be the decisions of other governments. If the cause is global contribution, then causation exists beyond the scope of governmental responsibility.

It barely requires formulation of a doctrine of quantum entanglement to prove that the effect of one object (government decisions approving emitting activities and failing to mitigate emissions overall) is reflected in a distant object, or a 'downstream' approach to understanding otherwise immense phenomena. The doctrine of quantum entanglement, however, provides the rationale for ignoring time and place to recognise that a classically categorised 'downstream' activity is necessarily implicated in local harms.

The temporal aspect of the incrementalist argument is that it has been many governments over many years. That may be so. But it is the government's continuing action in the face of clear knowledge about the consequences of that action that breaches the duty. At no point in the last three

64 Morton, *Hyperobjects* (n 20) 3.
65 Ibid 3, 12.
66 Ibid 27.
67 J T Houghton et al, *Climate Change 2001: The Scientific Basis* (IPCC, 2001) ch 1.
68 See, eg, *Xstrata Coal Queensland Pty Ltd v Friends of the Earth* [2012] QLC 13, [516].

decades (at least) can a government claim to have made these decisions ignorant of their effect.

I find that the government's actions contributed to the growth of the hyperobject in a way that was reasonably foreseeable by the government and its Ministers.

Harms Arising from the Hyperobject

While a system of systems, component systems are measurably part of the hyperobject. The bushfires and floods of 2020–22 constitute extreme weather events that have been forecast for decades.[69] They are not weather within its normal parameters.

I find that the harms experienced by each plaintiff arose from the hyperobject of climate change.

Defences

The doctrine of quantum entanglement applied to negligence permits the Court to link the effect of the defendants' actions to the harms suffered by the plaintiffs. Informed by current understanding of the interrelationship between objects and measurable effects regardless of time and space, it attributes liability to an actor where that actor knew of the effect of their actions on climate change, and the resultant implications for other legal subjects.

It is therefore no defence to argue that there was no single action by the defendant that is identifiably the cause of the harm. Nor is it a defence to attribute harm to an intervening action. For example, the living conditions of those experiencing family violence following the Lismore floods is not attributable to the perpetrator of the violence. The underlying social dysfunction is inextricably linked to climate change.[70]

I find that the defendants have not established a defence to the claim.

The plaintiffs' claim is upheld.

69 See generally, Intergovernmental Panel on Climate Change.
70 Caridade, Vidal and Dinis (n 29).

Chapter 7

The Case of Young People v Government of Ireland

Aoife Daly and Orla Kelleher

Commentary

Backdrop

There is arguably no more important issue than the climate crisis. It is an urgent threat to human rights globally. Children are, however, in a particularly vulnerable position for a number of reasons, including their relative vulnerability economically, physically, and politically (being disenfranchised); as well as the fact that they will, on average, live longer and therefore suffer far greater effects compared to the average adult. One significant threat among the many posed by the climate crisis is rising sea levels, particularly for island nations.

For this hypothetical judgment, we envisage a context on the island of Ireland in 2033 where over one-fifth of the landmass that existed in 2020 is now underwater due to rising sea levels.[1] This is due to the climate crisis. Villages such as Barna in Galway on the West Coast of Ireland have been particularly badly affected with significant levels of homelessness, as well as unemployment from lost land and resources, caused by the crisis. Children and young people have also been disproportionately affected for numerous reasons varying from trauma due to displacement, and lost income.

In 2022, Ireland's climate governance framework, which commits to reducing greenhouse gas ('GHG') emissions by 51% by 2030 relative to 2018 levels (a 44.5% reduction by 2030 relative to 1990 levels), and carbon neutrality by 2050, is patently inadequate to make a fair share contribution to the global temperature goals of limiting global heating to 1.5°C and well below 2°C as an outer limit.[2] Ireland has the second highest per capita emissions in the European Union (11.6 tonnes of CO_2 equivalent per capita

1 See <https://coastal.climatecentral.org/map/12/-6.2452/53.3362>.

2 *Paris Agreement*, opened for signature 22 April 2016, [2016] ATS 24 (entered into force 4 November 2016) art 2(1) ('Paris Agreement').

DOI: 10.4324/9781003389569-9

in 2020)[3] and has benefited from its past high emissions. The adoption of climate mitigation measures, which broadly align with the global average reduction rate for what needs to be done for an even chance (50%) of holding temperatures below 1.5°C, flouts international environmental law principles of equity and common but differentiated responsibilities and respective capacities (CBDR-RC).[4] The *Climate Action and Low Carbon Development Act 2015* (as amended) ('*Climate Act*') sets out a long list of factors that the government must have regard to in the making of climate policies but makes no reference to the interests of children or future generations.[5] The *Climate Action Plan 2021* barely mentions children, with the only substantive reference being a stated intention to establish a youth climate assembly to contribute views on the plan and a vague statement of intention to provide education to children on the environment. There is no reference to children's rights.

In this hypothetical case we envisage a group of children coming together in a similar fashion to the children in the *Juliana* case in the United States,[6] or in the *Sacchi* case[7] before the United Nations Committee on the Rights of the Child (the implementing body of the United Nations *Convention on the*

3 Central Statistics Office, 'Environmental Indicators Ireland 2022' (Web Page, 14 November 2022) <https://www.cso.ie/en/releasesandpublications/ep/p-eii/environmentalindicatorsireland2022/greenhousegasesandclimatechange/#:~:text=In%202020%2C%20Ireland%20had%20the,EU27%20average%20of%207.4%20tonnes>.

4 See, generally, Lavanya Rajamani et al, 'National "Fair Shares" in Reducing Greenhouse Gas Emissions within the Principled Framework of International Environmental Law' (2021) 21(8) *Climate Policy* 983.

5 Section 4(8) of the *Climate Action and Low Carbon Development Act 2015* (NI) (as amended) ('*Climate Act*').

6 Non-profit organisation Our Children's Trust began to file lawsuits across the United States in 2011, engaging children and youth as litigants. The applications were largely legally unsuccessful. One of these, the *Juliana* lawsuit, was mounted against the United States government in 2015 by 21 young people, who are arguing that the government is in breach of its atmospheric public trust obligations, and violating their constitutional rights to life, liberty and property, by failing to implement effective mitigation policies. The case was dismissed by the Ninth Circuit Court of Appeals in 2020 (*Juliana v United States* 947 F3d 1159 (9th Cir 2020). On 1 June 2023, Judge Aiken in the United States District Court for the District of Oregon granted the plaintiffs leave to file a second amended complaint in the matter. See Climate Case Chart, '*Juliana v United States*' (Web Page, 1 June 2023) <http://climatecasechart.com/case/juliana-v-united-states>.

7 In this application, the youth applicants claimed that states were violating their obligations under the *Convention on the Rights of the Child* due to the failure to mitigate climate change adequately. The application did not proceed to a full hearing due to the failure to exhaust domestic remedies. See *Sacchi v Argentina*, Petition Submitted under Article 5 of the Third Optional Protocol to the United Nations *Convention on the Rights of the Child* (23 September 2019).

Rights of the Child[8]). Included in the submissions of these young Irish plaintiffs would be stories similar to those presented in *Sacchi* – stories outlining the difficulties in their everyday lives due to the effects of the climate crisis.

The Legal Context

Ireland has a dualist system. In the hypothetical 'universe' in which this judgment is based, Ireland has incorporated rights instruments, treating them as compatible with its Constitution. This means that the two sources of law are considered concurrently by Irish courts. The exercise of considering this scenario is novel and exciting, as it permits us to examine what may occur if Ireland did indeed incorporate these instruments, and what this might mean in particular for children's rights and environmental rights. The reason that we have provided for the change in the Irish approach to international human rights law treaties is that, in our scenario, almost every other country in the world has by this future point in 2033 incorporated those treaties, and Ireland was under pressure to follow suit.

It is an exciting time for proponents of the human right to a healthy environment. In 2021, the Human Rights Council adopted a resolution recognising the right to a healthy environment.[9] This constituted the first time that a United Nations body recognised the right as a universal human right. In 2022, a resolution was also adopted by the United Nations General Assembly calling upon states, international organisations and business enterprises to increase efforts to ensure a healthy environment for all.[10] Since 2022, there has been a newly appointed United Nations Special Rapporteur for climate change. In addition, at the time of writing, the United Nations Committee on the Rights of the Child is drafting a general comment on the right of children to a healthy environment, acknowledging the position of prominence that this has right now, and the role of children in bringing the issue to the fore.

These developments are certainly consolidating the right of children to a healthy environment in the international arena. The United Nations *Convention on the Rights of the Child* is already a high-profile document, signed and ratified by almost every country in the world. The right of children to have their best interests given due consideration in matters affecting them is the most prominent right concerning children,[11] embedded in

8 *Convention on the Rights of the Child*, opened for signature 20 November 1989, 1577 UNTS 44 (entered into force 2 September 1990) ('CRC').

9 'The Human Right to a Clean, Healthy and Sustainable Environment', HRC Res 48/13, UN Doc A/HRC/48/13 (adopted 8 October 2021).

10 'The Human Right to a Clean, Healthy and Sustainable Environment', GA Res 76/300, UN Doc A/RES/76/300 (adopted 28 July 2022).

11 Laura Lundy et al, *The UN Convention on the Rights of the Child: A Study of Legal Implementation in 12 Countries* (UNICEF, 2012) 4.

the legal systems of most States. It is, therefore, potentially a very powerful principle on which to base claims relating to environmental harm. It is likely that it will be relied upon in future climate cases, at both national and international levels, as a base from which to argue that governments have legal duties to children in the climate crisis.

The Irish people approved by referendum in 2012 the enactment of a new provision of the Irish Constitution – Article 42A – which explicitly recognised the rights of children as separate from those of their parents. This was a significant step for Ireland, a country which many have argued provides undue deference to the rights of parents over their children, as opposed to the rights of children as individuals.[12] The recognition of the rights of children in the Constitution of Ireland could provide a strong basis on which to base claims relating to environmental harm. The United Nations *Convention on the Rights of the Child* may provide a useful elaboration on what the substance of children's rights actually is under the Irish Constitution.

At the time of writing, environmental constitutionalism has entered into a new and exciting phase in Ireland. While the Irish Supreme Court refused to recognise the existence of an unenumerated or 'derived' right to an environment consistent with human dignity or a right to a healthy environment in *Friends of the Irish Environment v Government of Ireland*[13] ('Case Ireland'), policy developments have overtaken this judicial finding. The Citizens' Assembly on Biodiversity Loss recently recommended that there should be a referendum to amend the Constitution with a view to protecting biodiversity.[14] The Assembly voted that the specific proposal to be put to the people should include substantive and procedural environmental rights for humans as well as nature rights. In the meantime, an underexplored dimension of environmental constitutionalism in both scholarship and practice is the potential for the nascent derived constitutional duty of vigilant and effective environmental protection in climate litigation in Ireland. One major contribution of the *Case of Young People v Government of Ireland* judgment is that it illustrates how the Irish Constitution, even as it currently stands, could be interpreted to secure more ambitious mitigation measures than those being currently delivered through Ireland's statutory carbon budgeting system. The judgment shows how the Irish Supreme Court could continue the emerging

12 Ursula Kilkelly and Conor O'Mahony, 'The Proposed Children's Rights Amendment: Running to Stand Still?' (2007) 10(2) *Irish Journal of Family Law* 19.

13 *Friends of the Irish Environment v Government of Ireland* [2020] IESC 49 ('Case Ireland').

14 The Citizens' Assembly, *Report of the Citizens' Assembly on Biodiversity Loss* (March 2023) 16 <https://citizensassembly.ie/wp-content/uploads/Report-on-Biodiversity-Loss_mid-res.pdf>.

dialogue between the highest or apex courts in climate cases[15] by drawing on the *dicta* of the German Constitutional Court on Article 20a of the German Basic Law,[16] to purposively interpret the equivalent Irish constitutional duty. It could also consider the poor record of Ireland on meeting climate targets, in contrast to many European Union countries such as Germany which met their 2020 targets (in our scenario we envisage they also do this for their 2030 and 2040 targets).

We expect, based on the evidence of the consequences of rising sea levels for Ireland, that Ireland is on course for serious changes in the coming two decades. This will likely include the loss of land and possibly even communities along our coast. We anticipate that, at the very least when this occurs, the Courts will give due regard to inadequate climate mitigation action and how this is an issue of constitutional rights. We hope that the Court will fully consider our obligations to children in Ireland, including those deriving from the level of international human rights law.

Judgment

Young People v Government of Ireland
Court: Supreme Court of Ireland
Judges: Daly and Kelleher JJ
Date: 31 July 2033

Background

Over one-fifth of the landmass of Ireland which existed in 2020 is now underwater due to the rising sea levels as a result of the climate crisis. In 2033, the economic, social, and environmental impacts of coastal flooding and erosion, which are being exacerbated by sea level rise and more extreme storm surges, are now acutely felt in coastal cities, towns, and villages including Dublin, Cork, Limerick, and Galway.[17] With 1.9 million (or 40%) of the Irish population living within five kilometres of the coastline and 40,000 living within

15 Anna-Julia Saiger, 'Three Hours from Paris to Karlsruhe: On the Close Reading of the Paris Agreement by the German Constitutional Court in Its Decision on Climate Change', *Völkerrechtsblog* (Blog Post, 1 May 2021) <https://voelkerrechtsblog.org/three-hours-from-paris-to-karlsruhe>.

16 In *Neubauer v Germany*, Bundesverfassungsgericht [German Constitutional Court], BVerfG, Order of the First Senate of 24 March 2021 – 1 BvR 2656/18 ('*Neubauer*').

17 Margaret Desmond, Phillip O'Brien and Frank McGovern, 'A Summary of the State of Knowledge on Climate Change Impacts for Ireland' (Environmental Protection Agency, Report No 223, 2017) 25-26 <https://www.epa.ie/publications/research/climate-change/EPA-RR-223_web.pdf>.

100 metres,[18] increases in coastal floods mean hundreds of families have already had to abandon their homes. The past decade has also seen a significant rise in unemployment from lost land and resources. Certain ways of life – such as those of fishing families – have been totally eroded.

The 12 plaintiffs are children and youth from all over Ireland and are aged between seven and 17 years. In their submissions, these young plaintiffs include descriptions of the difficulties in their everyday lives wrought by climate change. The families of five of the plaintiffs, who come from the coastal village of Barna in Galway, have had to abandon their family homes and businesses due to sea level rise and more extreme storm surges.[19] It has been well documented (as adduced through evidence presented by the plaintiffs at the oral hearing) that morbidity and mortality rates from skin cancer, overheating, water- and food-borne diseases, and respiratory illnesses have increased dramatically in Ireland in recent years. Each of the plaintiffs avers to adverse impacts of the climate crisis on their physical and mental well-being.

The plaintiffs are seeking a declaration from the court that the government is under a legal and constitutional obligation to increase the scale and speed of its domestic climate mitigation measures so as to make a fair share contribution to the global temperature goal of limiting global temperature to as close to 1.5°C above pre-industrial levels as possible.

In their submissions, the plaintiffs emphasise the failure of the government to implement the principle of the best interests of the child in the *Convention on the Rights of the Child*. Implementing this principle would have involved prioritising children's interests over unfettered economic growth. The plaintiffs also argue that the Irish government is abdicating its constitutional duty to ensure effective and vigilant environmental protection by failing to curb GHG emissions swiftly and sharply enough to limit global heating to as close to 1.5°C above pre-industrial levels as possible. The plaintiffs placed significant reliance on the 2028 *Summary for Policymakers of the IPCC's 7th Assessment Report*; this states with virtual certainty (99–100%) that global heating is already 1.3°C above pre-industrial levels and is likely to reach 2.9°C above pre-industrial levels by the end of the century, unless rapid, deep, and immediate reductions in all GHG emissions are achieved. The plaintiffs acknowledge that the Irish government through a combination of domestic mitigation measures and international carbon offsets is managing to live

18 Central Statistics Office, *Profile 2 Population Distribution and Movements* (Media release, 11 May 2017) <https://www.cso.ie/en/csolatestnews/pressreleases/2017pressreleases/pressstatementcensus2016resultsprofile2-populationdistributionandmovements>.

19 'County Council Accused of "Negligence" After It Ignores Flood Risk Warnings in Barna', *Galway Pulse* (online, 23 January 2022) <https://galwaypulse.com/2022/01/23/county-council-accused-of-negligence-after-it-ignores-flood-risk-warnings-in-barna>.

within the carbon budgets set under Ireland's *Climate Act*. The *Summary for Policymakers of the IPCC's 6th Assessment Report* stated that global CO_2 emissions need to be halved by 2030 relative to 2019 levels, and to reach net zero around 2050, in order to have an even chance (50%) of holding temperatures below 1.5°C.[20] The Irish carbon budgeting system adopted a global average reduction rate, that is, a 44.5% reduction in emissions by 2030 relative to 1990 levels (which, as stated above, was achieved) and climate neutrality by 2050 (which Ireland is on track to achieve).[21]

The plaintiffs do not seek to enforce the *Climate Act*, which they argue is manifestly inadequate for tackling the scale of the escalating climate crisis, but to instead challenge the constitutionality of the government's current approach to climate mitigation. The plaintiffs claim that the government's climate mitigation measures are inconsistent with a fair share contribution to limiting global heating to 1.5°C above pre-industrial levels. By adopting a carbon budgeting system based on the global average reduction rate, the Irish government's mitigation measures fail to reflect equity and the principle of common but differentiated responsibilities and respective capacities ('CBDR-RC').[22] It follows, the plaintiffs argue, that the government's current mitigation policies fall foul of the government's implied constitutional obligation to ensure vigilant and effective environmental protection as identified by the Irish High Court in *Brownfield Restoration v Wicklow County Council*[23] ('*Brownfield*'). The essence of the plaintiffs' case is that, to discharge its constitutional obligation to protect the environment, the government must 'do everything in its power'[24] to protect the plaintiffs from the adverse impacts of climate change. This translates into an individual and collective obligation on the Irish government to keep global heating as close to 1.5°C above pre-industrial levels as possible, through deep, rapid, and immediate emission reductions at the domestic level and international cooperation to ensure global emission reductions.

20 Jim Skea et al, *Climate Change 2022: Mitigation of Climate Change. Summary for Policymakers* (Intergovernmental Panel on Climate Change, 2022), 17, 23.

21 Sections 3(1) and 6A(5) of the *Climate Act* (n 5) ss 3(1) and 6A(5).

22 These principles are enshrined in articles 3(1) and 4(1) of the *United Nations Framework Convention on Climate Change*, opened for signature 4 June 1992, 1771 UNTS 107 (entered into force 21 March 1994) ('UNFCCC'); and articles 2(2), 4(1), and 4(3) of the Paris Agreement (n 2). These principles are also incorporated into the Irish *Climate Act* (n 5) via section 3(3) which requires the Minister for the environment/government to perform their functions in relation to carbon budgets, sectoral emissions ceilings, and climate policy planning in a manner consistent with article 2 of the UNFCCC and articles 2 and 4(1) of the Paris Agreement.

23 *Brownfield Restoration v Wicklow County Council* [2017] IEHC 456 ('*Brownfield*').

24 *Kolyadenko and Others v Russia* (2012) Eur Court HR [216]; *Budayeva and Others v Russia* (2008) Eur Court HR [175].

The government denies that the plaintiffs are entitled to the relief sought on a number of grounds. The government accepts the science of climate change and acknowledges that residents of Ireland, including the plaintiffs, are experiencing disruptions to their daily lives from climate-induced heatwaves, storms, and floods. However, the government disputes that an implied constitutional duty to ensure effective and vigilant environmental protection can be derived from the text or structure of the Irish Constitution. Furthermore, the government contests the plaintiffs' assertion that the Constitution demands more ambitious mitigation measures than those being delivered through Ireland's statutory carbon budgeting system. The government acknowledges the obligation to uphold the principle of the best interests of the child. It argues, however, that Ireland is not responsible for the climate harm suffered by the children in this case for two reasons. Firstly, Ireland's emissions are only partly to blame (the international community of States has together caused the climate to heat). Secondly, the best interests of children include consideration of how to best manage the economy of Ireland, to ensure the financial health of all citizens. The government argues that successive governments have done this to the best of their ability.

Children's Rights

Relevant Law

The United Nations *Convention on the Rights of the Child* constitutes the cornerstone for the rights and freedoms of children all over the globe. It sets out the thorough range of economic, civil, and political rights for all children. It has by now been signed, ratified, and /or incorporated into international law by every country in the world.

A provision of the Irish Constitution – Article 42A – was approved by referendum in 2012 by the Irish people. This provision explicitly recognises the rights of children and identifies children as a unique category of people whose rights must be protected. It recognises children's distinct attributes and vulnerabilities. Article 42A(1) states that '[t]he State recognises and affirms the natural and imprescriptible rights of all children and shall, as far as practicable, by its laws protect and vindicate those rights'.

In 2030, the Oireachtas (Irish houses of Parliament) incorporated the main United Nations human rights treaties into domestic law, as have most other jurisdictions. This was done via the *International Human Rights Law Standards Act 2030*, which states that these international human rights standards would be given equal standing as compared with the highest source of law, the Irish Constitution. When this occurred, added to Article 42A(1) was the wording: '[c]hildren in Ireland will have their rights vindicated in line with the requirements of the United Nations *Convention on the Rights of the Child*'. Today, in 2033, children's rights therefore have a strong basis in Irish national law, deriving from both the Irish Constitution and international human rights law standards which have been incorporated into Irish law.

It is necessary to establish whether the State has in this instance breached the rights of the young plaintiffs by, firstly, failing to mitigate the harm they have suffered due to the climate crisis (by giving insufficient consideration to their best interests/rights), and secondly, failing to provide remedies for those affected by it.

The Principle of the Best Interests of the Child

Although rights are drafted in vague terms, there are numerous sources through which definitions can be provided for those rights in particular scenarios. As well as domestic sources, comments of the United Nations Committee on the Rights of the Child and other United Nations sources now have strong persuasive guidance in our interpretation in this country of the nature of human rights.

All children have the right to have their best interests given due consideration in matters which affect them. This is, then, a crucial right in both the United Nations *Convention on the Rights of the Child* (Article 3) and also in domestic law systems (in Ireland that is via the Irish Constitution Article 42A) – indeed, globally it is the most prominent right concerning children. It applies to decisions of law and policy in relation to an individual child; or those relating to children as a group.

This includes in the area of environmental issues. The Special Rapporteur on Human Rights and the Environment emphasised in 2018 that states have special obligations to protect children from environmental harm, due to their relative vulnerability.[25] In 2023, the Committee on the Rights of the Child via General Comment No 26, on children's rights and the environment with a special focus on climate change, stated that:

> Environmental decisions generally concern children and the best interests of the child shall be a primary consideration in the adoption and implementation of environmental decisions affecting children. These include laws, regulations, policies, standards and guidelines, plans and strategies, budgets, international agreements and development assistance.[26]

25 John H Knox, *Report of the Special Rapporteur on the Issue of Human Rights Obligations Relating to the Enjoyment of a Safe, Clean, Healthy and Stable Environment*, UN Doc A/HRC/37/58 (24 January 2018) [38].

26 Committee on the Rights of the Child, *Draft General Comment No 26: Children's Rights and the Environment with a Special Focus on Climate Change* (9 December 2021) ('CRC Draft General Comment No 26') [52].

It was anticipated in 2022 that, if coastal rises continued at the rate of increase experienced at that time, a significant amount of Irish coastline will be submerged by 2030. A map produced by the organisation Coastal Climate Central[27] outlined the scenarios that would be faced in the coming ten years if sea levels rose even further to five metres above the high tide line. Many coastal towns in Ireland would be badly affected. Clearly, when it was established in 2020 that sea levels were rising, the impact of this on the rights and interests of children should have been considered. This became a particularly strong obligation on the State when the 2030 Act was passed.

Children's Rights Impact Assessments

It is now well established in international practice that children's rights impact analyses must be conducted in any processes likely to affect children. Indeed, this was the case by 2030. The Committee on the Rights of the Child has outlined the circumstances under which this should occur:

> Ensuring that the best interests of the child are a primary consideration in all actions concerning children (art. 3 (1)), and that all the provisions of the Convention are respected in legislation and policy development and delivery at all levels of government demands a continuous process of child impact assessment (predicting the impact of any proposed law, policy or budgetary allocation which affects children and the enjoyment of their rights) and child impact evaluation (evaluating the actual impact of implementation). This process needs to be built into government at all levels and as early as possible in the development of policy.[28]

The importance of environmental assessment by any State, and the obligation to particularly consider children's potential vulnerabilities (and participation), has also been emphasised by the Special Rapporteur on human rights and the environment. Any projects, laws, and policies should avoid any harm to human rights.[29] Children's rights impact assessments must be based on the *Convention on the Rights of the Child* as a framework. This means that assessments should be underpinned by the main principles of the *Convention on the Rights of the Child*, and on input from children themselves.[30]

27 See Climate Central, 'Costal Risk Screening Tool' *Climate Central* (Web Page, 2021) <https://coastal.climatecentral.org>.

28 Committee on the Rights of the Child, *General Comment No 5: General Measures of Implementation of the Convention on the Rights of the Child*, 34th sess, UN Doc: CRC/GC/2003/5 (27 November 2003) ('CRC General Comment No 5') section E.

29 Knox (n 25) [46].

30 Committee on the Rights of the Child, *General Comment No 14 (2013) on the Right of the Child to Have His or Her Best Interests Taken as a Primary Consideration*, 62nd sess, UN Doc: CRC/C/GC/14 (29 May 2013) [99].

The right of children for their best interests to be a primary consideration in policy-making requires that research be undertaken to ensure adequate information and analysis of the scope of a particular problem such as global heating (and the consequently rising oceans). The Committee on the Rights of the Child states that:

> [c]ollection of sufficient and reliable data on children, disaggregated to enable identification of discrimination and/or disparities in the realization of rights, is an essential part of implementation.[31]

It is an international obligation that data be collected across all ages, and throughout the jurisdiction – including in coastal communities. Given the impact of global heating on coastal communities highlighted in the Coastal Climate Central map, the State had an obligation to conduct specific research into the impact of the climate crisis on children in that particular demographic. Moreover, in such data-gathering exercises, data must include the views of children themselves as a crucial source.[32]

Ireland Context

While Ireland has managed to stay within its carbon budget for 2030, and reduced its emissions by 44.5% relative to 1990 levels, these measures have not been enough either individually or collectively to ensure or make a fair share contribution to limiting global heating to 1.5°C above pre-industrial levels. Conversely, many European Union countries such as Germany had far more ambitious emission reduction targets for 2020 and 2030 and managed to achieve them. Successive climate mitigation policies adopted under Ireland's *Climate Act* have failed to explicitly consider children's rights and interests, particularly the devastating impact that reaching 1.3°C of heating is already having on children's lives and futures. Ireland has clearly failed to uphold the standards by which the State has agreed to abide.

The Committee makes clear that foreseeable harm must be mitigated:

> Adaptation measures should target both short- and long-term impacts, such as those aimed at sustaining livelihoods and developing sustainable water management systems. Measures that are necessary to protect children's rights to life and health from imminent threats, such as extreme weather and floods, should include establishing early warning systems and risk notification and increasing the physical safety and resilience of infrastructure, including school, water and sanitation and

31 CRC General Comment No 5 (n 28) [48].
32 Ibid [50].

> health infrastructure, to reduce the risk of climate-related hazards. States should adopt emergency response plans such as provision of humanitarian assistance and access to food and water and sanitation.[33]

The Committee also emphasises that planning must include explicit consideration of the United Nations *Convention on the Rights of the Child*: 'When determining the appropriateness of their mitigation measures in accordance with the Convention, States should take into account the following criteria: (a) Mitigation measures should clearly indicate how they respect, protect and fulfil children's rights under the Convention'.[34]

Children have enjoyed processes in Ireland whereby their views were gathered, such as the National Youth Assembly on Climate in 2021 and the Children and Young People's Assembly on Biodiversity Loss in 2022. No explicit plan was put in place, however, to ensure the implementation of these views, nor was a children's rights impact assessment conducted. Furthermore, there has been no explicit consideration of the rights and interests of children who are suffering harm due to the effects of climate change, including rising sea levels.

A Constitutional Duty to Ensure Vigilant and Effective Environmental Protection

In addressing the issue of whether the Irish government's climate mitigation measures are compatible with its constitutional duties, the court shall structure its analysis around the following two questions. Firstly, is there an implied constitutional duty to ensure effective and vigilant environmental protection and, if so, is the duty implicated by climate change? Secondly, if the court identifies a duty and finds that it is implicated by climate change, do the government's climate mitigation measures constitute an unconstitutional dereliction of the government's duty?

A Constitutional Duty

The existence of an implied constitutional obligation to ensure vigilant and effective environmental protection is contested by the government. The government argues that the existence of constitutional environmental rights is *res judiciata*. It cites a finding of this Court in Case Ireland, where it was held that a constitutional right to a healthy environment was 'superfluous' if it could be encompassed within the right to life or bodily integrity, or

33 CRC Draft General Comment No 26 (n 26) [106].
34 Ibid [111].

'excessively vague and ill-defined' if it did extend beyond these rights.[35] The plaintiffs argue that the government has misconstrued the Supreme Court's *dicta* in Case Ireland, emphasising that the Court limited itself to a finding that 'as thus formulated' a *right* to a healthy environment could not be derived from the Constitution. The plaintiffs stress that this Court, the Irish Supreme Court, has not yet ruled on whether a constitutional *duty* to protect the environment can be implied from existing constitutional rights, obligations, and values. The Irish High Court in *Brownfield* previously held that an:

> implied constitutional commitment to intergenerational solidarity [reflected in the children's rights provisions and the directive principles of social policy of the Irish Constitution] militates against… a lax and forgiving approach to fundamental issues of stewardship of the environment in trust for future generations. Vigilant and effective protection of the environment is an implied constitutional obligation, to be laid at the door of private parties as well as the State.[36]

The Irish High Court makes a cogent case for the identification of a constitutional duty of vigilant and effective environmental protection. Specifically, the High Court identified a clear 'root of title in the text/structure of the Constitution'[37] from which such a constitutional duty could be derived. Such a duty could, of course, also stem from the constitutional values of dignity, freedom, and the democratic nature of the State, given that a healthy environment is a pre-condition for the realisation of each of these constitutional values.[38] It is the view of this Court that the constitutional duty of environmental protection must necessarily entail climate protection, as there cannot be a healthy environment in the absence of a safe and stable climate system.

Compatibility of the Government's Mitigation Measures with the Implied Constitutional Duty

In delineating the parameters of this implied constitutional duty, it is instructive to look to the judgment of the German Constitutional Court in *Neubauer v Germany*, in which a similar express constitutional obligation, Article 20a, featured prominently. Article 20a of the Basic Law provides that 'the State, mindful also of its responsibility towards future generations, shall protect the natural foundations of life and animals'. In *Neubauer*, it was held that Germany's emission reduction measures were incompatible with the State's

35 Case Ireland (n 13) [9.5].
36 *Brownfield* (n 23) [307].
37 Case Ireland (n 13) [8.6].
38 See the Preamble and Article 5 of the *Constitution of Ireland 1937*.

constitutional duty to protect against climate harm and to take climate action insofar as the emission reduction measures 'irreversibly offload[ed] major emission reduction burdens onto periods after 2030' without specifying reductions for the period after 2030.[39] The Court, referring to Article 20a, found that the global temperature goal of limiting global heating to well below 2°C is more than 'an expression of political will'; it must be understood as 'a specification of the climate action required under constitutional law'.[40] Given the global consensus that the safest guardrail is to limit global heating to as close to 1.5°C above pre-industrial levels as possible, it is difficult to see any valid reason why this temperature goal should not form an important benchmark for constitutionally required climate protection in Ireland.

The German Constitutional Court's explanation of how Article 20a operates in litigation, in the protection of future generations, and *vis-à-vis* competing constitutional rights, is also helpful. The Constitutional Court described Article 20a as a justiciable provision designed to bind the political process to favour environmental interests and safeguard future generations.[41] Article 20a does not take absolute precedence over all other constitutional rights, principles, and interests but must be balanced against them.[42] However, within this balancing process, the constitutional obligation to take climate action should be accorded increasing weight as the climate crisis worsens.[43] The German Constitutional Court made clear that Article 20a imposes limitations on the legislature's scope for decision-making and creates a 'special duty of care' towards future generations, particularly where there is scientific uncertainty about the possibility of serious and irreversible environmental impacts.[44] The Court indicates that Article 20a is concerned with how environmental burdens are shared out between different generations and therefore obliges the State to protect the natural foundations of life for future generations to ensure that they 'are not forced to engage in radical abstinence'.[45]

It is our view that the implied constitutional duty under Irish constitutional law should be seen to share these core characteristics, given this obligation is also designed to ensure intergenerational solidarity. In circumstances where the government's current carbon budgeting system is not rationally connected

39 See the Court's Press Release in English: Bundesverfassungsgericht [German Constitutional Court], 'Constitutional Complaints Against the Federal Climate Change Act Partially Successful' (Media Release, 29 April 2021) <https://www.bundesverfassungsgericht.de/SharedDocs/Pressemitteilungen/EN/2021/bvg21-031.htm>
40 *Neubauer* (n 16) [209].
41 Ibid [197].
42 Ibid.
43 Ibid.
44 Ibid [229].
45 Ibid [193].

to the +1.5°C temperature goal and ignores considerations of equity and CBDR-RC, it must be concluded that the government's current approach to mitigation falls short of what the implied constitutional duty of vigilant and effective environmental protection requires. The government has failed to proffer a satisfactory justification for unevenly distributing the emission reduction burden to such an extent between older generations and children and young people, including the plaintiffs, who are now suffering acutely from the impacts of the climate crisis. To quote the German Constitutional Court, the principle of proportionality dictates that '[o]ne generation must not be allowed to consume large portions of the CO_2 budget while bearing a relatively minor share of the reduction effort if this would involve leaving subsequent generations with a drastic reduction burden and expose their lives to comprehensive losses of freedom'.[46] It must be acknowledged that, having used up more than Ireland's fair share of the global carbon budget, the government has unjustifiably abdicated its constitutional duty. Children and young people in Ireland are already experiencing a comprehensive loss of freedoms and are therefore entitled to the declaratory relief sought.

Remedies

The Court declares that the government is under a legal and constitutional obligation to rapidly increase the scale and speed of its domestic climate mitigation measures so as to make a fair share contribution to the global temperature goal of limiting global warming to as close to 1.5°C above pre-industrial levels as possible. The government's current approach to climate mitigation falls significantly short of what is required under constitutional and human rights law. The Court acknowledges that it remains the prerogative of government to determine the sectors of the economy that must bear the greatest emission reduction burdens. This Court may nevertheless stipulate the minimum legal requirements for any updated climate mitigation policies. In updating its climate mitigation policies, the government must explain how it has taken into account equity and CBDR-RC in the formulation of its mitigation measures. The government must also explain how its updated mitigation measures constitute a fair share contribution, which is sufficient and collectively consistent with meeting the +1.5°C temperature goal. Finally, the government's updated climate mitigation policies must undergo a full and rigorous children's rights impact assessment to ensure the best interests of children and future generations are fully taken into account in the formulation of policy. If the government does not take swift action to give effect to the Court's findings within six months, the plaintiffs may reapply to this Court for injunctive relief.

46 Ibid [192].

Conclusion

For the reasons set out above, the Court finds in favour of the plaintiffs.

The extent and magnitude of the rise in sea levels has undoubtedly caused devastation for the plaintiffs, and indeed for numerous children in coastal communities in Ireland. The environmental harm is an urgent and systemic threat to children's rights globally. We must commend the efforts of these children and youth plaintiffs who have sought to hold successive governments to account for the failure to adequately consider their rights and interests, and the failure to mitigate the climate crisis.

Chapter 8

The Truth and Reparations Commission

Climate Reparations for the Anthropocene

Zoe Nay and Julia Dehm

Re Future Generations (2035) Truth and Reparations Commission of Oceania

Commentary

Like other judgment projects, this speculative practice of writing judgments for the Anthropocene is an 'exercise in the "art of the possible"'.[1] In donning our 'imaginative robes', we write to prefiguratively enact the types of normativities and legal adjudicative practices that will be needed to address the structural injustices of the Anthropocene.[2] In our judgment, we thus adopt an orientation to the future that maintains a hopefulness about the possibilities of drastic economic, political, and social transformation to mitigate climate change, even as we simultaneously recognise that historical accumulation of anthropogenic greenhouse gas emissions already in the atmosphere constrains the conditions under which the future will be collectively made.[3] We situate our judgment in the near future – 2035 – to foreground the urgency for bold policy intervention to break from currently hegemonic fossil-fuel intensive models of capitalist economic growth in what the Intergovernmental Panel on Climate Change (IPCC) has called the 'critical decade'.[4] In writing from

This chapter has benefited enormously from the valuable comments of Tim Lindgren and Erin Fitz-Henry at the University of Melbourne. We thank them for their careful and insightful feedback.

1 Erika Rackley, 'Why Feminist Scholars Should Write Judgments: Reflections on the Feminist Judgments Project in England and Wales' (2012) 24(2) *Canadian Journal of Women and the Law* 389, 392.
2 Amy J Cohen and Bronwen Morgan, 'Prefigurative Legality' *Law & Social Inquiry* (forthcoming).
3 See Julia Dehm, 'International Law, Temporalities and Narratives of the Climate Crisis' (2016) 4(1) *London Review of International Law* 167.
4 Intergovernmental Panel on Climate Change, *Global Warming of 1.5°C: An IPCC Special Report on the Impacts of Global Warming of 1.5°C Above Pre-Industrial Levels and Related Global Greenhouse Gas Emission Pathways, in the Context of Strengthening the Global Response to the Threat of Climate Change, Sustainable Development, and Efforts to Eradicate Poverty* (Report, October 2018) Summary for Policymakers 3–24 ('IPCC').

DOI: 10.4324/9781003389569-10

the vantage point of 2035, and providing a history of the future, we hope to make potential future trajectories that might currently seem unimaginable appear as real potentialities that could be collectively struggled for and enacted. While utopian, we have aspired to make these imagined future trajectories appear plausible, by highlighting how the legal developments we narrate are the product of an ongoing history of struggle and have their roots in already existing demands, campaigns, and social movement advocacy.

This hopeful but also critically realistic orientation we adopt to the future also characterises our engagement with law and the legal form. On one level, our objective is to crystallise utopian longings in judicial form; however, our speculative judgment simultaneously seeks to acknowledge how the judicial form itself is constraining and limiting, especially in how it is premised upon the authority of the judicial voice and posits a singular truth. Thus, while our judgment seeks to subvert and stretch the judicial form, by incorporating often excluded voices and drawing on literature and poetry, it ultimately shows that in order to enact a more expansive notion of reparative justice, we will also need to imagine different practices of judgment, of narration, and of truth-telling.

Situating the 'Truth and Reparations Commission'

There is a growing call for climate change reparations as a key aspect of climate justice. Such demands for some form of reparation or compensation for the impacts of climate change are not new but were for many years sidelined and marginalised.[5] Since the drafting of the *United Nations Framework Convention on Climate Change*[6] ('UNFCCC') in the early 1990s, small island states have called for an insurance mechanism to provide redress for loss and damage associated with climate change. Over three decades later, at COP27 in Sharm El Sheikh, the international community finally established a fund to finance loss and climate-related loss and damage in especially vulnerable developing countries. However, the climate regime still does not establish any obligations on those who have an oversized responsibility for causing the climate crisis to contribute funding to address climate-induced loss and damage or to provide climate reparations.

Contemporary discussions about climate reparations draw on long-standing demands for the repayment of ecological and climate debts.[7] Tendayi

5 Julia Dehm, 'Climate Change, "Slow Violence" and the Infinite Deferral of Responsibility for "Loss and Damage"' (2020) 29(2) *Griffiths Law Review* 220, 225–35.

6 *United Nations Framework Convention on Climate Change*, opened for signature 4 June 1992, 1771 UNTS 107 (entered into force 21 March 1994).

7 See, eg, Karin Mickelson, 'Leading Towards a Level Playing Field, Repaying Ecological Debt, or Making Environmental Space: Three Stories About International Environmental Cooperations' (2005) 43(1) *Osgoode Hall Law Journal* 137.

Achiume has stressed that '[r]eparations require addressing historic climate injustice, as well as eradicating contemporary systemic racism that is a legacy of historic injustice in the context of the global ecological crisis'.[8] Thus, as Maxine Burkett has identified, reparations are both backward- and forward-looking, as programmes 'justified by past harms and are also designed to assess and correct the harm and improve the lives of the victims into the future'.[9] Similarly, Olúfẹ́mi O Táíwò foregrounded how reparations must be a 'constructive project': in order to contend with the immense scale of injustice requires 'thinking more broadly about how to remake the world system'.[10] Writing with Beba Cibralic, they warn against conceptualising reparations as a one-off cash transfer or apology and instead propose an understanding of reparations as a 'systemic approach to redistributing resources and changing policies and institutions that have perpetuated harm'.[11] Building on this, Sarah Mason-Case and Julia Dehm have argued that, because climate change is inextricable from the 'unfinished business of decolonisation', solutions must entail the full decolonisation of international law.[12] Similarly, Sahar Shar has identified that to fully comprehend, let alone to start to repay, such a 'climate debt' requires moving beyond the confines of Western epistemologies or worldview given that '[t]he Global North owes a debt that we cannot fully comprehend with the cognitive tools given to us by Western worldviews'.[13] Responding to this demand, she suggests, requires an ethics of relationality and connectedness and a 'constellation of commitments' that has as their collective aim 'the profound transformation of the global system and national economies'.[14]

To explore the themes of reparative justice our speculative judgment grapples with, we considered it necessary to imagine a new institution – the

8 E Tendayi Achiume, 'Report of the Special Rapporteur on Contemporary Forms of Racism, Racial Discrimination, Xenophobia and Related Intolerance on Ecological Crisis, Climate Justice and Racial Justice', UN Doc A/77/549 (25 October 2022) 23.

9 Maxine Burkett, 'Climate Reparations' (2009) 10(2) *Melbourne Journal of International Law* 509, 522–3.

10 Olúfẹ́mi O Táíwò, *Reconsidering Reparations* (Oxford University Press, 2022) 1.

11 Olúfẹ́mi O Táíwò and Beba Cibralic, 'The Case for Climate Reparations', *Foreign Policy* (10 October 2020) <https://foreignpolicy.com/2020/10/10/case-for-climate-reparations-crisis-migration-refugees-inequality/>.

12 Sarah Mason-Case and Julia Dehm, 'Redressing Historical Responsibility for the Unjust Precarities of Climate Change in the Present' in Benoit Mayer and Alexander Zahar (eds), *Debating Climate Law* (Cambridge University Press, 2021) 170, 187; referencing Vasuki Nesiah, 'German Colonialism, Reparations and International Law', *Völkerrechtsblog* (Blog Post, 21 November 2019) <https://voelkerrechtsblog.org/german-colonialism-reparations-and-international-law>.

13 Sahar Shah, 'What is Owed Towards a Concept of Climate Reparations', *It's Freezing in LA!* (Blog Post, 14 October 2021) <https://www.itsfreezinginla.co.uk/post/what-is-owed-towards-a-concept-of-climate-reparations>.

14 Ibid.

fictional Truth and Reparations Commission of Oceania – inspired by Truth and Reconciliations Commissions established in the transitional justice context, most famously in South Africa after the end of Apartheid.[15] However, in contrast to transitional justice mechanisms, which seek to enable victims and perpetrators to live together again in the aftermath of atrocity through a process of truth-telling,[16] our envisioned Tribunal specifically does not include 'reconciliation' in its name or objectives. There can be no just co-existence with those who have done the most to cause the climate crisis, such as the 90 'Carbon Majors' who are responsible for two-thirds of greenhouse gas emissions since the Industrial Revolution.[17] The fossil fuel corporations, who have actively spread climate disinformation and promoted delay to allow them to continue to make profits of $2.8 billion daily[18] and benefit from ongoing public subsidies of almost $6 trillion in subsidies annually,[19] need to be brought under democratic control and publicly expropriated to provide compensation and financial support to those who are suffering from the impacts of climate change. Our imagined Truth and Reparations Commission of Oceania is a specialised body mandated to provide forms of compensation and redress to those within its geographical jurisdiction who are suffering from climate harms, and financially able to do so by redirecting the expropriated value generated from fossil fuel extraction to remedying the harms that those activities have caused.

Judicial Storytelling, Pluralising Voices, and Epistemic Humility

In an effort to resist and reimagine the constraints of judicial form, our judgment draws upon extra-legal literature and poetry, prioritising the voices and stories of Indigenous Australian and Pacific Islander writers. In doing so, we take seriously the traditions of storytelling in the region and reflect on the

15 South African Truth and Reconciliation Commission, 'The TRC Report' <https://www.justice.gov.za/trc/report/index.htm>.

16 See, eg, Valeria Vázquez Guevara, 'Living Together After Violent Conflict: Museum-Making as Lawful Truth-making' in Shane Chalmers and Sundhya Pahuja (eds), *Routledge Handbook of International Law and the Humanities* (Routledge, 2021) 379.

17 Richard Heede, 'Tracing Anthropogenic Carbon Dioxide and Methane Emissions to Fossil Fuel and Cement Producers, 1854–2010' (2014) 122 *Climatic Change* 229; Douglas Starr, 'Just 90 Companies Are to Blame for Most Climate Change, This "Carbon Accountant" Says', *Science* (online, 25 August 2016) <https://www.science.org/content/article/just-90-companies-are-blame-most-climate-change-carbon-accountant-says>.

18 Damian Carrington, 'Revealed: Oil Sectors "Staggering" $3Bn-a-Day Profits for Last 50 Years', *The Guardian* (online, 21 July 2022) <https://www.theguardian.com/environment/2022/jul/21/revealed-oil-sectors-staggering-profits-last-50-years>.

19 'Fossil Fuel Subsidies', *International Monetary Fund* (Web Page, 2022) <https://www.imf.org/en/Topics/climate-change/energy-subsidies>.

ways in which truth-telling practices could be transformed to allow for the recognition of multiple truths. We have also chosen to focus on representing the stories of children, youth, and future generations. Children and youth are disproportionately impacted by the adverse effects of climate change, from sudden-onset impacts such as floods, cyclones, and storm surges to slow-onset impacts such as food and water insecurity, forced displacement and migration, and disruptions to education. As the impacts of climate change increase in frequency and severity, future generations are inheriting a world increasingly hostile to human life. At the same time, children and young people have played a critical role in raising the profile of transformative approaches to climate justice and imbuing climate law with storytelling. Scholars have documented the nascent trend of youth-led climate litigation serving as a form of 'disruptive dissent',[20] through which children and youth seek to '*transform* norms, rules, regulations and institutions *within* existing political and economic structures'.[21] Beyond the courtrooms, Indigenous and Pacific youth have formed non-governmental organisations and grassroots movements to enhance action to address climate change. This can be seen in the normative force of the Pacific Island Students Fighting Climate Change, who appealed to the Government of Vanuatu to seek an Advisory Opinion before the International Court of Justice on matters of international law and climate change. Likewise, Australia's first Indigenous youth climate network, Seed, has brought together Aboriginal and Torres Strait Islander youth with the aim to pave the way for healthy, clean, and sustainable societies built on strong cultural and community connections. Our judgment draws inspiration from youth mobilisation movements in the region and takes seriously their calls for climate justice when considering what climate reparations could and should be.

Finally, our Anthropocene judgment highlights the limitations of existing settler-colonial law in responding to the injustices of climate change. There is a growing body of literature that highlights the way in which law has been complicit in the creation of Anthropocene conditions and the need for radically reimagining law and legality in and for the Anthropocene.[22] Building on this literature, we highlight the importance of multiple forms of knowledge to properly comprehend climate harms and thus complement global

20 Karen O'Brien, Elin Selboe, and Bronwyn M Hayward, 'Exploring Youth Activism on Climate Change: Dutiful, Disruptive, and Dangerous Dissent' (2018) 23(3) *Ecology & Society* 42.

21 Larissa Parker et al, 'When the Kids Put Climate Change on Trial: Youth-Focused Rights-Based Climate Litigation Around the World' (2022) 13(1) *Journal of Human Rights and the Environment* 64, 70 (emphases in original).

22 Julia Dehm and Usha Natarajan (eds), *Locating Nature: Making and Unmaking International Law* (Cambridge University Press, 2022); Margaret Davies, *Ecolegality: Legality, Life and the Normativity of Nature* (Routledge, 2022).

scientific modelling of climate impacts and harms with forms of knowledge that are situated and embedded and emerge from specific standpoints of marginalisation.[23] As Farhana Sultana suggests, a conceptual and abstract analysis needs to be 'interlace[d] with the fleshiness of climate, the pasts and presents in our bodies, minds, soils, kin – where the theory is in the flesh, and struggles form the basis of political consciousness and oppositional epistemologies against oppression in shared worlds'.[24] If the work of reparations is to be constitutive and (re)world-making, it needs to push beyond the confines of Western epistemologies or worldviews and create space for the alternative epistemologies and cosmologies that have been devalued in Eurocentric modernity. Such knowledges are produced not just through the grounded lived experience of climate impacts but are forged in the practice of resistance to climate coloniality.

Judgment

The three young representative plaintiffs from communities in Oceania disproportionately impacted by climate change are seeking acknowledgement and compensation for the impact of climate change on behalf of future generations. The three plaintiffs have all bravely shared their diverse experiences of climate loss with the Tribunal reflecting the different impacts of climate change across space, place, and time. While they have shared their stories with the aim of advancing both climate justice and climate reparations, we stress that these stories are not just a *means* to climate justice but that enacting climate justice demands that we 'listen more carefully to those most vulnerable to the ravages of climate change, such as Oceania's frontline communities'.[25] As Julian Aguon reminds us,

> we ... need more than facts to win. We need stories. And not just stories about the stakes, which we know are high, but stories about the places we call home. Stories about our own small corners of the Earth as we know them. As we love them.[26]

While we will outline the contours of their stories here, we urge all to read their stories in full, to grasp the gravity of the situation at hand. In the words

23 Farhana Sultana, 'The Unbearable Heaviness of Climate Coloniality' (2022) 99 *Political Geography* 102638.

24 Ibid 3.

25 Julian Aguon, 'To Hell with Drowning', *The Atlantic* (online, 1 November 2021) <https://www.theatlantic.com/culture/archive/2021/11/oceania-pacific-climate-change-stories/620570/>.

26 Ibid.

of Kathy Jetñil-Kijiner, to listen to the voices of those impacted by climate change is to 'remember / that beyond / the discussions / are faces'.[27]

Facts Submitted by the Plaintiffs

The plaintiffs are Amy MacLeary, a Wangan and Jagalingou woman from the area known as Queensland's Galilee Basin; Ciara Saunders, from Zenadth Kes, also known as the Torres Strait Islands; and Malika Gorrie, from the Marshall Islands. Amy has given testimony about the Wangan and Jagalingou Peoples' sacred connection to their country, and how they have protected their lands, water, people, history, and totems for thousands of years. In her evidence, Amy drew attention to the devastating impacts of the first and only coal mine in the Galilee Basin, the Carmichael coal mine, that was approved when she was six years old. Although the mine's approval was withdrawn by the Queensland government in 2024, due to ongoing resistance by the Wangan and Jagalingou Peoples, in its short period of operation the mine depleted the land, degraded sacred sites, and had cascading impacts on the Great Barrier Reef, surrounding wetlands and freshwater springs, and biodiversity in the region. Despite efforts to foster adaptation and repair climate impacts, Amy is acutely aware that some loss and damage cannot be repaired or financially compensated, including loss of ancestral lands and burial sites, biodiversity, and traditional knowledges. Due to the ongoing impacts of unjust climate change policies, Amy seeks reparations for current and future generations of Wangan and Jagalingou Peoples.

Ciara Saunders is a Gudamalulgal woman from the island of Boigu in Zenadth Kes, also known as the Torres Strait Islands of Australia and its surrounding seas. Torres Strait Islander Peoples have inhabited the islands of Zenadth Kes for over 70,000 years. As a consequence of climate change, Ciara has witnessed those islands being eroded and inundated by sea-level rise; she has experienced the impacts of biodiversity loss and coral bleaching on her communities' way of life; and she has been forced to contemplate what it would mean to be separated from her ancestral lands. Ciara emphasises that, to her and her peoples, Zenadth Kes is not merely territory but also essential to identity and cultural heritage. Thus, while climate finance is an essential first step to providing climate reparations, the loss and damage experienced in Zenadth Kes is not quantifiable or substitutable.

Malika Gorrie is a young woman from the Marshall Islands. In her statement, she conveyed her bittersweet experience of giving birth to her young daughter. She emphasised that the world her daughter is inheriting has been made more precarious by climate change – with food and water insecurity an

27 Kathy Jetñil Kijiner, 'Poem: 2 Degrees' (30 June 2015) <https://www.kathyjetnilkijiner.com/poem-2-degrees/>.

ongoing concern for Marshallese people. Malika emphasised that the adverse impacts of climate change do not occur in isolation – they compound histories of environmental extraction, exploitation, colonisation, and nuclear testing in the Marshall Islands. Locating loss and damage within longer histories of the Marshall Islands and recognising that the impacts of climate change will continue for generations to come, Malika calls on the Commission to establish an ongoing voice for young people in Oceania's climate decision-making processes. Malika and her fellow plaintiffs contend that a voice for young people is necessary to ensure future climate policies and plans work towards building a better world for generations to come – so their peoples can continue to grow up on their islands, to learn from their lands and seas, to develop deep connections with place, and to engender solidarity in Oceania.

In reviewing the plaintiffs' statements, the Commission is reminded of the words of novelist Octavia E Butler: 'The world is full of painful stories. Sometimes it seems as though there aren't any other kind and yet I found myself thinking how beautiful that glint of water was through the trees'.[28] The plaintiffs have clearly articulated the ways in which climate change has impacted them – as individuals, as women, as members of their communities, and as peoples of Oceania. Their stories provide an insight into the diverse and complex ways in which climate change has shaped the lives of young people and future generations, from sea-level rise shaping their lands to climate-related instability shaping their career choices and connections to place. Equally, it is important to recognise how the plaintiffs have shaped their worlds. They have emphasised their deep connections to place and their commitments to protecting their peoples, lands, and seas in the face of climate change, even though they and their communities have contributed least to its causes.

This Tribunal accepts the statements provided by the plaintiffs as authoritative evidence of the impacts of climate change in the region: their intimate knowledge of their lands and waters and their access to traditional stories that narrate ecological change mean they have an important epistemic viewpoint, from which this Tribunal can learn. Their accounts not only complement but deepen the facts that have been comprehensively established by climate science. The Tribunal recognised how the perspective of Indigenous peoples 'arise from memories, knowledges, histories, and experiences of oppression that differ from many of the nonindigenous scientists, environmentalists, and politicians who are prominent in the framing of the issue of climate change'.[29] Although it was known since at least 1990, that Island and coastal communities were among the most vulnerable to the adverse impacts

28 Olivia E Butler, *Parable of the Sower* (Headline, 2019) 248.

29 Kyle Whyte, 'Indigenous Climate Change Studies: Indigenizing Futures, Decolonising the Anthropocene' (2017) 55(1–2) *English Language Notes* 153, 153.

of climate change,[30] industrialised countries continued to pursue a hegemonic model of fossil-fuel intensive neoliberal capitalist growth that relied upon an extractive and exploitative relationship to the natural world. Countries of the Global North failed to provide adequate funds for adaptation and in international negotiations actively avoided the question of liability for the losses and damages associated with climate change.[31] When the planet reached the crucial 1.5°C warming limit in 2030, Oceania had long since been experiencing extreme heat events, food and water insecurity, more frequent and severe sudden-onset disasters, and slow and structural impacts, such as sea-level rise, ocean acidification, and the inequitable distribution of financial resources.

While, in retrospect, it is incomprehensible that it took until 2025 for countries around the world to galvanise the political will to take the urgent, scaled-up, ambitious action, thankfully the drastic steps taken in the mid-2020s have prevented temperature increase from exceeding 1.6°C to date and may still allow temperature increases to stabilise at 1.5°C provided transformative action continues to occur across all sectors and geographies. Yet even though worst-case scenarios have been avoided, we live in a climate-transformed world: the equatorial Pacific experiences 100 days of marine heat waves annually, destabilising subsistence-based fishing practices, marine ecosystems, and ways of life. Further, sea-level rise has displaced peoples – including some of the plaintiffs – from their ancestral lands, destroyed coastal burial sites, and contributed to soil salinisation, with cascading impacts across Oceania's Islands and Atolls.

The disproportionate impacts of climate change experienced in Oceania build upon colonial policies of dispossession, forced relocation, assimilation, environmental exploitation, and debt injustice. Indigenous scholars have described climate change as an 'experience of going *back to the future*': that is, as an 'intensified repetition of anthropocentric environmental change inflicted on Indigenous peoples via colonial practices that facilitate the capitalist industrial expansion'.[32] In the context of these cascading pressures, the peoples of Oceania have mobilised resources, knowledge, and networks of

30 Intergovernmental Panel on Climate Change, *Climate Change: The IPCC 1990 and 1992 Assessments IPCC First Assessment Report Overview and Policymaker Summaries and 1992 IPCC Supplement* (Report, 1990) 2.2.5 <https://www.ipcc.ch/site/assets/uploads/2018/05/ipcc_90_92_assessments_far_overview.pdf>.

31 Intergovernmental Panel on Climate Change, 'Summary for Policymakers' in *Climate Change 2022: Impacts, Adaptation and Vulnerability*. Contribution of Working Group II to the Sixth Assessment Report of the Intergovernmental Panel on Climate Change (Cambridge University Press, 2022) ('AR6 WGII') SPM.C.3.5.

32 Whyte (n 29) 156.

kinship and exchange to adapt to the impacts of the Anthropocene.[33] Yet, while recognising their agency and leadership in the climate crisis, we must not overlook the fact that Oceania would not be facing these existential threats if it were not for the actions of industrialised states and high-emitting corporations. Thus, it is both a legal and ethical necessity to listen deeply and intently to the stories of the plaintiffs in this hearing, to understand what 'reparations' mean for current and future generations, while acknowledging that some forms of loss and damage can never be completely repaired.

Reparations Under International Law

This Commission was established in 2030 pursuant to the *Truth and Reparations Commission of Oceania Act* 2030 (Cth). The Commission has two key functions: (i) to amplify stories of climate injustice in the region and (ii) to provide reparations for associated losses and damages. The Commission is made up of ten members, with representative membership from across Oceania. The Commission has the authority to make broad ranging orders to give effect to the climate record, including through financial compensation, but also non-financial orders. The establishment of the Commission followed important political developments in the preceding years, especially the coming into force of the global *Fossil Fuel Non-Proliferation Treaty* in 2027 and the action in Australia and countries around the world to phase out all fossil fuel subsidies and nationalise fossil fuel companies in the mid-to-late 2020s. The redirection of public funds that has previously been used to subsidise the fossil fuel industry, as well as the expropriation of fossil fuel companies, provided governments with billions of dollars of revenue to spend on climate mitigation and adaptation policies. These funds also enable the operations of this Commission, and other regional Commissions like it, and enable compensation payments to those who are suffering from the impacts of climate change.

Prior to the establishment of this Tribunal, and others like it in other regions of the world, there were limited legal avenues for redress or reparations for climate harms. Over the past five years of its operation the Commission has gradually expanded the concept of reparations. In the first case before the Commission, *Re Dahlms*, the Commission addressed the question of compensation for the loss and damage suffered during the Black Spring bushfires of 2029. In this landmark judgment, the Commission identified that reparations required that those who were impacted by these fires be returned to the *status quo ante* or, if not possible, provided with economic compensation for losses and damages. However, this focus on tangible

33 Tracey Banivanua Mar, *Decolonisation and the Pacific: Indigenous Globalisation and the Ends of Empire* (Cambridge University Press, 2016).

and quantifiable economic losses that characterised these earlier cases have been challenged in subsequent jurisprudence. In *Re Kennedy* [2031], which addressed the harms arising from the multiple flood events impacting the eastern States of Australia in 2030, the Commission recognised the need for compensation to account for the severe emotional and psychological stress of living through these extreme weather events.[34] This precedent was reaffirmed in *Re McIntosh* [2032], where the Tribunal was called on to consider a broader sense of emotional and physical harm than those arising from climate-induced disasters, such as the grief that accompanies adjusting to a climate-transformed world and the sorrow caused by solastalgia, the grief and melancholia of feeling homesick or dislocated because one's environment has been irreversibly transformed.[35] In *Re Agathy and Others* [2033], the Commission was required to consider what types of public memorialisation are necessary and how forms of public memorialisation might provide opportunity for healing and learning and provide acknowledgement of crime or suffering.[36] Likewise, in *Re Swan* [2034], the Tribunal had to grapple with the impossibility of repairing the broader cultural loss arising from the loss of land, and thus of the relationship with the land, and the other beings and stories that inhabited it. In that case, the Commission acknowledged the need to allow Indigenous elders and jurists to articulate how these harms were understood from within their own legal system and to recognise the role of ceremony in mourning these losses that cannot be repaired.

Reconsidering Climate Reparations

Building on the Tribunal's earlier jurisprudence, we are called on to move towards an 'Oceanic approach' to climate reparations for future generations. This Oceanic approach is inspired by the work of Epeli Hau'ofa, who described the ocean as a metaphor to transcend the territorial ways of imagining identities and instead focus on our interconnectedness and ever-moving expansiveness.[37] Taking an Oceanic approach to reparations means embracing fluidity. Accordingly, reparations should be determined reflexively and relationally, informed by the voices and stories of those affected.

34 Blanche Verlie, '"Climatic-Affective Atmospheres": A Conceptual Tool for Affective Scholarship in a Changing Climate' (2019) 33 *Emotion, Space and Society* 100623.

35 Glenn Albrecht et al, 'Solastalgia: The Distress Caused by Environmental Change' (2007) 15(1) *Australasian Psychiatry* S95.

36 In thinking about memorialisations and public commemoration as a key part of reparation, the Tribunal was influenced by similar judgments in the transitional justice space: see, eg, *Prosecutor v Nuon and Khieu (Judgment)* (Extraordinary Chambers in the Courts of Cambodia, Trial Chamber, Case No 002/19-09-2007/ECCC/TC E465, 16 November 2018).

37 Epeli Hau'ofa, *We Are the Ocean: Selected Works* (University of Hawaii Press, 2008).

Climate reparations must be both forward- and backward-looking. Reparations are a mechanism for taking responsibility for historical injustices, a crucial first step in pursuing climate justice, because, in the words of science-fiction writer NK Jemisin, '[h]ow can we prepare for the future if we won't acknowledge the past?'[38] But reparations must also be a 'constructive project' that takes a forward-orientated approach to dismantling systemic injustices. This entails eradicating the systemic racism and inequality that are the legacy of historic injustice, dismantling ongoing capitalist relations, and remaking the world so different social relations are possible. This requires a material redistribution of resources, a structural change in decision-making structures and political power, and a profound shift in how different lives and different voices are valued.

Firstly, it requires a systemic approach to redistributing resources in order to foster climate justice. Distributive justice has long since been recognised as a key pillar of climate justice. However, rather than simply redistribute existing wealth, it is necessary to radically transform the social, economic, and political structures that historically and in ongoing ways produce stark economic inequalities. The harm being addressed is not simply the unequal appropriation of atmospheric space by industrialised countries, but the fact that these countries unjustly benefited from these harmful acts to the detriment of others. Moreover, it is crucial to recognise how excessive historical emissions by countries of the Global North 'were constitutive in enabling the conditions for dispossession, violence, slavery, racial difference, and uneven wellbeing that generated – and continue to generate – stark asymmetries within and between countries'.[39] Therefore, as Gurminder K Bhambra and Peter Newell have shown:

> Understanding climate change in the context of colonial histories implies more than the payment for loss and damages experienced today as a result of accumulated emissions. It requires instead a broader recognition of how socially and regionally uneven concentrations of wealth, which has resulted in climate-changing emissions, were created in the first place.[40]

Therefore, we find that a transformative approach to reparations requires more than simply trying to return impacted peoples to the position they were in before the harmful acts: given the harmful acts of unequal historical emissions have produced a highly unequal international political and economic

38 NK Jemisin, *The Stone Sky* (Orbit, 2017) 216.
39 Mason-Case and Dehm (n 12) 174.
40 Gurminder K Bhambra and Peter Newell, 'More than a Metaphor: "Climate Colonialism" in Perspective' (2022) *Global Social Challenges Journal* 1, 6.

system, reparations require remaking the world differently. In envisioning a different global political economy, we affirm the vision put forward by science fiction writer Kim Stanley Robinson:

> To be clear, concluding in brief: there is enough for all. So there should be no more people living in poverty. And there should be no more billionaires. Enough should be a human right, a floor below which no one can fall; also a ceiling above which no one can rise. Enough is as good as a feast – or better.[41]

As Olúfẹ́mi O Táíwò has highlighted, resource redistribution is necessary but insufficient to provide meaningful climate reparations,[42] because resource redistribution, when deployed through traditional institutional and governance models, still locates power in the hands of the few. Therefore, a transformative approach to reparations also requires structural changes to decision-making structures and radical shifts in political power. Existing relations of power and authority are themselves a product of the history of unequal historical emissions and historical and ongoing processes of colonisation, resource extraction, environmental degradation, and exploitative capital accumulation they enabled. Therefore, in addition to the *ex-post* redistribution of resources to those disproportionately impacted by climate change – including and especially for Black, Indigenous, and other marginalised communities – there is a need to shift power and control over democratic processes to those disproportionately impacted communities, both within and beyond the formal political system.[43]

Finally, a profound shift in how different lives and different voices are valued is needed. This entails listening to the voices of affected communities and following their leadership to improve lives into the future. Moreover, it calls for the validation of 'epistemologies of the South', namely, the 'knowledges anchored in the experiences of resistance of all those social groups that have systematically suffered injustice, oppression, and destruction caused by capitalism, colonialism, and patriarchy'.[44] We recognise the vastness of Oceania and the diverse interconnections of resources, knowledge, and networks of kinship and exchange between those who inhabit the oceans and islands; this therefore requires not elevating one person or one voice but a multiplicity of (potentially conflicting) voices. Oceania's ingenuity in the face of

41 Kim Stanley Robinson, *The Ministry for the Future* (Orbit, 2020) 58.

42 See Olúfẹ́mi O Táíwò, 'Community Control and the Climate Crisis: Power, Governance, and Racial Capitalism' in Craig Calhoun and Benjamin Y Fong (eds), *The Green New Deal and The Future of Work* (Columbia University Press, 2022) 295.

43 Ibid 299.

44 Boaventura de Sousa Santos, *The End of the Cognitive Empire: The Coming of Age of Epistemologies of the South* (Duke University Press, 2018) 1.

environmental harm has been evident for decades, as has their leadership at local to global scales; however, for too long those who should have listened to these voices ignored, marginalised, and suppressed them. A transformative approach to reparations requires the full political and economic self-determination of all peoples within Oceania, as the precondition for collectively building a more equitable socio-ecological system for all. In the words of Kathy Jetñil-Kijiner, the peoples of Oceania 'deserve to do more than just / survive / we deserve / to thrive'.[45]

Towards a Youth Justice Forum on Climate Change

This Tribunal has held that a systemic approach to reparations that entails symbolic justice, the redistribution of resources, and the remaking of legal and institutional arrangements that have produced the intersecting harms of colonialism, racial injustice, and climate harms faced by these young plaintiffs is needed. It is convention that before concluding a judgment the Tribunal should make orders to give effect to its ruling. However, we hesitate to do so. We hesitate for two reasons. Our first hesitation relates to the content of such orders and the immense difficulties in articulating all the concrete and grounded measures that would be necessary to give effect to such a systemic conception of reparations. However, it is our second hesitation, which concerns the very act of making such orders, that is more fundamental. We worry that in the very act of making such orders we would reproduce the dynamics of climate coloniality that the transformative conceptualisation of reparations we have sought to articulate is seeking to undo.

We write this judgment as two white women, settlers living and working on the unceded land of the Wurundjeri peoples. In making a judicial ruling our voice is elevated above other voices, and thus over other conceptions of what reparations entails. Thus, the very act of legal pronouncement would reinforce the existing intersecting structural inequalities that have marginalised the voices of those who are most impacted by the climate crisis. We have, in our work on this Commission, sought to use our relative privilege to open these legal spaces to voices that have often been marginalised. However, we increasingly feel that in order to create space for the voices of those who have been most impacted by climate change, and to allow for the perspectives of those who have the greatest grounded and situated understanding of what reparations must mean in concrete, material terms, we might need to cede authority and simply listen. As the transformative account of reparations we have sought to articulate identifies, climate justice is not just about

45 Kathy Jetñil Kijiner, 'Dear Matafele Peinem' (23 September 2014) <https://www.kathyjetnilkijiner.com/united-nations-climate-summit-opening-ceremony-my-poem-to-my-daughter/>.

certain outcomes, but also the process by which we come to decide on these outcomes and the means by which we enact or realise them. Given our vision of reparative justice entails transforming not just the 'content of the decisions made at different levels of government' but also transforming 'who is empowered to make those decisions', it fundamentally requires greater democratic control over processes of decision making.[46] This thus necessarily requires creating space for greater democratic control over decisions about what reparations entail.[47]

A further hesitation we have is that when we speak as a judge or a tribunal member, we speak in the name of the law, and in making orders we posit law as a tool that can help achieve reparative climate justice. On the one hand, we recognise that climate litigation has important regulatory functions, while also providing a forum to amplify historically marginalised voices and 'humanise' climate impacts and risks.[48] In this way, the courtroom can potentially be harnessed as a platform to empower those most vulnerable to and least responsible for climate change to compel major emitters to take effective and equitable action in accordance with legal and moral obligations.[49] On the other hand, the foregrounding of law as the primary solution to climate injustice also entails a deliberate blindness to the many ways in which legal doctrines and frameworks are themselves implicated in settler-colonialism and the production of the Anthropocene conditions the world now faces. Legal regimes at the national and international levels have enabled the extraction, combustion, and circulation of fossil fuels globally, promoted an extractivist relationship with the natural world, and enabled unchecked capitalist accumulation and economic growth. To present law as offering the means to repair dangerously disavows law's complicity in structuring and enabling the very harms that now need redress. We remain cautiously hopeful about the possibility of dismantling the laws that have produced Anthropocene conditions, the 'legal imaginary *of* the Anthropocene', and enacting ecologically just lawful relations, or a 'legal imaginary *for* the Anthropocene'.[50] However, this requires a radical 'unmaking' of existing legal norms and doctrine, in

46 Táíwò (n 42) 305.

47 Ibid.

48 On the humanising role of climate litigation, see Jacqueline Peel and Hari Osofsky, 'A Rights Turn in Climate Change Litigation?' (2018) 7(1) *Transnational Environmental Law* 37, 67; Jacqueline Peel and Hari Osofsky, 'Transnational Climate Litigation: The Contribution of the Global South' (2019) 113(4) *The American Society of International Law* 679, 685.

49 See, eg, Maria Tigre, 'Climate Change and Indigenous Groups: The Rise of Indigenous Voices in Climate Litigation' (2022) 9(3) *Dezembro* 214; Sam Bookman, 'Indigenous Climate Litigation in Anglophone Settler-Colonial States: Context, Cases, and Caution', *Verfassungsblog* (Blog Post, 25 March 2022) <https://verfassungsblog.de/indigenous-climate-litigation-in-anglophone-settler-colonial-states/>.

50 Anna Grear, 'Legal Imaginaries and the Anthropocene: "Of" and "For"' (2020) 31 *Law and Critique* 351.

order to confront and uncover the problematic assumptions that lie at the heart of law, in order to give rise to a very different 'conception of law which acknowledges, and in fact emerges from, the entangled nature of existence'.[51]

The land we write from is imbued with other laws. For too long, Indigenous laws have not been acknowledged as laws, but instead treated as 'no more than oral stories, mere myths and fables'.[52] Despite the violence of colonialism, Indigenous elders and communities have 'struggle[d] to keep a body of First Nations law alive for future generations'.[53] Indigenous traditions have kept alive 'a view of law, which lives in all things and emanates love, caring and sharing, and respect for all things in the natural world' and 'supports the capacity of peoples to care for country and all our relations, including those we have with the animal, plant and broader natural world'.[54] We cannot speak about or make pronouncements about Indigenous laws: we have neither the authority nor training to do so. But fundamentally returning to and re-empowering Aboriginal authority in relation to the natural world is the only way forward: it is no exaggeration to say that the 'future of all lifeforms on earth' depends upon this.[55]

Finally, when we speak as a judge or tribunal member, we are required to speak with an authoritative voice and to make pronouncements about the veracity of facts and the reliability and credibility of witnesses. The very act of speaking in and through judicial voice is an act of consolidating a specific 'truth'. Such a conception of a singular truth does not speak to the lived realities of how knowledge is negotiated and produced across Oceania. As Epeli Hau'ofa writes:

> One of the more positive aspects of our existence in Oceania is that truth is flexible and negotiable, despite attempts by some of us to impose political, religious, and other forms of absolutionism. Versions of truth may be accepted for particular purposes and moments, only to be reversed when circumstances demand other versions; and we often accede to things just to stop being bombarded, and then go ahead and do what we want to do anyway.[56]

51 Margaret Davis, *Asking the Law Question* (Thomson Reuters, 4th ed, 2017) 497.
52 Irene Watson, 'Aboriginal Relationships to the Natural World: Colonial "Protection" of Human Rights and the Environment' (2018) 9(2) *Journal of Human Rights and the Environment* 119, 138.
53 Irene Watson, *Aboriginal Peoples, Colonialism and International Law: Raw Law* (Routledge, 2014) 12.
54 Ibid
55 Watson (n 52) 140.
56 Hau'ofa (n 37) 454.

Dismantling climate colonialism necessarily requires challenging the epistemic violence that underpins it and overcoming the coloniality of power. Epistemic justice requires epistemological and ontological shifts that foster pluriversality and recognises the multiplicity of knowledges.[57] Ultimately, repetitively remaking the world will require de-centering our own authority and ways of knowing in order to create space for plural epistemologies, plural ways of being in the world, plural laws, and a plurality of justices.

We thus cannot, in isolation, make orders about what will be necessary to enact the structural change necessary for climate justice. Instead, we seek to establish a deliberative and democratic process that will address these issues that foreground the perspectives and the voices of those most impacted. We therefore order the establishment of a democratic, deliberative, and participatory process that foregrounds the voices and perspectives of those most impacted by already existing and future impacts of climate change. This body, which shall be called the Youth Justice Forum on Climate Change, must be made up of representative groups of young people from Oceania, with special arrangements made to make sure the voices of Indigenous youth, Pacifika, and other marginalised, racialised, and ethnic communities are elevated. It shall be empowered to make decisions about what sort of future actions are necessary to bring about reparative justice to future generations who are living in a climate-transformed world.

In making these orders we not only conclude our judgment but take off our robes and step down from the authoritative position we have adopted as tribunal members. Adjudication has played an important role in seeking reparative justice in the Anthropocene; but it is now past time for a more deliberative, participative, and democratic political engagement about how we collectively build ecologically just futures.

57 Ulrich Oslender explains that pluriversality refers to the recognition that 'there are worlds out there (and have always been) that have historically been marginalized and suppressed by a Western cosmology and universalizing tendency that claimed a superior position for itself vis-à-vis those other worlds'. See, *Walter D Mignolo*, 'Foreword: On Pluriversality and Multipolarity' in Bernd Reiter (ed), *Constructing the Pluriverse* (Duke University Press, 2018) ix.

Chapter 9

How to Blow Up a Coalmine[1]: The Trial of the Waratah 7

Nicole Rogers

Commentary

This judgment is handed down at the tail end of a critical decade, a time when what Roy Scranton has described as the 'time lag between carbon dioxide increase and subsequent effects, between the wind we sow and the whirlwind we reap',[2] has become vanishingly small. The climate-related destruction of the early 2020s has proved to be the harbinger for a terrifying, never-ending succession of climate disasters, culminating most recently in Australia in the devastating megafires of the 2027/2028 Summer of Terror. While the description of that summer in the judgment may seem extreme and apocalyptic, such events are very plausible in the near future. Australians and their governments were forewarned by the Commissioners of the 2020 Royal Commission into National Natural Disaster Arrangements that the Black Summer of 2019/2020 'provided only a glimpse of the types of events that Australia may face in the future';[3] this sombre prediction has been realised in the Summer of Terror of the futuristic scenario.

In the first year of this decade, when the COVID pandemic first swept the world, we witnessed the collaboration of nation-States in a united response to a global emergency. Yet, when it comes to climate action, political leaders continue to prevaricate and defer commitments, engaging in the rhetoric of greenwash scathingly dismissed by youth climate activist Greta Thunberg in 2021 as 'blah blah blah'.[4] In a similar vein, Donna Haraway adopts science-fiction writer Kim Stanley Robinson's description of this period of human

1 The title of this chapter, and indeed the fact scenario, draw inspiration from Andreas Malm's book *How to Blow Up a Pipeline: Learning to Fight in a World on Fire* (Verso, 2021).

2 Roy Scranton, *We're Doomed. Now What?: Essays on War and Climate Change* (Soho Press, 2018) 321.

3 *Royal Commission into National Natural Disaster Arrangements* (Report, 28 October 2020) 22.

4 Quoted in Damian Carrington, '"Blah, Blah, Blah": Greta Thunberg Lambasts Leaders over Climate Crisis', *The Guardian* (online, 28 September 2021) <https://www.theguardian.com/environment/2021/sep/28/blah-greta-thunberg-leaders-climate-crisis-co2-emissions>.

DOI: 10.4324/9781003389569-11

history as 'The Great Dithering': 'a time of ineffective and widespread anxiety about environmental destruction' including 'violent climate change'.[5] Future ambitions of net zero targets necessitate immediate action, yet dubious practices of emissions accounting and carbon offsetting undermine even modest apparent declines in certain emissions.

In Australia, at the time of writing, the espousal of climate commitment on the part of a recently elected Labor government at federal level has not halted the spread of new coalmines and gas projects and the expansion of existing ones. Although the federal Minister for the Environment took the unprecedented step of refusing approval for Clive Palmer's central Queensland coal project in 2023,[6] the Commonwealth government subsequently rejected a Greens proposal that all new coalmines and gas projects be halted.[7] Australia continues to mine coal and rely upon coal exports,[8] with seeming disregard for the inevitable consequences of the burning of its coal in other parts of the world. The Queensland government has also supported and, in fact, encouraged the establishment of new coalmines.[9]

It is for these reasons that I have assumed, for the purpose of the judgment, that the trajectory of greenhouse gas emissions will continue to rise and that, by 2029, the world will be heading towards three or more degrees of global warming. I have also, albeit pessimistically, assumed that the Queensland government will eventually approve the Waratah coalmine in the event of a successful legal challenge, on the part of Waratah Pty Ltd, to the government's current refusal of the mine's environmental authority.[10]

5 Donna J Haraway, *Staying with the Trouble. Making Kin in the Chthulucene* (Duke University Press, 2016) 144–5.

6 Tess Ikonomou, 'Landmark Decision Sees Clive Palmer's Queensland Coal Project Rejected', *The Brisbane Times* (online, 8 February 2023) <https://www.brisbanetimes.com.au/national/queensland/landmark-decision-sees-clive-palmer-s-queensland-coal-project-rejected-20230208-p5cj0o.html>.

7 Katharine Murphy and Adam Morton, 'Greens Will Back Safeguard Mechanism Revamp If Labor Bans New Coal and Gas Projects', *The Guardian* (online, 15 February 2023) <https://www.theguardian.com/australia-news/2023/feb/15/greens-will-back-safeguard-mechanism-revamp-only-if-labor-bans-new-coal-and-gas-projects>.

8 Judith Brett, 'The Coal Curse: Resources, Climate and Australia's Future', *Quarterly Essay* (Issue 78, 2020).

9 For instance, in 2020, the Queensland government deferred the payment of mining royalties in order to ensure that Adani's controversial Carmichael coalmine could go ahead – a step described by an Australian Conservation Foundation spokesperson as 'roll[ing] out the red carpet for a mine that will soon become a stranded asset': Ben Smee, 'Queensland Strikes Royalties Deal with Adani Coalmine, Allowing It to Defer Payment', *The Guardian* (online, 1 October 2020) <https://www.theguardian.com/environment/2020/oct/01/queensland-strikes-royalties-deal-with-adani-coalmine-allowing-it-to-defer-payment>.

10 As discussed in the judgment, the Queensland government refused to issue the environmental authority in April 2023, after the President of the Queensland Land Court held that approving the mine would infringe numerous rights under the *Human Rights Act 2019* (Qld).

Activism and Violence

Given widespread climate inaction by governments, climate activists are adopting increasingly desperate strategies. There have been corresponding retaliatory moves by governments determined to stamp out such activities. In Australia, the enactment of anti-protest legislation by State governments,[11] the upscaling of charges, and even the reactivation of antiquated offences[12] have not deterred climate activists from engaging in disruptive activities designed to highlight the urgency of climate action and the enormity of the climate crisis.

The potential for punitive consequences is very real. In December 2022, climate activist Violet Coco was sentenced to a fifteen-month jail term after blocking one lane of the Sydney Harbour Bridge with a truck. She was also initially denied bail. The sentence was overturned on appeal, with the judge finding that the police had presented false evidence: namely, that her actions had impeded the passage of an ambulance. [13]

Most climate activist organisations, including Extinction Rebellion, espouse and adhere to a creed of non-violence. Even the recent spate of throwing foodstuffs at major artworks has not resulted in lasting damage to these masterpieces. The justification for such staged performances of simulated desecration and violation is quite straightforward: 'We need to break the mirage that everything is fine and shatter the illusion of normal life'.[14]

Not all contemporary climate activists, however, have renounced the use of violence against property. In 2022, youth activists sabotaged petrol pumps at two British service stations[15] and climate activist group Pipe

11 See, eg, *Summary Offences and Other Legislation Amendment Act 2019* (Qld), *Sustainable Forests Timber Amendment (Timber Harvesting Safety Zones) Act 2022* (Vic), *Roads and Crimes* Legislation *Amendment Act 2022* (NSW), and *Police Offences Amendment (Workplace Protection) Act 2022 (Tas).*

12 In 2022, after a three-minute protest in the public gallery of the Queensland parliament, 14 people were charged at the Speaker's request with disturbing the legislature: an offence which had not been used for 30 years: Joe Hinchliffe, Sean Ruse, and Michael McGowan, 'Violet Coco Is Not Alone: the Climate Activists Facing Jail', *The Guardian* (online, 10 December 2022) <https://www.theguardian.com/australia-news/2022/dec/10/violet-coco-is-not-alone-the-climate-activists-facing-jail>.

13 AAP and Jordyn Beazley, 'Climate Activist Deanna "Violet" Coco's 15-Month Jail Sentence Quashed on Appeal', *The Guardian* (online, 15 March 2023) <https://www.theguardian.com/australia-news/2023/mar/15/climate-activist-deanna-violet-cocos-15-month-jail-sentence-overturned-on-appeal>.

14 Quoted in Andreas Malm, 'History May Absolve the Soup Throwers', *The New York Times* (online, 20 October 2022) <https://www.nytimes.com/2022/10/20/opinion/just-stop-oil-soup-sunflowers-climate.html>.

15 Damien Gayle, 'Just Stop Oil Protesters Sabotage Petrol Pumps on M25 Motorway', *The Guardian* (online, 28 April 2022) <https://www.theguardian.com/environment/2022/apr/28/just-stop-oil-protesters-sabotage-petrol-pumps-on-m25-motorway>.

Busters damaged a pipeline intended to supply Heathrow Airport with aviation fuel.[16] Windows have been smashed. German climate movement Letzte Generation sabotages fuel pipelines.[17] The Tyre Extinguishers deflate tyres of fuel-guzzling SUVs in surreptitious night raids.[18]

In the United States, the prosecution and sentencing of Jessica Reznicek, who together with Ruby Montoya used a blowtorch to blast holes in parts of the Dakota Access Pipeline in 2016 and 2017, illustrates the punitive approach taken by judges to climate activists who damage property. These acts were, according to Reznicek and Montoya, undertaken with 'purity of heart'[19] and 'steady loving hands'.[20] Reznicek was nevertheless sentenced to 96 months in prison, and three years of supervised release, after the court applied a 'terrorism enhancement' clause. Her appeal against this sentence was unsuccessful.[21]

The fact scenario in the Waratah 7 judgment assumes an escalation of climate activist strategies in Australia: a preparedness to engage in acts of property destruction that target fossil fuel infrastructure. This sort of escalation and, in fact, more extreme acts of violence are anticipated in some works of climate fiction,[22] and in Andreas Malm's provocative text *How to Blow Up a Pipeline*.[23] Malm's book has inspired a 2022 feature film of the same title, in which a group of young people orchestrate an act of eco-sabotage against an oil pipeline.[24] After the Summer of Terror, Malm's call to arms resonates with the Waratah 7, who intentionally destroy parts of the mining infrastructure at the site of a controversial new coalmine: the Waratah coalmine.

The sentencing of Jessica Reznicek, the longstanding invocation of rhetoric of eco-terrorism by official agencies,[25] and the growing use of

16 Jack McGovan, 'Is Breaking Things the Best Way Forward for Climate Activists?', *Open Democracy* (online, 13 October 2022) <https://www.opendemocracy.net/en/climate-crisis-sabotage-property-destruction>.

17 Ibid.

18 Oliver Milman, '"Like a Public Shaming": A Night with the Eco-Activists Deflating SUV Tyres', *The Guardian* (online, 27 July 2022) <https://www.theguardian.com/environment/2022/jul/27/tire-deflators-suv-new-york-climate-crisis>.

19 Jessica Reznicek and Ruby Montoya, quoted in Malm (n 1) 96.

20 Quoted in ibid 100.

21 *USA v Reznicek* (unpublished) US 21-2548 (8th Cir, 2022).

22 See, eg, Kim Stanley Robinson, *The Ministry for the Future* (Orbit, 2020); *Last Light* (Stan, 2022); Sean Rabin, *The Good Captain* (Transit Lounge, 2022).

23 Malm (n 1).

24 See Peter C Baker, 'Will We Call Them Terrorists?', *The New York Times Magazine* (online, 5 April 2023) <https://www.nytimes.com/2023/04/05/magazine/how-to-blow-up-a-pipeline-terrorism-climate-change.html>.

25 See, eg, testimony on the United States environmental groups Earth Liberation Front and Animal Liberation Front, provided by John E Lewis, Deputy Assistant Director of the Federal Bureau of Investigation before the Senate Judiciary Committee, 18 May 2004: available at John E Lewis, 'Testimony: Animal Rights Extremism and Ecoterrorism', *Federal Bureau of*

counter-terrorism surveillance tactics against climate activists,[26] all point to the likelihood of prosecution under federal anti-terrorism legislation for Australian climate activists who engage in acts of property violence. I have chosen, instead, to stage the trial of the Waratah 7 in the Queensland District Court before a hypothetical judge, Judge Wood, and to limit the charges to property damage under the Queensland *Criminal Code*. This may downplay the criminal consequences for activists who engage in future acts of property destruction. However, the possibilities for raising the defence of necessity or, for that matter, extraordinary emergency in trials for terrorism offences are vanishingly small.

Here I want to highlight the awful choices that future activists confront, as another six years of inaction spell doom for the planet's climatic equilibrium, and how a courageous judge might respond when faced with the outcome of these choices.

Necessity and the Extraordinary Emergency Defence

The Waratah 7 have argued that the extraordinary emergency defence applies to their actions. This is a statutory version of the defence of necessity, also known as the choice of evils defence, widely invoked by climate activists[27] since the 2008 acquittal of the so-called Kingsnorth Six[28] but thus far with little success. Activists are arguing that it is necessary to break the law in order to avert the greater evil of climate change.

Broad questions around what Scott Veitch calls law and irresponsibility,[29] which can also be framed as law and thoughtlessness following Hannah Arendt's conclusions about Nazi bureaucrat Adolf Eichmann's contribution to the Final Solution,[30] arise in the context of these defences.

Investigation (Web page, 18 May 2004) <https://archives.fbi.gov/archives/news/testimony/animal-rights-extremism-and-ecoterrorism>.

26 Peter D Burdon, 'The Targeting of Environmentalists with State-Corporate Intelligence Networks' in Kirsten Anker et al (eds), *From Environmental to Ecological Law* (Routledge, 2020) 25.

27 See Nicole Rogers, 'Beyond Reason. Activism and Law in a Time of Climate Change' (2018) 12(2) *Journal for the Study of Radicalism* 157, 169–176; Lance N Long and Ted Hamilton, 'The Climate Necessity Defense: Proof and Judicial Error in Climate Protest Cases' (2018) 38 *Stanford Environmental Law Journal* 57.

28 Jonathon Mingle, 'The Climate Change Defense', *The New York Times* (online, 12 December 2008) < https://www.nytimes.com/2008/12/14/magazine/14Ideas-Section2-A-t-004.html>.

29 Scott Veitch explores 'the role of law and legal mechanisms in the production of suffering, and the role of legality in enforcing and legitimating this': Scott Veitch, *Law and Irresponsibility. On the Legitimation of Human Suffering* (Routledge-Cavendish, 2007) 10.

30 Donna Haraway writes that 'Arendt witnessed in Eichmann not an incomprehensible monster, but something much more terrifying – she saw commonplace thoughtlessness ... an astonishing abandonment of thinking': Haraway (n 5) 36.

Samuel Alexander and Peter Burdon have articulated the dilemma thus:

> When we look back on today, we will ask ourselves: Did we do enough? Were we complicit in a broken system? Should we have been so obedient given that we knew our gutless governments were leading us down a dead end?[31]

Judge Wood refrains from speculating in this vein but nevertheless acknowledges in her judgment that certain forms of law-abiding behaviour can be conducive not only to injustice but to great evil. This, in essence, is the conundrum that the defence of necessity and the extraordinary emergency defence are designed to address: there are situations in which complying with the law cannot be considered reasonable conduct.

Judicial Bias

In the final section of the judgment, Judge Wood addresses the hypothetical controversy generated by a progressive judge, with a background in radical environmental activism, sitting in judgment on climate activists. This controversy is not unprecedented; as Judge Wood comments, Judge Brian Preston encountered a version of this when he handed down his judgment in the Rocky Hill coalmine case.[32]

François Kunc, himself a sitting judge of the New South Wales Supreme Court, described the hostile industry and media reaction to Judge Preston's judgment thus:

> Headlines included 'Absurd Rocky Hill Decision Tarnishes Rule of Law', 'Mine Judge's Green Links', 'Did Environment Court Judge Veer from Justice to Morality', 'Activist Judge Shakes Climate Change World'. His Honour's speeches were combed through and much attention was given to his role in 1985 in establishing the NSW Environmental Defenders Office, which had acted for the successful opponents to the mine. Only some of the reports noted that no party had asked his Honour to recuse himself from the hearing.[33]

Justice Kunc wrote that '[w]hat is important is to see how quickly a judge becomes the object of allegations of bias and impropriety when the judge is said to have "overreached" into the political arena'.[34]

31 Samuel Alexander and Peter Burdon, 'Extinction Rebellion: Crisis, Inaction, and the Question of Civil Disobedience as Ecosocialist Strategy' in Leigh Brownhill et al (eds), *The Routledge Handbook on Ecosocialism* (Routledge, 2021) 242, 247.

32 *Gloucester Resources Ltd v Minister for Planning* (2019) 234 LGERA 257.

33 François Kunc, 'Current Issues' (2019) 93(4) *Australian Law Journal* 251, 253.

34 Ibid 253-4.

Conclusion

Justice Wood finishes her judgment by praising the defendants. Again, this is not unprecedented. In 2007, Victorian Deputy Chief Magistrate Jelena Popovic dismissed charges against anti-logging activist Holly Creenaune, describing her as a 'remarkable young woman' with 'passion, drive and ability' who has 'worked consistently and effectively in relation to improving our environment and maintaining the environment in a sustainable way so that it isn't further degraded'.[35] Increasingly, if largely anecdotally, United Kingdom magistrates and judges are expressing sympathy and support for the climate activists appearing before them.[36] In May 2023, a Western Australian magistrate stated that climate activist Violet Coco should be commended for her 'strong personal views', after she spraypainted four Woodside Energy logos on a Perth police station.[37]

Journalist George Monbiot has concluded that 'protesters condemned as criminals today will be the heroes of tomorrow. One of the greatest fears of those who wield power is that judges and juries will come to see them this way'.[38]

Judgment

District Court of Queensland

Citation: *Police v EY, LH, SM, PT, GW, OK and AD* [2029] QDC 21
Delivered on: 3 October 2029
Judge: Wood DCJ

The seven defendants in this matter, EY, LH, SM, PT, GW, OK and AD, are members of a climate action group that calls itself Last Generation Queensland (hereinafter 'Last Generation'). Last Generation was established in 2023 with the stated mission of halting the production of fossil fuels in the Galilee Basin in Queensland. The defendants have been charged under section 469(1) of the *Criminal Code 1899* (Qld) with wilfully and

35 Steve Butcher, 'Tree-Sitting Activist Wins High Praise from Judge', *The Sydney Morning Herald* (online, 22 December 2007) <https://www.smh.com.au/national/tree-sitting-activist-wins-high-praise-from-judge-20071222-gdru5r.html>.

36 George Monbiot, 'Today's Climate Activist "Criminals" Are Tomorrow's Heroes: Silencing Them in Court Is Immoral', *The Guardian* (online, 22 February 2023) <https://www.theguardian.com/commentisfree/2023/feb/22/climate-activist-criminals-heroes-truth>.

37 AAP and Nino Bucci, 'Magistrate Commends Activist Deanna "Violet" Coco for "Noble" Beliefs but Says She Went Too Far', *The Guardian* (online, 4 May 2023) <https://www.theguardian.com/australia-news/2023/may/04/deanna-violet-coco-to-face-court-after-allegedly-spray-painting-woodside-logo-on-perth-police-station>.

38 Monbiot (n 36).

unlawfully destroying property: namely, mining equipment and vehicles at a functioning coalmine site owned by Waratah Coal Pty Ltd, the Galilee Coal project (hereinafter 'the Waratah mine'). The prosecution alleges that on 12 February 2029, the defendants orchestrated a series of drone attacks on the site, thereby causing damage to infrastructure and property belonging to Waratah Coal Pty Ltd. The seven defendants have agreed to be tried together in relation to the alleged offences. They have all pleaded not guilty.

The defendants do not dispute any of the facts presented by the prosecution. They have argued, however, that the so-called extraordinary emergency defence under section 25 of the *Criminal Code 1899* applies to their conduct.

Section 469(1) of the *Criminal Code* states: Any person who wilfully and unlawfully destroys or damages any property is guilty of an offence which, unless otherwise stated, is a misdemeanour, and the person is liable, if no other punishment is provided, to imprisonment for 5 years.

Section 25 states: Subject to the express provisions of this Code relating to acts done upon compulsion or provocation or in self-defence, a person is not criminally responsible for an act or omission done or made under such circumstances of sudden or extraordinary emergency that an ordinary person possessing ordinary power of self-control could not reasonably be expected to act otherwise.

Activists and the Extraordinary Emergency Defence

Greg Rolles was the first Queensland climate activist to argue the extraordinary emergency defence in his trial, which took place in the Bowen Magistrates Court in 2019. He was unsuccessful;[39] on appeal, his sentence was reduced but his conviction was not overturned.[40] In the first half of this decade, Queensland magistrates were uniformly dismissive of similar attempts. The conflation of sudden and extraordinary in their reasoning proved to be one obstacle, albeit in my view a clear misreading of the section. In addition, this group baulked at the suggestion that acts of civil disobedience could constitute reasonable conduct.

Growing judicial tolerance of acts of civil disobedience committed by Queensland climate activists became apparent from 2025, when the first acquittal of a climate activist on the basis of the extraordinary emergency defence took place in the Brisbane Magistrates Court.

39 The magistrate concluded that Rolles's 'self indulgent, subjective belief that he was morally obliged to take personal responsibility to combat climate change does not excuse his unlawful behaviour': *Police v Rolles* (Bowen Magistrates Court, MAG-00225465/18(5), 28 May 2019) unpublished decision, 5.

40 *Rolles v Commissioner of Police* [2020] QDC 331.

Nevertheless, when protest activities involve law-breaking, opinion remains sharply divided on whether they are justifiable. The proceedings currently at hand are notable in that they constitute the first proceedings in which the legitimacy of acts of property destruction by climate activists has been considered in a Queensland courtroom.

Existence of an Extraordinary Emergency

The unique feature of the Queensland statutory defence lies in its invocation of emergency and not just any emergency, but either an extraordinary emergency *or* a sudden emergency. There is nothing sudden about human-induced global warming. Climate disasters that are occurring this century are, somewhat paradoxically, the consequence of historical emissions; the greenhouse gas emissions of today will cause the climate disasters of the future. The wording of the defence, however, clearly distinguishes between 'extraordinary' and 'sudden' emergencies. Either will suffice for the purpose of the section.

I turn therefore to the first question which must be addressed in relation to this defence: namely, whether an extraordinary emergency existed at the time that the offences took place.

In the twenty-first century, we have seen, in Australia alone, three decades of horrific fires. This decade was ushered in by the fires of Black Summer. The very palpable sense of emergency that prevailed during Black Summer, and indeed was officially acknowledged with various emergency declarations,[41] ebbed away with the rain events of February 2020, and the simultaneous arrival of a new global threat: the novel coronavirus which still, in its ever-multiplying variants, wreaks havoc on the global population. Flooding subsequently became the climate disaster of note but by the middle of the decade, Australia had descended again into drought.

The second half of this decade has been distinguished by fiery summers, with the worst of these to date being the Summer of Terror at the end of 2027 and the beginning of 2028. The catastrophic events of the Summer of Terror are indelibly etched on our memories. As the defendants have highlighted in their evidence, this was indisputably a national emergency, declared to be such by the Governor-General exercising powers given to her under the *National Emergency Declaration Act 2020* (Cth).[42] Members of the Australian Defence Force and voluntary firefighters were rapidly over-

41 See Nicole Rogers, *Law, Climate Emergency and the Australian Megafires* (Routledge, 2022) 24–5.

42 Enacted in response to a recommendation from the 2020 Royal Commission into National Natural Disaster Arrangements: *Royal Commission into National Natural Disaster Arrangements* (Report, 28 October 2020) Recommendation 5.1.

whelmed by the magnitude of the emergency; some lost their lives. Thousands of fire victims died while awaiting rescue. Others, attempting to flee across State borders in a search for refuge, found that certain States had closed their borders in a repeat of their pandemic response in 2020 and 2021.[43] Towns were levelled to the ground and ancient forests consumed in firestorms. Bunkers guaranteed to withstand megafires proved to be deathtraps for those who sought shelter inside them. Cities were cloaked in a toxic haze for many weeks; masked children, rushed by anxious parents to and from schools, collapsed from smoke inhalation. Many children and, indeed, many adults have suffered enormous trauma in losing their homes, and continue to suffer from anxiety and depression.

The prosecution accepts that the events of the Summer of Terror took place as set out in the evidence presented by the defendants. In their evidence, the defendants also reflected upon their personal experiences. 'For months, we were in constant fear for our lives.' 'All the familiar places of my childhood – now a barren, scorched wasteland'. 'The people of our town, including my family, have relocated to the city. Many of us are living in tents and substandard accommodation. But at least we can't get burnt out again there'. 'Our Country has lost its animals, our relatives,[44] its beating heart'. There is little doubt that the defendants believe that they are, personally, in the throes of an extraordinary emergency.

For the defence to be made out, however, there must be objective evidence of an extraordinary emergency, evidence which transcends personal beliefs. Here, the defendants have relied upon the scientific findings of the Intergovernmental Panel on Climate Change (hereinafter 'the IPCC') in its latest round of reports; these reports were tendered in evidence and the prosecution did not take issue with the material that appears in these documents. The Summary for Policymakers of its Synthesis Report for the Seventh Assessment Report, released in March 2029, sets out the current state of scientific knowledge in relation to climate change and its impacts. The IPCC draws upon calibrated language to indicate the level of confidence in each finding. In the Summary, the authors state that they are virtually certain

43 The closure of State borders to climate refugees is anticipated in Alice Robinson's *The Glad Shout* (Affirm Press, 2019), a work of climate fiction.

44 Here it is worth noting that celebrated Indigenous writer, Alexis Wright, reflecting on Black Summer, described it as a 'tragedy that silenced the songs and voices of the vast multitude of animals, our company, our relatives, that belong in the bushlands and forests': Alexis Wright, 'The Black Summer Bushfires Killed 3 Billion Animals. They are Our Relatives; They Deserve to Be Mourned', *The Guardian* (online, 1 April 2023) <https://www.theguardian.com/australia-news/2023/apr/01/alexis-wright-black-summer-bushfires-vigil-sydney-speech-3-billion-animals-killed-are-our-relatives-they-deserve-to-be-mourned>.

(99–100% certainty) that we are now, at 1.5 degrees of global warming,[45] locked into at least 2 degrees of global warming. Most of the world's current population of 8.4 billion people has experienced or will experience very serious and potentially life-endangering climate impacts in their lifetime. Food and water security cannot be guaranteed. Every terrestrial and marine ecosystem has been adversely impacted by climate change, with mass mortality events. The authors state that, in the absence of immediate, rapid, and dramatic global greenhouse gas reductions resulting in net zero CO_2 emissions, tipping cascades will be triggered which will take the world to 3 degrees of global warming or more (90–100% certainty). The authors state with virtual certainty that the urgency of climate action has never been greater.[46]

I am satisfied that there is sufficient evidence to establish the existence of an extraordinary emergency, extraordinary in its planetary scale and in its ramifications for every life form. I am also satisfied, in light of the evidence presented in the 2029 IPCC Synthesis Report, that this is an emergency that requires immediate action.[47] I shall therefore proceed to look at the second part of the defence.

Reasonable Conduct on the Part of an Ordinary Person

In Australia, the freedom to engage in protest activities is recognised as an important right. Justice Murphy of the High Court famously said, in relation to an Indigenous man seeking racial justice on a Queensland reserve, 'Mr Neale is entitled to be an agitator'.[48] Members of Last Generation believe that they are entitled to be agitators, and this may well be so.

In order to take advantage of the defence, however, the defendants must establish that agitators can, in certain circumstances, break the law. The extraordinary emergency defence enables law-breaking within certain parameters. The defendants must demonstrate that their acts of property destruction constitute reasonable conduct on the part of an ordinary person with ordinary power of self-control, in the face of the extraordinary emergency of climate change.

45 Should emissions continue to rise at the current rate, models suggest that 1.5 degrees of global warming will be reached in 2030.

46 Most of these conclusions can be found in the Summary for Policymakers of the Synthesis Report for the IPCC Sixth Assessment Report, although it is assumed that the world will be further advanced on the trajectory of global warming by 2029: Hoesung Lee et al, *Synthesis Report of the IPCC Sixth Assessment Report (AR6): Summary for Policymakers* (Report, Intergovernmental Panel on Climate Change, March 2023) <https://report.ipcc.ch/ar6syr/pdf/IPCC_AR6_SYR_SPM.pdf>.

47 In *Rolles v Commissioner of Police* [2020] QDC 331, at [40], Judge Rinaudo, referring to earlier authorities, held that 'an emergency for the purposes of s 25 must require immediate action'.

48 *Neale v the Queen* (1982) 149 CLR 305, 317 (Murphy J).

Mr Edwards, from the Office of the Director of Public Prosecutions, suggested to this Court that members of Last Generation are not ordinary people. His argument proceeded on the basis that the vast majority of Australians do not engage in acts of civil disobedience but are, rather, law-abiding citizens. Faced with the phenomenon of climate change, they continue to carry out their normal pursuits: going to work, looking after their families, and paying taxes. Many take steps to address their own carbon footprint by purchasing and driving electric vehicles, installing solar panels on their houses, and eating less meat. Some may write letters to their Member of Parliament, as befits a citizen in a representative democracy. This is, he argues, reasonable conduct on the part of ordinary people who are concerned about climate change. It is only the outliers, the exceptional rather than ordinary individuals, who feel compelled to break the law.

To my mind, there are two fundamental flaws in this argument. Firstly, behaviour which is commonplace, even behaviour which is law-abiding, does not necessarily constitute reasonable conduct. Nor does it necessarily constitute moral or ethical conduct: a point made with some force by philosopher Hannah Arendt when she recounted her experience as an observer of the trial of Adolf Eichmann in Jerusalem, in 1961.[49] Eichmann, a Nazi bureaucrat, could see nothing wrong with following the rules and orders of the Nazi regime, even though those rules and orders had genocidal outcomes. Is it reasonable to follow rules that cause genocide? Eichmann believed so and was at peace with his conscience. The Israeli State, Arendt and the global community have viewed his choices very differently.

Secondly, as the defendants pointed out, the actions outlined by Mr Edwards are not, of themselves, enough to achieve net zero CO_2 emissions. The defendants referred the Court to literature, in which it is claimed that the concept of individual carbon footprints is the product of an industry campaign to divert popular attention away from the need to halt fossil fuel production.[50] Whether or not that is the case, the 2029 IPCC Synthesis Report is very clear on this particular point: fossil fuel production must cease immediately.

The defendants further argued that it is reasonable conduct, on the part of ordinary people, to seek to protect their children from life-endangering harm. This is indisputably true. Every parenting instinct compels us to try to protect our children. This is no less the case when the threats are planetary in scale

49 Hannah Arendt, *Eichmann in Jerusalem. A Report on the Banality of Evil* (Penguin Books, 1994).

50 See, eg, Rebecca Solnit, 'Big Oil Coined "Carbon Footprints" to Blame Us for Their Greed. Keep Them on the Hook', *The Guardian* (online, 23 August 2021) <https://www.theguardian.com/commentisfree/2021/aug/23/big-oil-coined-carbon-footprints-to-blame-us-for-their-greed-keep-them-on-the-hook>.

and unprecedented in character. The burden of climate impacts will disproportionately fall upon today's children and the as-yet unborn, as renowned climate scientist James Hansen and his colleagues concluded in 2017.[51] There are, however, clear limits to what we can lawfully do in this regard. Parents do not have a legal licence to kill and destroy, on the basis that they are protecting their children. This particular argument does not advance the defendants' case.

The defendants relied, more convincingly, upon the exhaustion of legal avenues. This century, despite global awareness of the planetary consequences of uncurbed fossil fuel consumption and a sequence of intensifying climate disasters, Australian governments have approved an extraordinary number of new coalmines and coalmine extensions. The defendants have asserted that every possible step has been taken by the climate activist movement, and by themselves, to utilise available legal channels for challenging such projects in Australia. They point out, however, that legal challenges to these approvals have been largely unsuccessful.

The defendants went further with this line of argument, arguing that legal avenues have been exhausted in relation to the Waratah coalmine in particular. Here, of course, they were alluding to the events that followed the historic judgment of President Kingham of the Queensland Land Court, delivered on 25 November 2022.[52] President Kingham held that approving the Waratah coalmine would result in the violation of a number of human rights enshrined in Queensland's *Human Rights Act 2019*, including the right to life, the cultural rights of First Nations peoples, the rights of children, the right to property and to privacy and home, and the right to enjoy human rights equally.[53] Subsequently, on 3 April 2023, the Queensland Department of Environment and Science refused the environmental authority for the mine. This decision was, however, challenged by Waratah Pty Ltd in the Supreme Court of Queensland. In these judicial review proceedings, in 2024, the decision was set aside as unreasonable and the Queensland government then, and I might add despite a much-vaunted commitment to renewable energy, approved the coalmine. The Waratah coalmine could proceed. Construction began in 2024 and the mine began operating in 2027.

Finally, the defendants drew upon Swedish theorist Andreas Malm's arguments about the need for strategic acts of property destruction. The defendants argued that while indiscriminate acts of property violence cannot necessarily be justified as reasonable conduct, they were targeting a particular instrument of planetary destruction: fossil fuel infrastructure. The

51 James Hansen et al, 'Young People's Burden: Requirement of Negative CO_2 Emissions' (2017) 8(3) *Earth System Dynamics* 577.
52 *Waratah Coal v Youth Verdict* [2022] QLC 21.
53 Ibid [44].

correlation between the property destroyed, and the crisis which they are seeking to avert, is clear. As Malm argues in his highly influential text, *How to Blow Up a Pipeline*, it is the 'machines of the fossil economy [which] *are* the problem'.[54] The defendants also emphasised that they made certain, through drone surveillance operations, that there would not be injury to any person or sentient being in their targeted attacks, and this was in fact the case. As one of the defendants put it, 'we were protecting human and non-human lives by destroying fossil fuel infrastructure, not endangering them'.

I am persuaded by the defendants' arguments that their conduct was reasonable in the circumstances: circumstances in which the Waratah coalmine had commenced operation despite a judicial finding that this could result in 'foreseeable and preventable life-terminating harm';[55] and in which great care was taken by the defendants to ensure that the targeting of the mine's infrastructure did not endanger human lives. There is, furthermore, nothing to suggest that in carrying out the actions in question the defendants could not be viewed as ordinary people with ordinary power of self-control. On the contrary, ordinary people with ordinary power of self-control are as capable of responding to an emergency as are exceptional people with exceptional power of self-control.

Bias

As everyone sitting in this courtroom is aware, accusations of bias have been levelled at me by certain journalists and news outlets. One journalist went so far as to write: 'We cannot expect fair and impartial decision making from a woman who spent a significant number of months refusing to work at a proper job and instead sitting in a tree'.

The thrust of these accusations is that I have a personal stake in the outcome of the climate crisis which affects my capacity to adjudicate on the conduct of climate activists. It is true that I have a personal stake in the outcome of the climate crisis, but I am not alone in this; as one political scientist put it in 2020, we all have 'skin in the game' that is climate change politics.[56] In this quite extensive and potentially defamatory commentary, much is made of my past involvement in environmental protest movements. As is now popularly known, I participated as a young adult in anti-logging protests in Victoria, spending, at one point, 270 days tree sitting in a mountain ash tree to prevent it from being logged. In 2011, as a consequence of

54 Malm (n 1) 105 (emphasis in original).

55 *Waratah Coal v Youth Verdict* [2022] QLC 21 [512].

56 Jessica F Green, 'Less Talk, More Walk: Why Climate Change Demands Activism in the Academy' (2020) 149(4) *Dædalus, the Journal of the American Academy of Arts & Sciences* 151, 159.

these activities, I found myself in the Victorian Magistrates Court before Magistrate Singh. I was acquitted of the relevant charges and, unusually, the magistrate praised my commitment and social conscience, stating that she expected great things from me in the future. I reflected deeply upon her words and it was her expression of confidence in my abilities which led me to undertake a law degree. Upon its completion, I found work with the national Environmental Defenders Office as an environmental lawyer, before being appointed a judge.

I take some comfort from knowing that more senior jurists with formidable reputations have been subject to similar accusations. Judge Brian Preston, for many years the Chief Judge of the New South Wales Land and Environment Court and currently a member of the International Climate Court, established in 2026 to adjudicate over international climate disputes, broke new ground in 2019 by holding that a coalmine should not proceed on the basis of its projected climate impacts; he was subsequently vilified.

I find it abhorrent that media commentators do not make accusations of judicial bias against judges of conservative backgrounds, who regularly hand down judgments that entrench the status quo and contribute to the confluence of factors that are creating a climate emergency, but rather only against those judges who are courageous enough to reshape the common law to achieve social and environmental justice outcomes. It hardly needs saying, given that I took an oath on this matter at the time of my appointment, that I am fully cognisant of the need to administer justice without fear, favour, or affection.

I have exercised impartiality and objectivity in evaluating the evidence presented to me at the trial. I am persuaded that we are indeed in the throes of an extraordinary emergency and that, furthermore, the defendants' actions were those of ordinary people with ordinary power of self-control acting reasonably in all the circumstances. The defendants' actions did not result in injury to any person. I am persuaded by their evidence that they took all reasonable, in fact excessive precautions to ensure that no one was injured or hurt by their actions. In the absence of governmental intervention to halt fossil fuel production, it is reasonable for ordinary people to take steps to decommission fossil fuel infrastructure which is endangering not only the planet but, in addition, their own lives and the lives of their children.

I will now repeat to the defendants, with some paraphrasing, the words of Magistrate Singh during my 2011 trial: words which empowered and inspired me to continue the all-important task of protecting our planet, and its Earth systems.

You are remarkable, brave individuals who are prepared to take personal risks in following your conscience and working for the greater good. I am sure that, in future years, the important work that you accomplish in trying to avert the worst of the climate crisis will be widely acknowledged. I wish you all the best in your endeavours.

You are acquitted of all charges.

Chapter 10

Piccadilly Circus Water Lilies

A Judgment on Participation and Place Experience in Future Planning Decisions

Chiara Armeni

Commentary

Scenario

It is 2100. Neither the net zero targets nor the nationally determined contributions under the Paris Agreement[1] adopted in the 2020s significantly changed the emission trajectory up to 2030. Global mean surface temperature has risen by around 3°C relative to 1850–1900.[2] Over the last century, European cities have been intensely hit by climate change. Among them, London has experienced a sea-level rise of one metre above 1995–2014 levels.[3] The call made in the early 2020s to move the Thames Barrier 17 kilometres towards the sea in order to reduce the flooding risk to a tolerable level was left unanswered by decision-makers.[4] As a result, by 2096 most of central London was partially submerged, and its residents, businesses, and institutions were forced to relocate to higher areas on the outskirts of the city. The High Court and the Court of Appeal of England and Wales were also relocated from the Royal Courts of Justice to Sutton House in the borough of Hackney.

While London's landmark sites are mostly underwater, they remain important places of memory, belonging, attachment, and identity for Londoners. Although they still connect people to a pre-climate change past, the way individuals and communities experience them has changed. Archaeological diving, canal boating, water gardening, water sports, or fishing are now

I am grateful to Maria Lee (University College London) and Ned Westaway (Francis Taylor Building) for their precious comments on earlier drafts.

1 *Paris Agreement*, opened for signature 22 April 2016, [2016] ATS 24 (entered into force 4 November 2016).
2 This scenario builds on the IPCC SSP2-4.5 scenario. See Intergovernmental Panel on Climate Change, *Climate Change 2021: The Physical Science Basis* (Cambridge University Press, 2021) Cross-Chapter Box 1.4, 'The SSP Scenarios as Used in Working Group I (WGI)'.
3 Projection based on Sea Level Projection Tool – NASA Sea Level Change Portal scenario based on the IPCC SSP2-4.5 in London in 2100, at <https://sealevel.nasa.gov/ipcc-ar6-sea-level-projection-tool>.
4 Jim W Hall, Hamish Harvey, Lucy J Manning, 'Adaptation Thresholds and Pathways for Tidal Flood Risk Management in London' (2019) 24 *Climate Risk Management* 42–58.

DOI: 10.4324/9781003389569-12

aggregating new communities of interest around these iconic places, reshaping old and creating new place experiences. Marine biodiversity has emerged where traffic and street lights once dominated the urban landscape. Several CO_2-neutral cooling beaches have been created, making Londoners feel in the Mediterranean Riviera while still viewing the tip of Big Ben in the distance.

This scenario was inspired by Robert Graves and Didier Madoc-Jones's 'Piccadilly Circus Water Lilies' presented at the Museum of London exhibition *Postcards from the Future* (October 2010–March 2011).[5] The exhibition showed the impact of climate change on 14 landmark views of London through climate change projections juxtaposed with contemporary visual images. In 'Piccadilly Circus Water Lilies', Graves and Madoc-Jones represented Piccadilly Circus as 'a haven of calm as water levels rise ever higher, [where] [w]ind turbines rotate gently in the breeze while fish swim around the water lilies'.[6] The artists used water lilies growing around wind turbines as a bucolic image of the future Piccadilly Circus. While I build on their overall representation, I refer to 'Water Lilies' as the new floating wind turbines *themselves*, which present as the central object of this judgment.

The Facts of the Case

On 7 January 2098, Wind of London Power Ltd applied for development consent to construct and operate 100 small-scale floating wind turbines in central London's shallow waters. The Secretary of State for the Future – responsible for granting consent – appointed a technical Examining Authority and a deliberative Permanent Citizens' Planning Assembly for the technical and democratic evaluation, as required by the new *Planning and Place Act 2097*.

In their joint report, the Examining Authority and the Permanent Citizens' Planning Assembly considered that while environmental and visual impact of the project would have been minimal and acceptable, its impact on people's place experience was significant and outweighed the overall benefits of the development. These examining bodies recommended that the Secretary of State refuse consent.

The Secretary of State disagreed for two reasons. First, she considered that climate change had irreversibly transformed central London's placescape into a 'non-place', an interstitial space of transit that 'cannot be defined as relational, or historical, or concerned with identity'.[7] Second, loss of place inevi-

5 GMJ, 'Postcards from the Future: Wish You Were Here?', *Postcards from the Future* (Web page) <http://www.postcardsfromthefuture.com>.

6 GMJ, 'Piccadilly Circus Water Lilies', *Postcards from the Future* (Web page) <http://www.postcardsfromthefuture.com/work-1/project-two-9mb57>.

7 Marc Augé, *Non-places: Introduction to an Anthropology of Supermodernity*, tr John Howe (Verso, 1995) 77.

tably modified the legal characterisation of people's practical, sensorial, and emotional experience as *place* experience and the weight to be given to it for the decision. The Secretary of State concluded that the presumption in favour of development of shallow-water flexible wind energy installations embedded in the National Policy Statement on Net Zero Energy Infrastructure outweighed the project's impact on non-place experience.

The non-governmental organisation Our Future Place challenged the Secretary of State's decision before the High Court. The challenge focused on her categorisation of the site as a non-place and on her assessment of the weight to be given to people's place experience. However, in her judgment, Ms Justice Alderman upheld the Secretary of State's reasoning and decision.

Our Future Place appealed against the order of Ms Justice Alderman dismissing the judicial review of the Secretary of State's decision. The Appellant argued that the Secretary of State

a) misinterpreted the definition of 'place' and 'place experience' under the *Planning and Place Act 2097* and
b) misapplied the reinforced requirement for considering deliberative participation in decision-making as it failed to consider 'place experience' in and of the proposed site as 'important and relevant' considerations for the decision under the *Planning and Place Act 2097*.

The Court of Appeal agreed with the appellant, quashed the Secretary of State's decision, and remitted the matter to them for reconsideration.

Relation with Current Deficiencies in Law and Analysis

Planning law is 'the most self-evidently future-oriented of all the legal fields'.[8] Planning is fundamentally about caring for the future, 'anticipating and preparing for coming challenges and exercising agency to purposively steer change for the common good'.[9] This fictional judgment deals with imagining the space for place experience and the role of public participation in future planning decisions through the lenses of renewable energy infrastructure in England. In so doing, it addresses two deficiencies in contemporary English planning law.

First, it addresses the limited recognition of place experience in the consent process for large-scale offshore wind energy projects in England. These are considered Nationally Significant Infrastructure Projects and

8 Elen Stokes, 'Futurescapes of Planning Law: Some Preliminary Thoughts on a Timely Encounter' in Maria Lee and Carolyn Abbot (eds), *Taking English Planning Law Scholarship Seriously* (University College London Press, 2022) 157, 158.
9 Andy Inch, 'Planning for the Future?' (2021) 2 (3) *Planning Theory & Practice* 341.

their decision-making process is regulated by the *Planning Act 2008*. Place experience is a complex notion that can be understood as 'an indiscernible combination of actions (for example shopping, playing, walking, cycling, bird-watching, fishing), sensory perceptions (for example the smell, sound or view of the landscape) and emotional connections (for example place attachment, sense of place, place identity)'.[10] Although an important concern in people's perception of and reaction to infrastructure development, the *Planning Act 2008* does not define or consider place experience as an 'important and relevant matter' for decision-making.[11] The challenge of dealing with place experience in planning decisions under the Nationally Significant Infrastructure Projects regime derives primarily from its 'comfortable reliance on technical assessment and a strong policy commitment to climate mitigation' to justify decisions.[12] These technical and policy elements often reduce the weight of other reasons – such as place-based values and experiences – for the decision.

The judgment deals with this deficiency by imagining a clear role for place experience in the legal agenda. The imaginary *Planning and Place Act 2097* gives central stage to place in planning decisions. It does so by (a) explicitly defining the notion of 'place' and 'place experience' as legally relevant and (b) recognising that place experience can be considered as an 'important and relevant matter' for the decision. In so doing, the Act locates place experience solidly within the legal planning framework to enable decision-makers to take it into account as a material consideration for a decision.

Building on this imagined legal framework, the judgment is concerned with the legal interpretation of 'place' and the impact of climate-induced changes upon it. The key point made in the judgment is that even when it is legally institutionalised, the dense notion of 'place' exceeds technical categorisations. 'Place' is a moving and evolving concept that adapts to and changes with the physical and societal environment and people's experiences of it. From this perspective, the judgment offers a reading of planning law and society as co-constructed in a post-climate change future by taking experiences and changes seriously into account in future regulatory decisions.

Second, this future judgment deals with the challenges of public participation in planning decisions. This is essential as it is through public consultation that concerns for place experience enter the decision-making process. English planning law institutionalises the right to participate in planning decisions, mostly through public consultation in the planning process combined with the right to participate in environmental assessment at strategic and project

10 Chiara Armeni, 'Place, Participation and Planning Law in a Time of Climate Change' in Lee and Abbot (n 8) 111.
11 *Planning Act 2008* (UK) s 104.
12 Ibid.

levels.[13] However, in a climate change context, the tensions between technical expertise and participation often lead to an erosion of participatory rights.[14] Planning decisions on wind energy Nationally Significant Infrastructure Projects in England are a good example of these challenges.[15]

This judgment's contribution is to insist on the role of procedural rights to participate in planning decisions. In its imagined future, participation is reinforced through new institutional arrangements – the imaginary Permanent Citizens' Planning Assembly – and through reinforced regulatory provisions – namely, the imaginary statutory obligation to define place experience by reference to the results of the Permanent Citizens' Planning Assembly and its recognition as important and relevant for the decision.

The judgment emphasises the input of participation in the decision-making process and allows broader – qualitative, emotional, cultural, and so on – considerations to at least be able to shape planning decisions in a post-climate change society. By requiring that place experience is given weight as an important and relevant matter, this future judgment strengthens the voice of participants claiming a practical, emotional, or sensorial relationship with a 'place'. This broader, place-sensitive, and participatory approach to future planning decisions is an invitation to imagine a climate-neutral and more democratic future through law.

Judgment

Neutral citation Number: [2100] EWCA Civ 23
Case No: 4583
IN THE COURT OF APPEAL (CIVIL DIVISON)
ON APPEAL FROM THE HIGH COURT OF JUSTICE
(PLANNING COURT)
THE HONORABLE MS JUSTICE ALDERMAN[16]
[2099] EWHC 12 (Admin)

Royal Courts of Justice
2 And 4 Homerton High Street, Hackney E9 6JQ
Date 7/12/2100

Before
Lady Justice Montgomery-Taylor

13 In an English context, *The Environmental Assessment of Plans and Programmes Regulations 2004* (UK) (SI 1633); *The Town and Country Planning (Environmental Impact Assessment) Regulations 2011* (UK) (SI 1824).

14 Chiara Armeni and Maria Lee, 'Participation in a Time of Climate Crisis' (2021) 48 (4) *Journal of Law and Society* 549; Maria Lee, 'The Aarhus Convention 1998 and the Environment Act 2021: Eroding Public Participation' (2023) 86(3) *Modern Law Review* 756.

15 Armeni (n 10).

16 The names in this judgment are inspired by Naomi Alderman, *The Power* (Penguin Book, 2017).

Lady Justice Monke
and
Lady Justice Cleary

Between:
R. (on application of Our Future Place) Appellant

– and –

Secretary of State for the Future Respondent

Ms Tatiana Moskalev QC (instructed by Our Future Place) for the Appellant
Mr Olatunde Edo QC (instructed by the Government Legal Department) for the Respondent

Hearing Dates: 1 December 2100

Approved Judgment

Lady Justice Montgomery-Taylor, Lady Justice Monke, and Lady Justice Cleary

Introduction

With permission granted by Montgomery-Taylor LJ, the appellant, Our Future Place, appeals against the judgment ('the Judgment') of Alderman J dated 17 November 2099 in which she upheld the decision dated 12 November 2098 ('the Decision') of the Secretary of State for the Future ('the Secretary of State') to grant development consent to the construction of the Piccadilly Circus Water Lilies project ('Water Lilies') by Wind of London Ltd.

The decision was taken following the rejection by the Secretary of State of recommendations made to her by the Examining Authority and the Permanent Citizens' Planning Assembly who had been appointed by her to report on this application and who recommended that consent be refused ('the Examining Authority and Permanent Citizens' Planning Assembly Recommendations').

As set out in the judgment of Alderman J, the development proposed by Wind for London Ltd comprises the construction and operation of up to 100 floating wind turbine generators of a generating capacity below 50 GW each. Each infrastructure comprises a turbine adapted for a downwind configuration, a floating platform deployed in shallow waters integrating mooring system and all the required control systems and wi-fi smart connection to the energy grid.

The site to which the application for development consent related covers an area of one km^2 from the old road junction of Piccadilly Circus (northern point) to Embankment point (southern point) on the old shore of what used to be the River Thames. The Circus used to connect Piccadilly, Regent Street, Shaftesbury Avenue, the Haymarket, Coventry Street (onwards to Leicester Square), and Glasshouse Street which have been submergences for the last four years. This area is a shallow-water area (38–50 metres deep).[17]

This shallow-water floating wind turbine development project is a Nationally Significant Infrastructure Project (NSIP) under the *Planning and Place Act 2097*.

The Issues Before the Court

This appeal focuses on the statutory interpretation of 'place' and weight to be given to 'place experience' concerns expressed by the deliberative Permanent Citizens' Planning Assembly.

The first legal issue before the court is whether the Secretary of State misinterpreted sections 104(7) and (8) of the *Planning and Place Act* in its interpretation of the definition of 'place' and 'place experience'. The question before the court is whether the changes caused by climate change upon statutorily defined elements of the notion of 'place' are to be viewed as transforming the area proposed for the development into a 'non-place'. This issue requires the court to assess whether the definition of place under the *Planning and Place Act* demands a strict interpretation (where place is a fixed category that once changed loses its legal status) or an evolutive interpretation (where place can be radically transformed while maintaining legal status).

The second legal issue is whether the Secretary of State misapplied her obligations under section 104(8) of the *Planning and Place Act* by not giving substantive weight to place experience concerns expressed by the deliberative Permanent Citizens' Planning Assembly as important and relevant matters for the decision.

The Legislative and Policy Framework

Planning Consent Process

Under section 33 of the *Planning and Place Act*, siting floating wind turbine infrastructure in shallow water requires a development consent order

17 Kun Xu et al, 'Design and Comparative Analysis of Alternative Mooring Systems for Floating Wind Turbines in Shallow Water with Emphasis on Ultimate Limit State Design' (2021) 219 *Ocean Engineering* 108377. See also prototypes at X30 platform: floating wind scale prototype in the Canary Islands; X1Wind, 'A Disruptive Floating Wind System to Change the Paradigm of the Wind Energy Industry' (Web Page, 2023) <x1wind.com>.

from the Secretary of State. The consenting process is triggered by the acceptance of the application by the National Infrastructure Planning Unit at the Planning Inspectorate for Net Zero Energy.

Once the application is formally accepted for examination, two examining bodies are appointed: an Examining Authority and a Permanent Citizens' Planning Assembly. Under section 98(1), the technical Examining Authority is composed of a three-person panel of professional planners appointed by the Secretary of State. Under section 98(2), the deliberative Permanent Citizens' Planning Assembly is composed of 500 members randomly selected by the Planning Inspectorate for Net Zero Energy according to criteria of demographic representation and the procedure set in section 99. The Assembly is formally appointed by the Secretary of State. To be selected as an Assembly member, participants must be above the age of 14 and have a plausible practical, emotional or other justifiable connection with the proposed development site.

Under section 100, participation rights can be extended to all legal or moral persons interested in contributing to the decision-making process upon request to the Planning Inspectorate for Net Zero Energy before the first meeting of the Permanent Citizens' Planning Assembly.

The examination lasts 12 weeks. After the examination is concluded, the examining bodies have a duty to produce a joint report including the summary of their examination process, the views expressed in the examination, their joint findings, and conclusions, and the final recommendations to the Secretary of State as to whether consent should be granted or refused.

The Secretary of State decides on the application by order within six months from the closure of the examination.

Section 102(2) provides that:

In deciding the application, the Secretary of State must have regard to:

(a) any national policy statement and marine policy documents that have effect in relation to development of the description to which the application relates;
(b) the recommendations of the examining bodies;
(c) technical and participatory evidence provided by the examining bodies in the context of their examining process;
(d) any impact assessment provided by the applicant; and
(e) any other matters which the Secretary of State thinks are both important and relevant to the Secretary of State's decision.

Section 103 provides that:

> The Secretary of State must decide the application in accordance with any relevant National Policy Statement, except to the extent that the Secretary of State is satisfied that doing so would:

(a) lead to the United Kingdom being in breach of its international obligations;
(b) lead to the Secretary of State being in breach of any duty imposed on them under any enactment;
(c) be unlawful under any enactment; or
(d) outweigh the benefits of the proposed development.

The decision on consent or refusal is subject to judicial review within six weeks from its notification.

Place-Related Provisions

Section 104 of the *Planning and Place Act* specifies that, in addition to the statutory requirements for impact assessment provided for Nationally Significant Infrastructure Projects, a Place Impact Assessment must be carried out for any development proposed in an area that qualifies as a 'place'.

Section 104(7) defines a 'place' as:

> any environmental locus in and through which individual and group actions, experiences, intentions and meaning are drawn together spatially and recognised individual and collective worth, value and significance.[18]

Section 104 (8) provides that:

(a) 'place experience' means a combination of individual or collective actions and routines, sensory perceptions, and emotional connections triggered by being and feeling in a place;
(b) 'place experience' must be defined on a case-by-case basis by reference to the results of deliberative participation and is capable of being an important and relevant matter for the decision.

Section 104 (9) also defines 'place impact' as:

> any direct or indirect effect on the way that elements of a place defined in section 104 (7) combine together and how people experience it.[19]

18 Definition built on David Seamon, 'Place Attachment and Phenomenology: The Synergistic Dynamism of Place' in Lynne C Manzo and Patrick Devine-Wright (eds), *Place Attachment: Advances in Theory, Methods and Applications* (Routledge, 1st ed, 2014) 11, 11.

19 Definition built on Ministry of Housing, Communities and Local Government, *National Design Guide* (2021) 14 <https://www.gov.uk/government/publications/national-design-guide>.

National Policy

The *National Policy Statement on Net Zero Energy Generation 2097* has an effect in relation to this application and therefore Section 104(1) of the *Planning and Place Act* applies.

The policy context is important here. The United Kingdom is well beyond the national energy transition. Since 2055, all non-renewable energy infrastructure has been decommissioned and no new fossil fuel energy installation has been given planning consent. The whole of the United Kingdom's energy supply today comes from wind, tidal, solar, hydrogen and other new energy sources, such as gravitational energy and re-nuclear energy. But the transition has had a significant impact on urban and rural land planning and management. In England, physical space for wind energy infrastructure is running short. In London, the relocation of households, businesses, and institutions meant that competition for land and its multiple uses increased dramatically.

As a result of reduced availability of land and a century of public opposition to onshore wind farms, the *National Policy Statement on Net Zero Energy Generation* established a moratorium on new onshore wind energy installation in England.[20]

This moratorium is balanced by a presumption in favour of development of large-scale offshore and small-scale shallow-water wind energy turbines in permanently flooded and submerged areas as they are 'presumed to have a minimum environmental and visual impact and to satisfy the continuous need for net zero carbon energy'.[21]

The Relevant Decisions and the Appeal

The Examining Authority and Permanent Citizens' Planning Assembly Report and Recommendations

In their joint report to the Secretary of State, dated 15 August 2098, the Examining Authority and Permanent Citizens' Planning Assembly stated that:

> After consideration of the range of potential impacts that would be likely to arise if the proposed development were to be built and operated, these Examining Bodies conclude that the proposed development shows minimal and acceptable environmental impact as well as Seascape, Landscape, and Visual Impact. The principles of the proposed development are in general in conformity with the National Statement on Net Zero Energy Infrastructure. We further conclude that, subject to the

20 United Kingdom, *National Policy Statement on Net Zero Energy Generation 2097* [5.1].
21 Ibid [5.2].

> scheme restrictions, the controls and mitigation are adequate. However, significant adverse impact will remain on place experience associated to the proposed site.

The examining bodies further note that:

> The deliberative participation exercise conducted by the Permanent Citizens' Planning Assembly produced substantive evidence of the significant place experience related to the area. Most of the participants claimed that the development site is 'the wrong place' and the development is 'not fitting' with their place experience. This includes practical (for example, archaeological diving, canal boating, water gardening, water sports or fishing, beach activities), sensorial (for example, enjoyment of the view and landscape), and emotional (for example, sense of identity, memory, attachment to the place) experiences which are now aggregating new communities of interest around these iconic places.
>
> It is clear to us that despite a radical change of landscape and physicality due to climate change-induced sea level rise, this space sees 'an interaction of individual bodily routines rooted in a particular environment that may become an important place of interpersonal and communal exchange, meaning, and attachment'.[22] These interactions should be considered evidence of place experience and are to be given due regard in the decision.

The examining bodies concluded that:

> In the light of concerns for the substantive impact on place experience, we conclude that the impact of the project upon individual and collective place experience outweighs its benefits in terms of energy generation.

Based on their conclusions regarding the criteria set out by the *Planning and Place Act* and all relevant and important matters, including the merits of the case for the proposed project, the examining bodies recommended that 'the case for granting development consent for the proposed project is not made'.

The Decision

In her decision letter, dated 12 November 2098, the Secretary of State disagreed with the examining bodies' recommendations. She considered that:

22 Seamon (n 18).

> Climate change has irreversibly transformed the character and identity of central London and consequently its ability to be qualified as a 'place' under the criteria of section 104 (7) of the *Planning and Place Act*. Although the result of public participation shows that individuals and groups attributed value to it, from a technical viewpoint the proposed area for the development could not be qualified as a place, having lost its defining character.
>
> In today's Anthropocene society, many of our familiar places have radically been transformed or have been lost due to climate change. Because of flooding and sea level rise, most of London's places have become non-places: interstitial spaces of transit between places that 'cannot be defined as relational, or historical, or concerned with identity'.[23] Non-places are 'spaces that are not themselves anthropological places and which [...] do not integrate earlier place'.[24] While there used to be a limited number of such non-places (for example, airports, railways, leisure centres, or shopping malls), we see today more places turning into non-places due to extreme climate impact. The physical environment is essential to the construction of meaning.[25] This means that if the physical environment is lost or radically changed, so is its meaning and experience changing places into non-places. Despite public concerns for those spaces, non-places do not matter to planning law. While the *Planning and Place Act* gives legal recognition to places and their experience of attachment, identity, and memory in the decision-making process, it does not extend this legal recognition to the experience of non-places in a post-climate crisis world.

As a result, the Secretary of State argued that the policy need for shallow-water flexible wind energy installations of the type of the Water Lilies outweighs the project's impact on non-place experience.

The Judgment

On 20 December 2098, the organisation, Our Future Place, challenged the decision before the High Court. In her judgment of 17 November 2099 Alderman J dismissed the claim, holding that:

> Whereas the examining bodies' report recognised individual and collective place-based values in relation with the proposed development, the

23 Augé (7) 77.
24 Ibid 78.
25 Richard C Stedman, 'Is It Really Just a Social Construction? The Contribution of the Physical Environment to Sense of Place' (2011) 16(8) *Society & Natural Resources* 671.

loss of place due to climate change transformed this area into a neutral, anonymous site. This new context does not require the Secretary of State to consider concerns related to potential impact on people's experiences in this site as important and relevant matters for the decision.

She found that the Secretary of State acted within the law in deciding that concerns for impact on place experience of the site could not outweigh the policy need for wind energy in this case.

On 3 January 2100, Our Future Place appealed against the order of Alderman J dismissing judicial review of the Secretary of State's decision. The Appellant argues that the Secretary of State:

1. misinterpreted section 104(7) and (8)(a) of the *Planning and Place Act 2097* in relation to the notions of 'place' and 'place experience'; and
2. misapplied the reinforced requirement for considering deliberative public participation in planning decision-making as she failed to consider 'place experience' in and of the proposed site as 'important and relevant considerations for the decision' under section 104(8)(b) of the *Planning and Place Act 2097.*

Discussion and Conclusions

(i) *Did the Secretary of State misinterpret the notions of 'place' and 'place experience' under section 104(7) and (8)(a) of the* Planning and Place Act 2097*?*

The essential argument put forward – as in the court below – is that the Secretary of State erred in the interpretation of the notion of 'place' under section 104(7) of the *Planning and Place Act.*

The Secretary of State argued that climate change-induced sea level radically transformed the characteristics of the development site area as a place. This transformation irreversibly changed the place into a non-place. As a result, the experiences of the areas – presented in particular in the results of the Permanent Citizens' Planning Assembly – cannot be considered as place experiences as this area cannot fit the standard of section 104(7) of the *Planning and Place Act.*

We cannot accept this interpretation. We agree that climate change has radically transformed many places, including those in the proposed development site. However, we cannot accept that the area can be considered a non-place as a space that 'cannot be defined as relational, or historical, or concerned with identity' purely based on an objective and technical assessment. Transformation and transition are inherent features of a post-climate crisis society. While some places have changed and stabilised, others are in continuous mutation, together with the ways in which people experience them.

Climate change is progressively transforming our familiar places not only by profoundly affecting their morphological and ecosystem characteristics but also through changing people's experience of them. This does not necessarily mean that places are lost. Section 104(7) is to be interpreted in a broad, evolutive, and inclusive way to account for the radical – positive and negative – transformation that climate change has brought to urban places and individual and collective experiences of them. These changes should be considered inherent in the definition of 'place' in a post-climate change society, rather than a non-statutory criterion categorising it as a non-place that is outside the scope of the *Planning and Place Act*.

In our view, the Secretary of State's interpretation of 'place' and 'place experience' amounts to a legal error. The idea of 'non-place' finds no support in the statutory language and the Secretary of State applied it erroneously. The Secretary of State's conclusions that the development site constitutes a non-place show that she has erred in interpreting the relevant provisions of the *Planning and Place Act*. This results in circumvention of the necessary focus on place required under the *Planning and Place Act*.

We conclude therefore that on this issue the appeal should succeed.

(ii) *Did the Secretary of State misapply the requirement for considering the result of deliberative participation in the case-by-case assessment of place experience under section 104(8)(b) of the* Planning and Place Act 2097*?*

> How people experience places and infrastructure within them are important questions for planning decisions. People's experience of places is reflected in the public participation process for planning consent, where wind energy infrastructure is often framed by participants as 'out of place' – as a threat or disruption to the place as it would be seen to 'industrialise' an otherwise rural place that is considered for its natural beauty and remoteness.[26]

This close relationship between participation and place experience is reflected in the *Planning and Place Act*, which provides that the assessment of place experience is to be made on a case-by-case basis by reference to the results of deliberative participation. This reinforced requirement for public participation is implemented by the establishment of the Permanent Citizens' Planning Assembly and the statutory obligation to define place experience by reference to the results of the Permanent Citizens' Planning Assembly, and its ability to be important and relevant for the decision.

26 Armeni (n 10) 112.

This means that the extent to which place experience claims are recognised as evidence for the decision should not be limited by technical and policy arguments. Section 104(8)(b) of the *Planning and Place Act*

> opens up discussion on the space for lay public place experience claims in the participation process and the role of law in shaping that space. It engages more widely with the opportunities and challenges of public participation in planning decisions.[27]

The Secretary of State based her assessment of place impact purely upon technical assessment of quantitative factors (specifically, environmental impact and Seascape, Landscape, and Visual Impact). She excluded the deliberative assessment of place experience as the 'combination of individual or collective actions and routines, sensory perceptions, and emotional connections triggered by being and feeling in a place' from her decision.

In our view, this amounts to a legal error. By failing to fully consider the wider meaning of place experience and the results of the Permanent Citizens' Planning Assembly in the assessment of place experience, the Secretary of State misapplied section 104(8)(b).

We conclude therefore that on this issue the appeal should succeed.

Results of the Appeal

For the reasons set out above we allow the appeal.

We quash the Decision of the Secretary of State and we remit the matter to them for reconsideration in the light of this judgment.

Finally, we record our gratitude to the parties for their helpful written submissions, and to the Counsels for their thoughtful oral submissions and for responding with patience and care to the multitude of questions posed by the Court during the hearing.

27 Ibid 115.

Part III

Postcolonial Justice

Chapter 11

The Problem with Cooperative Action Problems

Conceptions of Agency and the Understanding of Environmental Crises

Oscar Davis, Bindi Bennett, and Kelly Menzel

Introduction

In both social and political domains, global environmental threats are often cast as problems of cooperative action. This raises several problems. Cooperative action problems frame crises in terms of the need to overcome individual or corporate concerns about selfhood, dignity and well-being. Typically, overcoming individual self-concern is attempted through social and legal mechanisms such as regulation, incentivization, nudging and so on. Cooperative action problems are not necessarily soluble, as framed. They are also demotivating; they discourage individual responsibility. They seek to manipulate individuals and corporations, discouraging moral development. In this chapter, we argue that contemporary environmental crises require something more radical; without a deep and widespread re-evaluation of deep-rooted assumptions about our relationship to our environments, regulative efforts to address environmental crises are threatened with comprehensive failure. We should take seriously the possibility that contemporary environmental problems are thoroughly resistant to technical solutions, either regulatory, economic or technological.

The conclusion of this line of thought is that it is a persistent mistake to frame environmental crises as existential crises. We need a new way of thinking about them. There is a pressing need for a renewed understanding of the relationship between the human and natural worlds, something environmental philosophers have been championing for many decades, but one that helps us frame environmental crises as urgent moral crises, directed not at future generations of humans but at something more fundamental still. Drawing on the philosophical underpinnings of Indigenous Australian worldviews, and recent work in environmental philosophy, we set out the characteristics of an ethic which sees environmental threat in just this way. We call this an account of extended moral agency and we illustrate how this central conceptual and practical feature of Australian Aboriginal ontology and morality construes our interconnectedness with the environment as the starting point of ethical deliberation. We then explore how this account

DOI: 10.4324/9781003389569-14

provides a conceptually rich articulation of our individual responsibility towards global environmental crises, one that overcomes our framing of the emergency as an existential one.

The Problem with Cooperative Action Problems[1]

Our response to climate crisis is a moral concern as much as it is a political, technical, and scientific concern, yet its moral features are underexplored. The existing moral literature on human responsibility in climate crises predominantly frames our responses as problems of cooperative action that require new mechanisms for bolstering collective solutions[2] or are otherwise new imaginings of the problems themselves.[3] In these cooperative action problems, overcoming individual self-concern is typically attempted through turning the commons into a public good, or otherwise enclosing the commons into a private good. These conventional solutions are grounded in the establishment of social and legal institutions. Cooperative action problems are not necessarily soluble, as framed. We need other and better ways of conceptualising human responsibility for climate crisis on the level of the individual. We require a framework for cultivating an environmental conscientiousness – the kind of beliefs and attitudes towards the natural world conducive to a genuine and personal response to climate emergency.

Natural and Social Relations

The climate crisis is leading to the irreversible damage of terrestrial, marine and cryospheric ecosystems. Yet, in Australia, a large mining corporation demolished a 46,000-year-old sacred Aboriginal site in order to expand their iron ore mine in the Pilbara region, and iron ore remains Australia's largest export.[4] In recent months, from the approval of the Willow Project in Alaska to the report that China permits the equivalent of two new coal power plants

1 We understand collective action problems here as the social dilemmas arising from group members choosing individual profit, immediate satisfaction and their own long-term interests over collaboration at the expense of group interests. Examples of this in society include resource depletion, overpopulation and global warming.

2 Ángel Galán-Martín et al, 'Time for Global Action: an Optimised Cooperative Approach Towards Effective Climate Change Mitigation' (2018) 11(3) *Energy & Environmental Science* 572.

3 Amy Poteete, 'Elinor Ostrom, Governing the Commons: The Evolution of Institutions for Collective Action' in Martin Lodge, Edward C Page, and Steven J Balla (eds), *The Oxford Handbook of Classics in Public Policy and Administration* (Oxford University Press, 2015) 529.

4 Calla Wahlquist, 'Rio Tinto Blasts 46,000-Year-Old Aboriginal Site to Expand Iron Ore Mine', *The Guardian* (online, 26 May 2020) <https://www.theguardian.com/australia-news/2020/may/26/rio-tinto-blasts-46000-year-old-aboriginal-site-to-expand-iron-ore-mine>.

a week, we can say with *very high confidence* that the 'world will continue to warm, the damage will increase, and the global response will be inadequate'.[5]

Our goal here is twofold. Firstly, we are in search of an environmental ethic that satisfactorily frames what is at stake on the level of the individual. Further to this, we are in search of a theory which evades the following trilemma. We require a theory in which (1) our evaluation is properly action-guiding, a theory which is practical; (2) a theory in which evaluations are not anthropogenic, and therefore arbitrary; and (3) a theory which is not anthropocentric, that is, it is not based on human interests and concerns.[6]

One response to this threefold challenge begins with the question, *are the rights or interests of beings in the natural world neglected?* According to proponents of this view, we should extend our consideration towards the rights and interests of organisms in the natural world through the rational extension of our inherited evolutionary predisposition to sympathise with others.[7] At some point in our evolutionary past, the interests and rights we counted as significant in our moral deliberation moved beyond immediate family to communities, from communities to societies, and from societies to nations. The history of cultural norms can be understood as having become more radically impartial in just this way. In most parts of the world, there have been generational shifts in our attitudes towards women's roles in society, what constitutes reasonable punishment in systems of justice, or even how we raise our children. We can learn from the adoption of the standpoint of others, supposedly including other nonhuman creatures, that we can *expand the circle* of our considered interests when forming moral judgments and making decisions if only we are committed to maximising what is good. Carter writes that 'unless there is a morally relevant difference which separates humans from nonhuman animals, it seems that it would be inconsistent of us not to count in our moral calculations their welfare alongside that of humans'.[8]

However, the notion that we ought to, for example, consider the interests and rights to life of the flora and fauna of the Pilbara when deciding whether to expand an iron ore mine appears to us to be problematic. How, for instance, can we balance the rights and interests of one species over another?

5 Elizabeth Kolbert, 'The U.N. Issues a Final Warning on the Climate—and a Plan', *The New Yorker* (online, 20 March 2023) <https://www.newyorker.com/news/daily-comment/the-un-issues-a-final-warning-on-the-climate-and-a-plan>. See Hoesung Lee et al, *Synthesis Report of the IPCC Sixth Assessment Report. Summary for Policymakers* (Report, Intergovernmental Panel on Climate Change, March 2023).

6 This trilemma is raised in: Damian Cox, 'On the Value of Natural Relations' (1997) 19(2) *Environmental Ethics* 173, 175.

7 See Peter Singer, *The Expanding Circle: Ethics, Evolution, and Moral Progress* (Princeton University Press, 1981) and Alan Carter, 'Hume and Nature' in Hugh LaFollette (ed), *Ethics in Practice* (Blackwell Publishing, 3rd ed, 2007) 691.

8 Carter (n 7) 696.

We could introduce criteria such that native species are always valued more than invasive species. Also, we could add that the interests of more intelligent species like octopi and dolphins are weighed more heavily than less intelligent species, like penguins or sloths. But this is a slippery slope of arbitrary value attribution, as our criterion selection seems to depend ultimately upon what *we* deem as valuable. This means that, on the one hand, theories grounded in this notion succumb to the above trilemma – they are both anthropogenic and anthropocentric. On the other hand, since the environmentalists and the mining corporation have fundamentally different values and goals, the debate framed this way tends to become intractable.

A second response poses the question, *is the natural world intrinsically valuable?* Proponents of this view argue that our actions are right when they tend to 'highly regard' the intrinsic value of the natural world and that we must therefore 'preserve the integrity, stability, and beauty of the biotic community'.[9] Value, understood in this way, is an objective property of a thing, in much the same way that colours or textures are things that a property possesses.[10] That certain states of affairs are intrinsically valuable means that they possess 'value independently of any positive effects that it might have on other valued things. Thus, the intrinsic value of an object is the value that it would retain even if all its instrumental value were to be somehow extinguished'.[11] The test is to imagine that the thing in question is the only thing in the universe and then to consider, *would you prefer the existence of this universe over its nonexistence?* Our response supposedly reveals our intuitions about the intrinsic value of that thing. The deep ecology movement is an extension of this intrinsic value-orientated view.[12] Broadly construed, deep ecologists endorse a 'biospheric egalitarianism' that sees all living things as having intrinsic value. In each formulation here, the ethicist's manoeuvre seeks to enable us to determine which parts of the nonhuman world have value, with intuition as our arbiter.

The most compelling reaction to the difficulties inherent in translating intrinsic evaluation of nature into practical action is to ground practical action in the present and future welfare of the human community. But this means that our best justification for considering and acting in the interests of the nonhuman community is 'an enlightened self-interest…We value nature

9 Aldo Leopold, *A Sand County Almanac: With Other Essays on Conservation from Round River* (Oxford University Press, 1981) 131.
10 These ideas are rooted in Moore's work: GE Moore, *Principia Ethica* (Cambridge University Press, 1993) <http://fair-use.org/g-e-moore/principia-ethica/s.12>.
11 Cox (n 6) 174.
12 Andrew Brennan and Norva YS Lo, 'Environmental Ethics' in Edward N Zalta and Uri Nodelman (eds), *The Stanford Encyclopedia of Philosophy* (2021) <https://plato.stanford.edu/entries/ethics-environmental>.

for its contribution to human welfare'.[13] This is not a genuine ethical stance towards nature. This is an ethical stance towards the human community and future generations, in which nature plays an instrumental role. Acting to preserve the *value* of the 'integrity, stability, and beauty of the biotic community' is a morally good thing, but it is a reduction of our relations to the natural world that expects benefits of some kind to flow to us. Outside of the variety of metaethical concerns with an account of the good grounded in our intuitions, this theory also falls prey to the aforementioned trilemma – it is both impractical and anthropogenic. Choosing when and why to attribute *intrinsic value* to features of the natural world threatens to descend into an impractical value relativism.

In summary, all three responses to the axiological challenge of valuing nature fail to produce adequate, motivating perspectives. Instead, let us reconsider the idea that treating the environment in the way that we do, means that we are failing in our responsibility to the environment and to us as humans. The question is not, why are these behaviours and attitudes towards the environment morally wrong? We propose that the question ought to be, what kind of relationship with the environment adequately embodies moral concern for it? This does not frame the moral problem of climate crises as an existential threat. Instead, the moral problem of climate crises emerges from our profoundly defective moral relationship with the natural world. This requires that we re-evaluate what we are in relation to nature and what nature is in relation to us. This sort of evaluation is at the very foundation of First Nations' philosophies around the world.

Taking Relations Seriously

The resources we utilise in our thinking about moral matters can inhibit or distort moral deliberation in such a way that moral responsibility is diminished or excused. Here, we have so far acknowledged that the subjugation of the natural world is propagated in environmental theories that are fundamentally anthropocentric and anthropogenic. By shifting our moral attention to the quality of relations, as opposed to subjects themselves, we are drawing out this limitation in the moral norms and concepts relevant to articulating our shared moral responsibility in environmental crises. This expands the moral project of environmental ethics from the attempt to prescribe where humans and nonhumans stand in the global community to a description of what membership in the global community involves.

The best or most morally relevant description of our relationship to the natural world requires an environmental philosophy that is open to engaging in cultural and empirical dialogue. Christensen writes that we require a

13 Cox (n 6) 178.

'nuanced, in-depth, and comprehensive knowledge and understanding of our moral lives, but [environmental] philosophy with its focus on abstraction and generality seems in many ways to be ill-suited to acquiring such knowledge'.[14] This would explain, Christensen argues, why environmental philosophers have tended to ground their theories in intuitions or abstractions such as the existence of value-instantiating properties. A more systematic reflection upon our responsibilities in the face of climate crises is radically incomplete without proper attention to the kind of beings that we are and the environments in which we live. This requires at least that environmental ethics draws upon cultural and empirical perspectives. Culturally, an Australian Indigenous Knowledge perspective on our moral responsibilities towards the environment is practical here because it is open to this kind of enquiry in the form of the Dreaming. Empirically, environmental ethics requires the understanding of the complex and cascading relationships underlying biotic diversity and ecosystem functioning – an understanding of relational ecologies.

The Dreaming, or moral memories of place, blend together law and lore with scientific and historical knowledges. Morality, on this view, is a description of the past or possible imaginings of the future, it is a guiding principle for reality as both a scientific and moral description, and it is embedded within place. It underlies the delineation of totemic species, it defines the structure of kinship relations, and it describes when, where, and how to carry out ecological maintenance rituals.[15] Autonomous regard and extended moral agency do more than just shift the focus of our ethical attention to the quality of relations; they serve as a conceptual reminder that moral deliberation involves offering descriptions of the particularities of our moral lives and taking responsibility for the descriptions that we offer – as these descriptions ultimately guide our understanding of what we ought to do.

A common theme among First Nations' philosophies is that ethics is relational in that it ensures the continued existence of the other.[16] This account is foundationally environmental, as the ontological ground for many Indigenous communities involves a web of responsibilities that emerges from the land and informs the interactions between agents in the social and natural worlds. Graham argues that this is a fundamental axiom underpinning Aboriginal Australian worldviews; she writes that '[t]he land, and how we

14 Anne-Marie S Christensen, *Moral Philosophy and Moral Life* (Oxford University Press, 2021) 168.

15 Mary Graham, 'Some Thoughts About the Philosophical Underpinnings of Aboriginal Worldviews' (2008) 45 *Ecological Humanities*.

16 At least among Indigenous Australian and American worldviews, there is a shared perspective of a 'principled approach to sharing resources in a way that leaves none without, and yet allows it to re-generate for the future'. See Bradley Bryan, 'Property as Ontology: On Aboriginal and English Understandings of Ownership' (2000) 13(1) *Canadian Journal of Law and Jurisprudence* 23.

treat it, is what determines our human-ness ... the relation between people and land becomes the template for society and social relations'.[17] In Indigenous Australian ways of being, doing, and knowing, land is 'the beginning of ethics [and] that which confers responsibility'.[18] This view overturns the dominant anthropocentric and anthropogenic accounts of value, as it shifts our moral attention to the nature of the responsibilities we have within our relationships. Graham construes these relations as grounded in autonomous regard:

> The concept of autonomous regard, as an integral component of an Aboriginal ethic, is a way of gazing at each other from a distance. This distance is necessary for respect, for recognition of sovereignty, of limits, and also the existence of the self in the other, because, although distant, the self is not starkly separate from the self of the other.[19]

The concept of relationality means that the goodness of our relationship with nature, rather than nature itself, is the primary concern of environmental ethics from an Indigenous Knowledge perspective. When we can speak to the nurturing of a relationship, we are in a better position to articulate our shared responsibility, making this holistic approach more descriptive of our place in the world. On this view, building good relations everywhere involves tending to and critically re-evaluating the nature of the web of relationships we already exist in.

In the face of environmental crises, taking relations seriously is more properly action-guiding because it requires that we rethink our relationship with the world and the conceptual tools we rely upon in our moral deliberation. Midgley emphasises this in her discussion of how, in many ways, plumbing and philosophy are quite similar. She writes that both philosophy and plumbing are activities that are carried out beneath the surface of our fairly complex and elaborate cultural systems, systems which sometimes go wrong:

> In both cases, this can have serious consequences. Each system supplies vital needs for those who live above it. Each is hard to repair when it does go wrong, because neither of them was ever consciously planned as a whole.[20]

17 Graham (n 15). See also, Mary Graham, 'Aboriginal Notions of Relationality and Positionalism: A Reply to Weber' (2014) 4(1) *Global Discourse* 17, 18.

18 Simone Thornton, Gilbert Burgh, and Mary Graham, 'Place-Based Philosophical Education: Reconstructing "Place", Reconstructing Ethics' (2021) 17 *Childhood and Philosophy* 1, 10.

19 Ibid 13.

20 Mary Midgley, 'Philosophical Plumbing' (1992) 33 *Royal Institute of Philosophy Supplements* 139.

The discovery that our philosophical plumbing in environmental philosophy is contingent upon anthropocentric and anthropogenic concepts is akin to the discovery of a serious leak. We have offered a normative response here which recognises how cultural and empirical descriptions can enrich our understanding of the particularities of our relationship with the world.

Relations that are exploitative, oppressive, or instrumentalising are bad relationships independent of their conduciveness to produce good or bad effects. If taking relations seriously means re-evaluating the relationship between the natural and social worlds, then this requires that we abandon the search for some prescriptive account of our relationship with nature and instead search for the best or most morally relevant descriptions of that relationship. These descriptions involve actively making ourselves aware of and becoming morally responsive to the preferences, threats and sensitivities of members of our communities, which necessarily extends into the natural. It is this relational capacity of environmental ethics from an Indigenous Knowledge perspective that we call, extended moral agency.

Intertwined with the cultural description of our moral responsibilities towards the environment is the empirical description. Advances in biology can better inform moral understandings of the particularities of our relationships with the world and ought, therefore, to play a foundational role in our moral deliberation. Ecology, for example, teaches us about the fragile connections between biodiversity and ecosystem functioning. The loss of just a few species has significant long-term impacts on the sustainability of ecosystem functioning. Our communities are deeply interconnected and our moral theorising must properly acknowledge these cascading relationships. Evolutionary biology can also better inform our foundational moral concepts such as individuality and what it means to be a moral agent – concepts upon which our moral relations with social and natural communities are largely based. Advances in understandings of the ontological development of life forms teach us that the individual is not a single and autonomous entity, but a collection of symbiotic organisms which co-exist and co-evolve:

> Thus, animals can no longer be considered individuals in any sense of classical biology...Our bodies must be understood as holobionts whose anatomical, physiological, immunological, and developmental functions evolved in shared relationships of different species. Thus, the holobiont, with its integrated community of species, becomes a unit of natural selection whose evolutionary mechanisms suggest complexity hitherto largely unexplored.[21]

21 Donna Haraway, *Staying with the Trouble: Making Kin in the Chthulucene* (Duke University Press, 2016) 334.

Taking this view seriously allows us to see and value the connections between all life forms without descending into anthropocentric or anthropogenic theory. Biological descriptions such as these remind us that critically re-evaluating the nature of the web of relationships we already exist in is the first step to building good relations everywhere.

Indigenous Solutions to Complex Problems

Indigenous people have learned some important lessons in creating sustainable collective solutions.

1. Involve people with 'lived experience': Indigenous people have long been saying *nothing about us without us*. In Australia, we developed solutions to common problems over 80,000 years so we know where the system has let people down, where it can be improved and what solutions worked in the past and are likely to succeed again.
2. Diversity: understanding things through intersectional and diverse lenses includes a wide range of perspectives. This knowledge is not static, it has come from collectives before us and therefore we can learn from both the successes and failures of these to direct us to continued learnings and solutions.
3. We over me: changing our values and actions to be a collaborative together that reminds people why they are collaborating, what they want to achieve for their lifetime and the lifetimes of others, and how they can do it together and stay anchored to this purpose. This means fostering those with a vision for change, those who can adapt and influence change, and those who expand the change and increase our knowledge. This must be de-Westernised. A collective does not necessarily have boundaries, although it will have clear activities and lines of responsibility and accountability to each other.
4. Storytelling: our stories unite, motivate, and enable others. Stories can be used to understand what we are trying to achieve collectively and what values we should hold.[22]

Conclusion

It is a persistent mistake to frame environmental crises as existential crises. We need a new way of thinking about them. There is a pressing need for a renewed understanding of the relationship between the human and natural

22 For example, the Dreamtime story of Tiddalick the frog, who drained the rivers dry due to his insatiable thirst. It was only when the other creatures collaborated to make Tiddalick laugh that the water was restored.

worlds, something environmental philosophers have been championing for many decades, but one that helps us frame environmental crises as urgent moral crises, directed not at future generations of humans but at something more fundamental still. Drawing on the philosophical underpinnings of Indigenous Australian worldviews, and recent work in environmental philosophy, we set out the characteristics of an ethic which sees environmental threat in just this way. We call this an account of extended moral agency, and we illustrate how this central conceptual and practical feature of Australian Aboriginal ontology and morality construes our interconnectedness with the environment as the starting point of ethical deliberation. We then explore how this account provides a conceptually rich articulation of our individual responsibility towards global environmental crises, one that overcomes our framing of the emergency as an existential one.

The insights of Indigenous worldviews around the world have existed and been practised for thousands of years – they are not novel approaches to our moral reality. But such theories count as possible imaginings that are not reducible to the dominant Western perspectives already in place. This sort of enquiry leads us to other and better ways of thinking about our relationship with the world and our understanding of environmental crises. We have the capacity to nurture collective solutions, together.

Chapter 12

A Voice, Truth and Treaty Thought Experiment

Robert Cunningham

Commentary

Introduction

The case of *Indigenous Custodians v Biodata Extraction and Excavation Ltd* [2037] HC 13 is the culmination of several important developments, throughout the 2020s and 2030s, in relation to 'Aboriginal Australians', 'Australian First Nations Peoples', or 'Aboriginal and Torres Strait Islanders'.[1] Those developments include the evolution of Voice, Truth and Treaty initiatives flowing from the 2017 *Uluru Statement from the Heart*. The latter flowed from a Referendum Council, which was established on a bi-partisan basis in 2015 by the Australian Commonwealth Government.

As a corollary of the Referendum Council, there were, in 2016 and 2017, 13 'First Nations Constitutional Dialogues', which sought to ensure Aboriginal decision-making lies at the heart of a Constitutional reform process. From the inception of the 2017 *Uluru Statement from the Heart* it took several more years – and the election in 2022 of a new Australian Federal Labor Government – for a 'Voice' referendum to occur.

Building upon a successful Voice Referendum in September 2023, subsequent Truth and Treaty initiatives developed relatively swiftly. By 2025, the Makarrata Truth Commission had been established, making way for the ensuing 2026 *Makarrata Treaty*.

The *Native Information Title Act 2028* (Cth) embeds several Voice recommendations, including *sui generis* Traditional Knowledge rights of Aboriginal Australians. These rights are concerned with the cultural customs,

1 The correct and appropriate phraseology used with respect to 'First Nations Peoples of Australia and the Torres Straight' remains a matter of conjecture and controversy. The phrase 'Aboriginal Australians' is used through this judgment given this was the language deployed in *Love v Commonwealth of Australia; Thoms v Commonwealth of Australia* (2020) 270 CLR 152 ('*Love & Thoms*'). In that case, Edelman J (as his Honour then was) said: 'The very words "Aboriginal" and "indigenous", ab origine or "from the beginning", enunciate a historical, and original, connection with the land of Australia generally' (at 314).

DOI: 10.4324/9781003389569-15

practices, language, and lore and provide a springboard for present and future social, political, and economic development of Aboriginal Australians. The *Native Information Title Act 2028* (Cth) also embeds significant aspects of the *Makarrata Treaty*, including an 'Aboriginal Australian Confirmation Framework' – based on the notion of pan-Aboriginality – alongside *sui generis* sovereign Traditional Knowledge rights over 'cultural customs, practices, language, and lore of First Nations Peoples in Australia'.

The factual matrix of *Indigenous Custodians v Biodata Extraction and Excavation Ltd* flows from the related High Court matter of *Australasian Information Environmentalism Alliance v Commonwealth of Australia* [2037] HC 12.[2] Although the resolution in the said case has some bearing on *Indigenous Custodians v Biodata Extraction and Excavation Ltd*, the latter nevertheless requires adjudication given the Indigenous Custodians are seeking an 'enduring injunction' against all biological data mining activities of the Second Defendant, Biodata Extraction and Excavation Ltd, that conflict with the *Native Information Title Act 2028* (Cth).

The over-arching issue in *Indigenous Custodians v Biodata Extraction and Excavation Ltd* is whether the biological data mining activities of Biodata Extraction and Excavation Ltd conflict with *Native Information Title Act 2028* (Cth). To resolve this issue, it is necessary, from the outset, to determine whether the 'Indigenous Custodians' have standing, which is resolved by assessing the application or otherwise of the *Makarrata Treaty* Aboriginal Australian Confirmation Framework. It is then, in turn, necessary to determine the extent to which those aspects of the *Makarrata Treaty* embedded within the *Native Information Title Act 2028* (Cth) conflict with the *Biological Data Mining Act 2025* (Cth); and to the extent there is any such conflict, which statute/provision prevails.

Makarrata Treaty Aboriginal Australian Confirmation Framework

While controversy underpinning defining 'Aboriginality' should not and cannot be ignored, as *Indigenous Custodians v Biodata Extraction and Excavation Ltd* makes clear, defining Aboriginality is an essential aspect of self-determination. Elucidating the *Commonwealth v Tasmania*[3] ('*Tasmanian Dam*') definition of Aboriginality outlined by Deane J in that case,[4] and referred

2 Robert Cunningham, 'Information Environmentalism and Biological Data: A Thought Experiment' in Nicole Rogers and Michelle Maloney (eds), *Law as if the Earth Really Mattered: The Wild Law Judgment Project* (Routledge Press, 2017) 355.

3 *Commonwealth v Tasmania* (1983) 158 CLR 1 ('*Tasmanian Dam*').

4 Ibid 274 (Deane J).

to by Bell J in *Love & Thoms*, as well as [21] of *Indigenous Custodians v Biodata Extraction and Excavation Ltd*, is insightful:

> In the *Tasmanian Dam Case*, Deane J proposed the meaning of the term 'Australian Aboriginal' as 'a person of Aboriginal descent, albeit mixed, who identifies himself as such and who is recognized by the Aboriginal community as an Aboriginal'. This was in the context of s 8(2)(b) of the *World Heritage Properties Conservation Act 1983* (Cth), which referred to Aboriginal sites having particular significance to 'the people of the Aboriginal race'. His Honour inclined to the view that the reference was to the Australian Aboriginal people generally rather than to any particular racial sub-group.[5]

There are two significant points flowing from Deane J's 'Australian Aboriginal' definition in *Tasmanian Dam*. First, the definition encompasses the concept of 'community recognition' as the third limb as part of a tripartite test, implicitly embedding self-determination. Second, the definition implicitly alludes to the notion of 'pan-Aboriginality', which has become a vital notion in the context of the 2026 *Makarrata Treaty* and, in particular, the Aboriginal Australian Confirmation Framework therein.[6] That framework reinforces the majority decision of *Love & Thoms* by indicating that 'the legal concept of Aboriginality, at its core, recognises that there is a unique group of Australians, Aboriginal Australians, who are descendants of the original inhabitants of this country and who identify as such and are accepted as such'.[7]

The 2026 *Makarrata Treaty* – agreed one hundred years after Unaipon called for the establishment of an Aboriginal state – established a framework process of Aboriginal Australian confirmation.[8] This is crucial because the Australian Commonwealth Government had made it clear, in the lead-up to the agreement, that the ambition was a singular *Makarrata Treaty*. Thus, it was necessary to facilitate a self-determination process that established an

5 *Love & Thoms* (n 1) 190-1 (Bell J).

6 Note Keane J in his Honour's minority judgment in ibid, at 224:

> The provisions of the World Heritage Act with which Deane J was concerned related to the protection and conservation of identified property 'of particular significance to the people of the Aboriginal race'. The description by Deane J of 'Australian Aboriginal' was put forward to give context to the operation of the World Heritage Act. *It is important to appreciate that it was not propounded as a test of membership of the body politic of the Commonwealth of Australia* (emphasis added and footnotes omitted).

7 Ibid 282 (Gordon J) (footnotes omitted).

8 'The Uluru Statement from the Heart Frequently Asked Questions', *The Uluru Statement* (Web Page, 2023) <https://ulurustatement.org/education/faqs/>.

appropriate pan-Aboriginality working definition to facilitate clarity about who might be affected by the rights and responsibilities embedded within the *Makarrata Treaty*. As *Indigenous Custodians v Biodata Extraction and Excavation Ltd* reinforces, it is imperative any such approach is grounded in principles of self-determination.

In exploring the legal notion of Aboriginality, Edelman J in *Love & Thoms* emphasises Aboriginal people's 'fundamental spiritual and cultural sense of belonging to Australia' and that this identity 'constitutes them as members of the Australian political community'.[9] And he further states that:

> Aboriginal people *belong* to Australia and are essential members of the community which constitutes the body politic of the nation state (emphasis added).[10]

It is this 'belonger' concept, which underpins the *sui generis* status of Aboriginal Australians.

While Gordon J indicated in *Love & Thoms* that it was *not* necessary to 'chart the outer limits of the [legal] concept [of Aboriginal Australians]',[11] since the 2026 *Makarrata Treaty* it has become increasingly important to do just that. Although *Indigenous Custodians v Biodata Extraction and Excavation Ltd* certainly does not complete the task of 'charting the outer limits', it does contribute to delineating the parameters as well as underscoring the practical significance of the (pan-Aboriginality) Aboriginal Australian Confirmation Framework.

Co-existence of Aboriginal Australian Sovereignty

As was envisaged, the 2026 *Makarrata Treaty* has become a springboard for significant explorations concerning the co-existence of Aboriginal Australian sovereignty with the Crown (or Commonwealth of Australia). As Keane J reinforced at [193] in *Love & Thoms*:

> The point is that the basis of the relationship between a sovereign State and its people is a function of political and legal considerations.[12]

It was perhaps for this reason – that sovereignty imports political and legal considerations – that Nettle J in *Love & Thoms* indicated it was necessary 'to stop to inquire' so as 'to re-examine the essentials of alienage and the nature

9 *Love & Thoms* (n 1) 287 (Edelman J).
10 Ibid 398 (Edelman J).
11 Ibid 282 (Gordon J).
12 Ibid 225 (Keane J).

of an Aboriginal person's relationship to the Australian polity'.[13] *Indigenous Custodians v Biodata Extraction and Excavation Ltd* has done precisely so. That is, the case has 'stopped to inquire' and, as a corollary, has found any controversy associated with Aboriginal Australians determining their own membership on the basis of self-determination or asserting the co-existence of sovereignty is grounded upon false claims.

Conclusion

By granting the orders sought by the Indigenous Custodians in *Indigenous Custodians v Biodata Extraction and Excavation Ltd*, the Australian High Court has reinforced principles of self-determination and asserted the co-existence of Aboriginal Australian sovereignty. In this way, the decision lays the foundation for further progress with respect to the Voice, Truth and Treaty initiatives flowing from the 2017 *Uluru Statement from the Heart* – a statement from the oldest living First Nations culture on earth.

Judgment

Indigenous Custodians v Biodata Extraction and Excavation Ltd [2037] HC 13

Facts

1. A summary of the relevant facts concerning *Indigenous Custodians v Biodata Extraction and Excavation Ltd* – which in turn relate to the application of the *Native Information Title Act 2028* (Cth) – is partially outlined in *Australasian Information Environmentalism Alliance v Commonwealth of Australia & Ors* [2037] HC 12.[14]
2. The relevant facts in *Australasian Information Environmentalism Alliance v Commonwealth of Australia & Ors* are rehearsed below to the extent that such is necessary to provide a rationale for the present judgment.
    a. In the late 2020s, to further its objectives of conserving the physical environment, Australasian Information Environmentalism Alliance commissioned a nationwide 'interactive biological art project' called Nature-Scape, which continues to operate throughout various National Parks and other nature reserves in Australia.
    b. According to the relevant promotional material, the aim of Nature-Scape is 'to leverage "citizen science" by co-joining conservationists with technological developments to secure positive environmental outcomes'.

13 Ibid 248 (Nettle J).
14 Cunningham (n 2) 355–72.

c. Nature-Scape was initially advertised (and Trade Marked) as a 'fully-immersive, multi-senses experience using hologramaphics to deepen humanity's relationship with, and understanding of, nature'.
d. Australasian Information Environmentalism Alliance and the Nature-Scape project sought to enhance appreciation and understandings of nature by harnessing the latest Human-Nature-Interface technology. Specifically, Australasian Information Environmentalism Alliance provided participants of Nature-Scape with a particular type of Human-Nature-Interface device called 'uber-sonic', which amplified and enhanced the human senses of sound, taste, smell, touch, and sight as well as influencing how those senses interact with nature.
e. The uber-sonic device was typically worn on the external body as a piece of 'tech-clothing' (akin to the Apple Body Sensory Attire 4S) or worn as 'cyber-binoculars' (via the Google Human Nature Interface Specs 3.1).
f. Relevantly, the use of a Nature-Scape uber-sonic device was fully dependent upon access to, and use of, the physical National Parks and other nature reserves used by Australasian Information Environmentalism Alliance from time to time, as well as the biological data inherent therein.
g. As Nature-Scape participants are immersed in the physical environment through bush-walking and other nature-based adventure activities, the uber-sonic device automatically collects and processes biological data of the surroundings. This information is in turn deployed by Australasian Information Environmentalism Alliance to advance conservation of the relevant area.
h. Prior to Nature-Scape, Biodata Extraction and Excavation Ltd had been relying on its Biological Data Mining Licence, which had been granted by the Commonwealth of Australia in 2026. Among other things, this licence allowed Biodata Extraction and Excavation Ltd to 'exploit biological data resources from National Parks and other nature reserves within Australia'. Biodata Extraction and Excavation Ltd has subsequently profited by selling biological data usage licences, mainly to large pharmaceutical and industrial biotechnology and nanotechnology companies based in Germany and the United States of America.
i. The legal dispute culminated when Biodata Extraction and Excavation Ltd sent a 'cease and desist' letter to Australasian Information Environmentalism Alliance. This correspondence stated that

> Biodata Extraction and Excavation Ltd has an exclusive Biological Data Mining Licence over the biological data within particular National Parks and other nature reserves in Australia and therefore Australasian Information Environmentalism Alliance does

not possess the access and/or use rights required to continue the Nature-Scape project.

j. In response, after several failed attempts to negotiate a mutually satisfactory resolution, Australasian Information Environmentalism Alliance initiated legal proceedings in the Federal Court of Australia seeking two primary orders:
   i. a revocation of the Biological Data Mining Licence granted by the first defendant, the Commonwealth of Australia; and
   ii. an 'enduring injunction' against all biological data mining activities of the second defendant, Biodata Extraction and Excavation Ltd that conflicts with the *Information Commons Rights Act* (2029).

k. This Court granted the orders sought by Australasian Information Environmentalism Alliance.

3. *Australasian Information Environmentalism Alliance v Commonwealth of Australia & Ors* involved the interpretation and application of two significant legislative initiatives developed in the 2020s, namely: the *Biological Data Mining Act 2025* (Cth) and the *Information Commons Rights Act 2029* (Cth).
4. In the present case, the Plaintiffs, being the Indigenous Custodians, seek an 'enduring injunction' against all biological data mining activities of the defendant, Biodata Extraction and Excavation Ltd, that conflict with the *Native Information Title Act 2028* (Cth).
5. Accordingly, it is now the task of this Court to determine the extent to which the biological data mining activities of Biodata Extraction and Excavation Ltd – and more specifically the *Biological Data Mining Act 2025* (Cth) – conflict with the *Native Information Title Act 2028* (Cth).
6. After alluding to the legislative foundations of the *Native Information Title Act 2028* (Cth) and the notion of Traditional Knowledge, the judgment will traverse:
   a. *Locus standi* as it relates to the Aboriginal Australian Confirmation Framework embedded within the *Makarrata Treaty*; and
   b. Whether there is any inherent conflict between the *Biological Data Mining Act 2025* (Cth) and *Native Information Title Act 2028* (Cth); and, if so, how this conflict is to be resolved?

Native Information Title Act 2028 *(Cth) and Traditional Knowledge*

7. The *Native Information Title Act 2028* (Cth) seeks to provide legislative support for *sui generis* Traditional Knowledge rights over the cultural customs, practices, language, and lore of Aboriginal Australians. The

legislation was enacted in 2028 after being initially recommended by The Voice[15] in January 2026.

8. The explanatory memorandum of the *Native Information Title Act 2028* (Cth) outlines the legislative aim of 'leveraging the present Native Title system relating to physical land and sea so as to secure future benefit-sharing flowing from cultural customs, practices, language, and lore of First Nations Peoples'. The *Native Information Title Act 2028* (Cth) gives effect to the legislative aim in a variety of ways, including 'veto rights of Aboriginal Australians over the moral and economic rights concerning cultural customs, practices, language and lore'.
9. The bundle of rights inherent within *Native Information Title Act 2028* (Cth) is often referred to using the umbrella phrase 'Traditional Knowledge'.[16] That phrase is adopted in this judgment.
10. The general background underpinning the significance of Traditional Knowledge to Aboriginal Australians is well-outlined in Federal Court of Australia cases such as *Milpurrurru v Indofurn Pty Ltd* (1994) 54 FCR 240 (*Carpets Case*), *Yumbulul v Reserve Bank of Australia* (1991) 28 IPR 481, and *Bulun Bulun v R & T Textiles* (1998) 86 FCR 244 (*T-shirts Case*).[17]

Standing and *Makarrata Treaty* Aboriginal Australian Confirmation Framework

11. The first substantive step in this matter involves determining whether the Indigenous Custodians have the requisite standing to bring this matter before this Court. This necessarily involves an evaluation of the relevant history alongside the Aboriginal Australian Confirmation Framework, which is embedded within the *Makarrata Treaty 2026*.
12. As a corollary of Australia's historical legacy, the issue of Aboriginality confirmation is not without controversy. In *Eatock v Bolt* ('*Eatock*'), Bromberg J put it this way:

15 The establishment of the Voice as a First Nations advisory body for the Commonwealth Parliament was one of the recommendations of the *Uluru Statement from the Heart*. At the time of writing, the outcome of a 2023 constitutional referendum on the Voice remains unknown.

16 The World Intellectual Property Organization indicates that while there is not yet an accepted definition of 'Traditional Knowledge' at the international level, it can be described as 'knowledge, know-how, skills and practices that are developed, sustained and passed on from generation to generation within a community, often forming part of its cultural or spiritual identity': see World Intellectual Property Organization (Web Page) <https://www.wipo.int/tk/en>.

17 See, also, Erin Mackay, 'Indigenous Traditional Knowledge, Copyright and Art – Shortcomings in Protection and an Alternative Approach' (2009) 32(1) *University of New South Wales Law Journal* 1.

> It is a notorious and regrettable fact of Australian history that the flawed biological characterisations of many Aboriginal people was the basis for mistreatment, including for policies of assimilation involving the removal of many Aboriginal children from their families until the 1970s. It will be of no surprise that a race of people subjected to oppression by reason of oppressive racial categorisation will be sensitive to being racially categorised by others. I accept that to be the case in relation to Aboriginal Australians.[18]

13. Justice Bromberg went on in *Eatock* to highlight the controversy of defining Aboriginality by quoting from an extract in the final *Report of the Royal Commission into Aboriginal Deaths in Custody*, as reproduced in a 2003 report of the Australian Law Reform Commission:

> No area of research and commentary by non-Aboriginal people has such potential to cause offence as does that which attempts to define 'Aboriginality'. This determination of non-Aboriginal people to categorise and divide Aboriginal people is resented for many reasons, but principally, I suspect, because the worst experiences of assimilation policies and the most long term emotional scars of those policies relate directly to non-Aboriginal efforts to define 'Aboriginality' and to deny to those found not to fit the definition, the nurture of family, kin and culture. To Aboriginal people there appears to be a continuing aggression evident in such practices.[19] (footnotes omitted)

14. Despite the controversy concerning Aboriginality definitions and the related issue of Aboriginal Australian confirmation – or perhaps because of it – a critical aspect of self-determination is for Aboriginal Australians to delineate membership, on their own terms. In other words, there is no issue more fundamental to the concept of self-determination than that of membership.
15. The history of Aboriginal Australian policies has been outlined by many scholars.[20] Australian Commonwealth Government policy positions

18 *Eatock v Bolt* (2011) 197 FCR 261, 301 ('*Eatock*').

19 Ibid. Cited by Mortimer J in *Helmbright v Minister for Immigration, Citizenship, Migrant Services and Multicultural Affairs (No 2)* (2021) 287 FCR 109 ('*Helmbright (No 2)*'), at 136.

20 See, eg, Megan Davis, 'Sovereignty and the First Australians' in Martin Hinton and John M Williams (eds), *The Crown: Essays on Its Manifestation, Power and Accountability* (University of Adelaide Press, 2018) 19; and Laura Rademaker and Tim Rowse, *Indigenous Self-Determination in Australia: Histories and Historiography* (ANU Press, 2020).

have included 'genocide', 'protection', 'assimilation', 'self-determination', and 'reconciliation'.[21]

16. The self-determination era concerning First Nations Peoples in Australia can be traced back to the establishment of the Aboriginal Tent Embassy set up outside Parliament House (now Old Parliament House) on 26 January 1972. Nearly a decade prior, in 1963, the Bark Petition was presented to Commonwealth Parliament and the Gurindji People walked off Wave Hill Station.[22]
17. The process of defining Aboriginality, and the related issue of Aboriginal Australian confirmation, is a legal question initially confronted by this Court in *Tasmanian Dam*[23] and subsequently *Mabo v Queensland (No 2)*[24] ('*Mabo (No 2)*') and *Love & Thoms*.[25]
18. In *Love & Thoms*, the question arose as to whether a person who satisfied the definition of an Aboriginal Australian could be considered an 'alien' under section 51(xix) of the Australian Constitution.
19. Given the *sui generis* relationship Aboriginal Australians have to the body politic of the Commonwealth of Australia, the question of Aboriginal Australian confirmation has particular significance. It is notable that the question of inclusion and exclusion is not one confined to Aboriginal Australians, having occupied the minds of those constructing the Australian Constitution. It is for this reason section 51(xix) of the Australian Constitution provides the Commonwealth power to make laws with respect to 'naturalization and aliens'. As stated by Chief Justice Keifel in *Love & Thoms*:

> At Federation it was well recognised that an attribute of an independent sovereign State was to decide who were aliens and whether they should become members of the community.[26] (Footnotes omitted)

21 Davies (n 20).

22 Robert French, 'The Voice – A Step Forward for Australian Nationhood' (Conference Paper, Exchanging Ideas Symposium, Judicial Commission of New South Wales, New South Wales Bar Association, and New South Wales Law Society, 4 February 2023) [18]. Note that the Bark Petition was built upon many historical attempts of Aboriginal Australians to seek social justice. Other examples include the Petition to Queen Victoria in 1846 in relation to exiled Tasmanian Aboriginal people on Flinders Island concerning an agreement made with Colonel Arthur, the Petition from Residents of Maloga Mission in 1881 seeking greater freedom from missionary control, and the Petition from Wurundjeri Elder William Barak to the Victorian Government opposing the Aboriginal Protection Bill in 1886.

23 *Tasmanian Dam* (n 3).

24 *Mabo v Queensland (No 2)* (1992) 175 CLR 1 ('*Mabo (No 2)*').

25 *Love & Thoms* (n 1).

26 Ibid 171 (Kiefel CJ).

20. The common law has made clear on numerous occasions that while the Australian Commonwealth Government can seek to delineate the parameters of 'Australian citizenship', it is the role of this Court to determine the constitutional meaning of 'aliens'. As emphasised by Gibbs CJ in *Pochi v Macphee*:

> [T]he Parliament cannot, simply by giving its own definition of 'alien', expand the power under s 51(xix) to include persons who could not possibly answer the description of 'aliens' in the ordinary understanding of the word.[27]

21. In the first formal judicial consideration of the matter in Australia, Deane J in *Tasmanian Dam* proffered the following tripartite formulation of Aboriginality:

> a person of Aboriginal descent, albeit mixed, who identifies himself as such and who is recognized by the Aboriginal community as an Aboriginal.[28]

22. Dean J's tripartite formulation was enunciated in the context of section 8(2)(b) of the *World Heritage Properties Conversation Act 1983* (Cth), which referred to Aboriginal sites having particular significance to the 'people of the Aboriginal race'.[29]
23. Although Deane J indicated in *Tasmanian Dam* that the determination of Aboriginality is a question of fact, the more recent High Court of Australia case of *Love & Thoms* highlighted the inquiry underpinning that determination involves a question of law: namely, what is the appropriate legal test of Aboriginality?
24. The appropriate legal test of Aboriginality was an issue also traversed in the Federal Court matter of *Helmbright v Minister for Immigration, Citizenship, Migrant Services and Multicultural Affairs (No 2)*[30] ('*Helmbright (No2)*'). Drawing heavily upon *Love & Thoms*, Mortimer J (as her Honour then was) sought to navigate the nuances between Deane J's tripartite formulation of Aboriginality in *Tasmanian Dam* and Brennan J's formulation in *Mabo (No 2)*. The latter reading:

> [m]embership of the indigenous people depends on biological descent from the indigenous people and on mutual recognition of a

27 *Pochi v Macphee* (1982) 151 CLR 101, 109 (Gibbs CJ).
28 *Tasmanian Dam* (n 3) 274 (Deane J).
29 Ibid 275 (Deane J). See, also, Bell J at 190–1 in *Love & Thoms* (n 1).
30 *Helmbright (No 2)* (n 19).

particular person's membership by that person and by the elders or other persons enjoying traditional authority among those people.[31]

25. As Mortimer J indicated in *Helmbright (No 2)*, there are several controversies flowing from the nuances between Deane J's *Tasmanian Dam* formulation of Aboriginality vis-à-vis Brennan J's *Mabo (No 2)* formulation. Those controversies are partially addressed through the following sub-inquiries:

    a. What is the extent to which connection to land and waters is required to satisfy Aboriginal Australian confirmation? [32]
    b. Whether Aboriginality affects:
        i. The political relationship between an individual Aboriginal Australian and the body politic of the Commonwealth of Australia; and, if so, what is the nature of that influence?[33]
        ii. The political relationship between Aboriginal Australians collectively and the body politic of the Commonwealth of Australia; and, if so, what is the nature of that influence?[34]

31 *Mabo (No 2)* (n 24) Brennan J at 70. See also Bell J at 191 in *Love & Thoms* (n 1). Note, however, the minority judgment of Keane J in *Love & Thoms*, when reflecting upon Brennan J's *Mabo (No 2)* Aboriginality formulation, stating at 224:

> In this passage, Brennan J was concerned to explain the basis on which an individual indigenous person may come to share in the communal rights of a particular indigenous group to a particular territory. Brennan J was plainly not seeking to describe the political relationship between an individual indigenous person and the body politic, being the Commonwealth of Australia, much less the relationship between all indigenous people collectively and the body politic.

32 *Mabo (No 2)* (n 24) 70 (Brennan J). See also, eg, Mortimer J's reflection in *Helmbright (No2)* (n 19) at 179 where her Honour (as she then was) said:

> The reality which is well-established across Australia is that Aboriginal and Torres Strait Islander groups all express a connection to particular land and waters, and that may well be why, even if Deane J's approach in the *Tasmanian Dam Case* is taken to s 51(xix), for the purposes of deciding alienage a Court may be unable to go to a wide understanding of 'community' such as that expressed in *Shaw v Wolf*, which is effectively unconnected to any land and waters. Given the emphasis by the majority in *Love/Thoms* on connection to country as underpinning the belonging of Indigenous people to the Australian polity, it may be necessary to retain in any test some link to land and waters in the approach to the description of the 'community' of which a person is a member.

33 See, eg, Keane J in *Love & Thoms* (n 1) 224.
34 Ibid.

26. To resolve these lines of inquiry, it is necessary to address what is a more fundamental question: the extent to which Aboriginal Australians can claim co-existence of sovereignty alongside the Commonwealth of Australia. It is to this issue the judgment now turns.

Co-existence of Aboriginal Australian Sovereignty

27. The *Uluru Statement from the Heart* referred to Aboriginal and Torres Strait Islander tribes as being 'the first sovereign Nations of the Australian continent and its adjacent islands'. It went on to state:

> This sovereignty is *a spiritual notion: the ancestral tie between the land, or 'mother nature', and the Aboriginal and Torres Strait Islander peoples who were born therefrom, remain attached thereto, and must one day return thither to be united with our ancestors. This link is the basis of the ownership of the soil, or better, of sovereignty.* It has never been ceded or extinguished, and co-exists with the sovereignty of the Crown. (emphasis added)

28. The perspective of sovereignty elucidated in the *Uluru Statement from the Heart* was advanced because the Commonwealth Parliament and this Court did not, when that Statement was made in 2017, recognise the sovereignty of Aboriginal Australians. *Mabo (No 2)*, for instance, explicitly rejected the sovereignty of Aboriginal Australians. As Gageler J said in *Love & Thoms*:

> That rejection has meant that, unlike the 'Indian Tribes' recognised in the Constitution of the United States, Aboriginal and Torres Strait Islander societies have never been treated constitutionally as 'distinct political societies' or as 'domestic dependent nations' the members of which have owed 'immediate allegiance to their several tribes'.[35]

29. The common law rejection of Aboriginal Australian sovereignty prior to the 2026 *Makarrata Treaty* was also emphasised by Keane J in *Love & Thoms* – when referring to prior cases such as *Coe v The Commonwealth*[36] and *Yorta Yorta*:[37]

> The assertion of a claim to [Aboriginal Australian] sovereignty has been rejected on the few occasions on which it has been articulated. Thus, in *Coe v The Commonwealth* (1979) 24 ALR 118, Mason CJ said:

35 Ibid 201 (Gageler J) (footnotes omitted).
36 *Coe v The Commonwealth* (1979) 24 ALR 118.
37 *Yorta Yorta Aboriginal Community v Victoria* (2002) 214 CLR 422.

> *Mabo [No 2]* is entirely at odds with the notion that sovereignty adverse to the Crown resides in the Aboriginal people of Australia. The decision is equally at odds with the notion that there resides in the Aboriginal people a limited kind of sovereignty embraced in the notion that they are 'a domestic dependent nation'.

Similarly, in *Yorta Yorta Aboriginal Community v Victoria* [2002] HCA 58, Gleeson CJ, Gummow and Hayne JJ said at [44]:

> [W]hat the assertion of sovereignty by the British Crown necessarily entailed was that there could thereafter be no parallel law-making system in the territory over which it asserted sovereignty. To hold otherwise would be to deny the acquisition of sovereignty and ... that is not permissible.[38]

30. Yet, as reinforced by Nettle J in his Honour's majority judgment in *Love & Thoms*, the position enunciated by *Mabo (No 2)*, which served to reject the sovereignty of Aboriginal Australians, leads to 'some difficulties':

> as espoused in Blackstone's day, territories governed by a sovereign could be acquired only by conquest or cession, and thereafter remained subject to the body of law earlier in force until alteration by the acquiring sovereign. On the other hand, territories which were not governed by a sovereign before the Crown acquired sovereignty could be acquired by settlement and thereupon receive English law, both statutory and unenacted, so far as applicable to the situation of the settlers and condition of the infant colony. But the application of that doctrine to the territory of the Australian colonies has given rise to 'some difficulties', which have been attributed to an 'incongruity between legal characterisation and historical reality', or between 'theory [and] our present knowledge and appreciation of the facts'.[39]

31. Regardless, the common law position concerning Aboriginal Australian sovereignty has now been ultimately annulled with the enactment of the 2026 *Makarrata Treaty*, which essentially embraced the majority position in *Love & Thoms*.
32. There are significant points underscored by the majority judgment of Edelman J in *Love & Thoms*, which established the foundation for the *Makarrata Treaty*, namely that:

38 *Love & Thoms* (n 1) 226–7 (Keane J).
39 Ibid 249–50 (Nettle J) (footnotes omitted).

a. Aboriginal people *belong* to Australia and are essential members of the 'community which constitutes the body politic of the nation state' (emphasis added).[40]
b. Any test of Aboriginality is not set in stone, 'particularly as an approach to determining Aboriginality as the basis for those fundamental ties of political community in Australia which are not dependent upon membership of a particular sub-group'.[41]

33. In underscoring the 'belong' concept, Edelman J stated in *Love & Thoms* that:

> The antonym of an alien to the community of the body politic cannot be a 'citizen'. It is a 'belonger' to the political community.[42]
> ...
> The sense of identity that ties Aboriginal people to Australia is an underlying fundamental truth that cannot be altered or deemed not to exist by legislation in the same way that changing legislative definitions of citizenship cannot alter the fundamental truth underlying identity that is shaped by the core combined norms that metaphorically tie a child to Australia by birth and parentage.[43]

34. Accordingly, Aboriginal Australians – based on principles of self-determination – are empowered to delineate who is included and who is excluded from the definition of Aboriginal Australian. As the plurality held in *Love & Thoms*, once a person is deemed 'Aboriginal', they cannot be deemed an 'alien' of Australia:

> Notwithstanding the amplitude of the power conferred by s 51(xix) it does not extend to treating an Aboriginal Australian as an alien because, despite the circumstance of birth in another country, an Aboriginal Australian cannot be said to belong to another place.[44]

35. Reflecting on the majority decision in *Love & Thoms*, Nettle J – his Honour being part of that majority – had this to say in the subsequent matter of *Chetcuti v Commonwealth*:

40 Ibid 290 (Edelman J).
41 Ibid 317 (Edelman J).
42 Ibid 288 (Edelman J) (footnotes omitted).
43 Ibid 314 (Edelman J).
44 Ibid 190 (Bell J) ((footnotes omitted).

> Admittedly, each member of the majority in *Love* expressed his or her reasoning to some extent differently. But common to all was the essentiality of the common law's recognition of the membership of an Aboriginal society continuously united in their acknowledgment of their ancient laws and customs deriving from before the Crown's acquisition of sovereignty over the Australian territory, and the inherent inconsistency of that fact with the permanent exclusion from Australia of a member of such an Aboriginal society. That is what some members of the majority characterised in terms of 'belonging', and that is what all members of the majority concluded put such Aboriginal persons beyond the ordinary understanding of 'alien'.[45]

36. As Nettle J enunciated in *Love & Thoms*:

> under the common law of Australia, an Aboriginal society retains an identifiable existence so long as its members are 'continuously united in their acknowledgement of laws and observance of customs' deriving from before the Crown's acquisition of sovereignty, and such may be inferred from 'subsidiary facts' of a social, cultural, linguistic, political or geographical kind.[46]

37. Mortimer J (as her Honour then was) reflected upon Nettle J's *Love & Thoms* sentiments in *Helmbright (No2)* indicating that what Nettle J was seeking to underscore was 'the survival of a society or community, and with a broader ongoing and well-recognised spiritual connection to country, which...existed at the time of the assertion of European sovereignty, and has not been extinguished even if [native] "title" has been extinguished'.[47]
38. Although the 'belonger' concept was used to validate a position that Aboriginal Australians could not be an alien of Australia, the concept has become equally applicable when considering the question of the co-existence of Aboriginal Australian sovereignty. That is, the *sui generis* nature of Aboriginal Australians relationship with the body politic of the Commonwealth of Australia is such that the co-existence of Aboriginal Australian sovereignty can no longer be denied. This is particularly so post-*Makarrata Treaty*.

45 *Chetcuti v Commonwealth* (2020) 272 CLR 457, 471–2 (Nettle J).

46 *Love & Thoms* (n 1) 253 (Nettle J). This passage is quoted by Mortimer J in *Helmbright (No 2)*. But the authority for the proposition ultimately flows from *Sampi v Western Australia* (2010) 266 ALR 537 [77] (North and Mansfield JJ).

47 *Helmbright (No 2)* (n 19) 149–50 (Mortimer J).

39. As reflected by The Honourable Robert French in a post-judicial speech, which was made in 2023 well prior to the 2026 *Makarrata Treaty*:

> Authority over land and waters within the First Nations legal framework and within the colonising legal framework are capable of co-existence just as tradition law and custom are capable of co-existence of the kind reflected in native title agreements. [48]

40. Ultimately, the source of Aboriginal Australian sovereignty flows from 'the ancestral tie between the land and its people' or, put another way, because Aboriginal Australians 'belong' to the Australian continent.
41. In the same way that Native Title has, since its inception, always co-existed alongside 'the colonising legal framework', the notion of Aboriginal Australian sovereignty can also co-exist alongside the sovereignty of the Commonwealth of Australia.

Conclusion

42. The 2026 *Makarrata Treaty* embeds an Aboriginal Australian Confirmation Framework, which is grounded in principles of self-determination. The rationale for the framework – which was essentially built upon the majority decision in *Love & Thoms* – is to facilitate a singular *Makarrata Treaty* between Aboriginal Australians and the Commonwealth of Australia.
43. In *Indigenous Custodians v Biodata Extraction and Excavation Ltd*, the Indigenous Custodians have satisfied the framework through means of self-determination. It is not for this Court to interfere with the implementation of the Aboriginal Australian Confirmation Framework and, accordingly, there is nothing before this Court to undermine the position of standing with respect to the Indigenous Custodians.
44. To the extent there is any conflict between the *Biological Data Mining Act 2025* (Cth) and those aspects of the Makarrata Treaty embedded within the *Native Information Title Act 2028* (Cth), the latter must prevail. This is because the co-existence of Aboriginal Australian sovereignty – alongside the Commonwealth of Australia – contains relevant and specific *sui generis* Traditional Knowledge rights over 'cultural customs, practices, language, and lore of First Nations Peoples in Australia'.
45. In this instance, Biodata Extraction and Excavation Ltd's assertion of 'exclusive biological data mining rights within National Parks and other nature reserves in Australia' must yield to the co-existing sovereign rights of Aboriginal Australians. This is because the specific *sui*

48 French n (22) [35].

generis Traditional Knowledge rights associated with the relevant biological data existed prior to any rights that flow from the colonising legal framework.

46. The Court orders:
    a. The Indigenous Custodians have standing to bring this action;
    b. An 'enduring injunction' against all biological data mining activities of the Defendant, Biological Extraction and Excavation Ltd, that conflict with the *Native Information Title Act 2028* (Cth); and
    c. The Defendant, Biological Extraction and Excavation Ltd, bears the costs of the proceeding.

Chapter 13

The Disillusion of International Law

Jo Bird and Greta Bird

Commentary

Prologue

This chapter was difficult to research and write. As white, female lawyers trained in Western law with its emphasis on 'benchmark man' as its central character,[1] we come to International Environmental Law and are confronted. Confronted by the yawning gap between our visceral knowledge gained as environmental activists and the promises paraded as environmental laws. In writing, we tap into the material of our bodies, material that holds feelings and experiences. We have envisaged a future scenario with a Chimera as a central character in the narrative. The Chimera is a metaphor for the dissolution of physical boundaries; boundaries that Western man has theorised to protect himself against the chaos of the biosphere. Calls for the decentering of the 'human' raise the questions: How is this to be achieved? Can Western law discover the answer and implement the changes required? Or will humanity be extinguished?

A monumental task confronts us. We are acutely aware of the laws of First Nations that come from the land and are based on a relational philosophy. Here care for Country is a primary responsibility. The land is not a commodity whose resources can be plundered but is full of living ancestors. If *ruwe* (Mother Earth) is respected, *ruwe* will provide what is necessary for life.[2] First Nations laws have been ignored or disrespected in most parts of the planet. This is demonstrated by their woeful treatment in the United Nations.

The writers assert that a future for human, and other, life on the planet is only possible if Western law is capable of humility. Western 'man' and Western laws are not the pinnacle of a mythical hierarchy. First Nations'

1 Margaret Thornton, *The Liberal Promise: Anti-Discrimination Legislation in Australia* (Oxford University Press, 1990) 20–22.

2 Irene Watson, *Aboriginal Peoples, Colonialism and International Law: Raw Law* (Routledge, 2015) 20.

DOI: 10.4324/9781003389569-16

relationship to land and care for Country offers guidance to those who show respect. The writers acknowledge, in particular, the inspiration provided by the activism and scholarship of Irene Watson.[3] The writing took place on the lands of the Bundjalung and Kulin Nations.

The chapter is organised in three sections. The first is the climate crisis and the failure of international environmental law to address fundamental structural problems in the legal system. The second is the work of philosophers and myth makers who are struggling to decentre the human. Third, we draw on First Nations Knowledges to create an alternative to a judgment, a collaborative response to the climate crisis through yarning circles.

The biosphere that sustains life on Earth does not end at State borders. As Anzaldua puts it, 'the skin of the earth is seamless. The sea cannot be fenced, el mar (the ocean) does not stop at borders'.[4] Halting climate change and other ecological catastrophes requires cooperation at local and planetary levels. However, the cooperative forces of nation-states and multinational corporations, the central powers crafting international law, are failing. A law that affirms the life of the planet is required.

Our future scenario envisages the 'end point' (death) of international law in its current form, a law that embeds environmental 'action' within the very systems that violate 'nature' and are the causes of ecological destruction.

We imagine a future two hundred years from now, where the effects of climate change, biowarfare, and nuclear disaster have led to an environment mostly uninhabitable for human or other life. In order to survive, Western scientists have by biological experimentation created a being, a Chimera, that is genetically both human and tree. This collapse of physical boundaries is a metaphor for the decentering of the human that must take place in international law, which continues to inscribe a relationship of domination of Earth and other species by humans. We do not write a judgment in the traditional sense. We see Indigenous ways of caring for Earth as the ways of the future. Therefore, our 'judgment' of the climate crisis is based on deep listening and yarning.

The story of the chimera speaks of human beings who found their 'civility' in processes and laws that separated them from the environment. They therefore destroy all that they see as 'not human', and in doing so, they destroy the very conditions for the existence of themselves as human beings. It is only at this point that they realise both their kinship with all life, and the centrality of Indigenous law in sustaining all life on the planet.

Our scenario draws upon new materialist/critical environmental law's insistence on a decentering of the (Western) subject of law. It also draws on

3 Irene Watson is a First Nations scholar and Pro Vice Chancellor at University of South Australia.

4 Gloria Anzaldua, *Borderlands/La Frontera: The New Mestiza* (Aunt Lute Books, 1987) 3.

Indigenous critiques of international law as Western law and the call by Irene Watson for non-Indigenous peoples to exit the colonial matrix of law.

There is a risk that decentering the human opens us to the accusation of being 'anti-human'. It is not the place of this chapter to offer a detailed critique of international human rights discourse. However, it can be said that human rights reinscribe or leave unchallenged a particular relationship of human beings to land, one that inscribes humanity at the centre, and the environment as object or resource of (over)consumption. At best, the environment is an object of 'protection'. The environment, from our perspective, does not exist with the primary purpose of maintaining human life, but has its own value, and value to all life on Earth, a value that preceded and will outlive the Anthropos.

International Environmental Law: Background

The 1960s was a period when much change occurred in the West, especially in connection with the environment. Youth were rebelling against the status quo, through music, art, fashion, dance and political movements. A catalyst for environmental awareness was the publication of Rachel Carson's *Silent Spring*. This book, about industrial pesticides, led to political action on pollution in the United States and beyond.[5]

International environmental law in the West began in Sweden. Sweden was experiencing high levels of air pollution and initiated the *Stockholm Declaration*,[6] the first international environmental law instrument. A duty to prevent transboundary pollution emerged. It was established that the actions of one state within its own territory could cause harm within the sovereign territory of another.[7]

In 1992 more than 100 heads of state met in Rio de Janeiro, Brazil, for the first international 'Earth Summit' convened by the United Nations. There was a palpable sense of excitement around the globe, especially among young environmentalists, that a blueprint could emerge to provide a way to work on global climate issues.

During the Summit, *Agenda 21* was adopted. This was a declaration by world leaders to prevent damage to the global environment and reduce poverty and diseases. The agenda sounded positive, but the aim was to put

5 Rachel Carson, *Silent Spring* (Houghton Mifflin, 1962). Carson drew attention to the detrimental effects of agricultural pesticides on birds.

6 *Report of the United Nations Conference on the Human Environment*, UN DocA/CONF 48/14/Rev 1 (1973, adopted 16 June 1972) ('*Stockholm Declaration*').

7 *Trail Smelter Arbitration (USA v Canada)) (Awards)* (1983) 3 RIAA 1907. The case remains the only international law case concerning air pollution. See Elena McCarthy, 'International Regulation of Transboundary Pollutants: The Emerging Challenge of Ocean Noise' (2001) 6(2) *Ocean and Coastal Law Journal* 257.

'sustainable development' at the forefront and encourage free trade. Critics were alarmed that this was similar to earlier agendas.

Indeed, the achievements of the Earth Summit fell below the expectations of climate activists. But notably, the *Framework Convention on Climate Change*,[8] coming out of the Summit, led to the *Kyoto Protocol*[9] and the *Paris Agreement*.[10] Some positive change was achieved.

Following the Earth Summit, people were optimistic that international environmental law would expand. The International Court of Justice opened an environmental chamber. However, in 13 years, the chamber was never used and was dissolved in 2006. There are very few specifically international environmental law cases.[11] One of these few cases relates to Vanuatu, whose statehood is under threat by rising sea levels.[12] After a 'herculean effort' Vanuatu secured the support of the majority of members of the United Nations General Assembly for an advisory opinion by the International Court of Justice into state responsibility for climate change.[13]

The *Framework Convention on Climate Change* offers some hope for international environmental agreement. However, it is still constrained within the architecture of global capitalism. Its main mechanisms are to reduce emissions through the 'greening' of markets that brought about the crisis in the first place.[14] Legally binding treaties are one of the most important sources of international law; however, states must consent to them to ensure that they are 'universal' agreements. Little progress has been made; scientific data demonstrates that there is a year-on-year increase in CO_2 emissions.[15]

8 *United Nations Framework Convention on Climate Change*, opened for signature 4 June 1992, 1771 UNTS 107 (entered into force 21 March 1994).

9 *Kyoto Protocol to the United Nations Framework Convention on Climate Change*, opened for signature 11 December 1997, 2303 UNTS 162 (entered into force 16 February 2005).

10 *Paris Agreement*, opened for signature 22 April 2016, [2016] ATS 24 (entered into force 4 November 2016).

11 Most have predominantly economic dimensions, see for example, *Certain Phosphate Lands in Nauru (Nauru v Australia)* [1992] ICJ Rep 240 and *Pulp Mills on the River Uruguay, Argentina v Uruguay* [2010] ICJ Rep 14.

12 Sovereignty-over-territory requires that a state have a territory recognised at international law, no matter how small. The smallest state is the Vatican.

13 This does not involve the re-opening of the environmental chamber of the International Court of Justice. In April 2023, the United Nations General Assembly supported a resolution, spear-headed by Pacific Youth, for the International Court of Justice to give an Advisory Opinion that may be a 'turning point in climate justice': 'UN resolution billed as a turning point in climate justice', *UN Environment Program* (Media Release, 31 March 2023) <https://www.unep.org/news-and-stories/story/un-resolution-billed-turning-point-climate-justice>.

14 See for example, Ryan Gunderson, Diana Stuart, and Brian Peterson, 'Ideological Obstacles to Effective Climate Policy: The Greening of Markets, Technology and Growth' (2018) 42(1) *Capital & Class* 133.

15 International Energy Agency, *CO_2 Emissions in 2022* (IEA Publications, March 2023) 3.

Sustainable Development

Environmental declarations and conventions have to fit within the framework of sustainable development, which presents itself as having humanitarian aims such as ending world poverty and hunger. However, in reality, the United Nations' much-lauded Sustainable Development Goals ('SDGs') are more about preserving global capitalism and the freedoms of multinationals. For Celine Tan, it is the hegemony of Northern states that 'underwrite the rules of multilateral engagement' that maintain deep asymmetries in the international economic order.[16]

The concept of state sovereignty has been supplanted by a neo-liberal economic order that favours the activities of multinational corporations with their resource extractive activities.[17]

The *General Agreement on Tariffs and Trade 1947*[18] was premised on the neo-liberal promise that free trade and unrestricted foreign investment will generate growth and alleviate poverty.[19] The World Trade Organization, established in 1995, was based on a similar philosophical premise.

The first principle of the *Rio Declaration* provides that 'human beings are at the centre of concerns for sustainable development',[20] and the second principle contains a reminder that states possess 'the sovereign right to exploit their own resources' according to their own national laws.[21] Sustainable development, now a pivotal concept in international environmental law,[22] bears some similarity with intergenerational equity given that it envisages 'development that meets the needs of the present' without compromising the needs of future generations.[23] We ask: how is this care for future generations to be achieved?

One answer is offered by Cardinal. He argues that we must not be constrained by the discipline of international law's epistemic biases, but instead should embrace resistance. Resistance should refuse a 'dialogue under the

16 Celine Tan, 'The New Biopower: Poverty Reduction Strategy Papers and the Obfuscation of International Collective Responsibility' (2011) 32(6) *Third World Quarterly* 1039, 1045.

17 For a critique of state sovereignty and its contribution to world governance see Achille Mbembe, *Necro-Politics* (Duke University Press, 2016).

18 *General Agreement on Tariffs and Trade 1947*, opened for signature 30 October 1947, 64 UNTS 187 (entered into force 1 January 1948).

19 Gillian Triggs, *International Law: Contemporary Principles and Practices* (LexisNexis Butterworths, 2nd ed, 2011) 740.

20 *Rio Declaration on Environment and Development*, UN Doc A/CONF 151/26 (vol I), adopted 14 June 1992, Principle 1 ('Rio Declaration').

21 Ibid Principle 2.

22 The concept of 'sustainable development' arose from 1987's Brundtland Report, see World Commission on Environment and Development, *Our Common Future (Brundtland Report)*, UN Doc A/42/427 (4 August 1987).

23 Rio Declaration (n 20) Principle 3.

rules of the dominant power because this power does not seek dialogue, but rather domination'. Therefore, we are called to seek 'an ethic of epistemic justice'.[24]

'Development' is a sacred pillar of international law, much like the colonial era concept of sovereignty-over-territory. It is an aspect of sovereignty, in that it is 'the sovereign right to exploit', and to consent to others to exploit, the resources within one's territory. Both the nation-state and the multinational foreign investor are at the core of this concept whose skeleton must be defended at all costs, or the corpus of international law would dissolve.

Connected with 'development' is the idea that the 'undeveloped' will be lifted out of poverty through the continued exploitation of 'the natural world'.[25] Development is theorised as liberatory and humanising for impoverished individuals living within the 'post' colonial state. The right to development is inscribed as a human right within which 'all human rights and fundamental freedoms can be realized'.[26] Development, in this way, is represented as synonymous with humanitarianism. It presents humanity with the question: do we choose Earth or human rights? 'Growth' is the mantra of neo-liberalism, and there is a risk that sustainable development is primarily a green mask for global capitalism.

The Exclusion of Indigenous Peoples

Despite 'decolonisation', First Nations, the only peoples with a non-destructive environmental law system based on respect for the ecosystem as a spiritual and material being, continue not to be recognised as lawmakers and participants in the 'universal' united nations system. The *United Nations Declaration on the Rights of Indigenous Peoples*[27] ('UNDRIP') is a non-binding and watered-down version of what was originally hoped for.[28] It denies sovereignty to Indigenous peoples, as this might threaten the skeleton of colonial nation-states such as Australia, Canada, the United States, and New Zealand. Article 46 of UNDRIP provides that 'nothing in this declaration shall be construed as authorizing or encouraging any action that would

24 Pierre-Alexandre Cardinal, 'Resistance and International Law; Decoloniality and Pluritopic Hermeneutics' (2016) 1(1) *Inter Gentes* 40, 51.

25 Antony Anghie, *Imperialism, Sovereignty and the Making of International Law* (Cambridge University Press, 2005) 310. Anghie writes: 'colonialism was central to the constitution of international law'.

26 *United Nations Declaration on the Right to Development*, UN Doc A/RES 41/128 (adopted 4 December 1986).

27 *United Nations Declaration on the Rights of Indigenous Peoples*, GA Res 61/295, UN Doc A/RES/61/295 (2 October 2007, adopted 13 September 2007).

28 Watson (n 2) 90–95.

dismember or impair...the territorial integrity of Sovereign and independent States'. As Irene Watson asserts:

> The UNDRIP enables business as usual, the power of states to do as they wish with our lands, resources, lives and laws ... Under the UN, what is achievable in terms of self-determination is limited by the power of the states to contain First Nations as state captives of colonialism.[29]

Mythic and Philosophical Positions Which Challenge International Law Thinking by Decentering the Human and Locating the Plant's Importance

While Eurocentric international law centres the human as divided from the natural world, mythology, philosophy and art speak of alternative forms of relationships between human and nature.

Representations of the human-plant in Western art and literature are not new. Corporealities that are both human and plant have existed in both Western and non-Western myth, epistemology and art throughout millennia. In choosing the human-tree chimera to write our own speculative myth, we explore/draw upon three such dramas/narratives.

In Ancient Greek mythology is the story of Daphne who is transformed into a laurel tree; her hair becomes leaves, her arms branches, her heart encased in bark, as she begs to escape Apollo.[30] The story epitomises the association of women with nature and passivity; both are resources, the property of men.[31]

In the science-fiction imagination of John Wyndham, the triffids are not the passive tree woman Daphne, whose leaves are harvested. They are flesh-eating plants, with human powers such as locomotion and intelligence, that emerged from the anxieties of post–World War II and a shrinking British empire. The plant-beings were created in a lab in the Soviet Union, and by accident escaped. Once humans lose the power of sight after being blinded by a meteorite shower, the triffids attack and kill human beings. The protagonist in the book, Bill Mason, makes the comment: 'Just what did happen is anyone's guess now. But one thing I'm quite certain of – that somehow or other we brought this lot down on ourselves'.[32]

In the 2021 sculpture *Nude*, by Hugh Hayden, the human form is skeletal, representing its impermanence and mortality. It is dead, as the brittle branches that it is a part of. The death of plant and human are represented as

29 Ibid 148.

30 Ovid, *Metamorphoses*, tr Stephanie McCarter (Penguin Classics, 2022) 525–52.

31 Apollo tells Daphne, 'If you cannot be my woman, you shall be my tree, out of which I will make laurel wreaths for my hair'. There are various feminist interpretations of *Metamorphoses* which are beyond the scope of this chapter.

32 John Wyndham, *The Day of the Triffids* (The Modern Library, 2003) 205.

one. It is notably headless and faceless without a human identity. Has it been beheaded in a violent murder, or does its wooden appearance indicate the decay and barrenness that is also being experienced by plant life? The human form is constructed of bald cypress, with branches emerging out of its spine, leg and arm bones. The human form is at the centre of the plant and almost indistinguishable from it. In an uncertain future, will humanity and its laws be stripped naked or will human law suffer the same fate of extinction in the face of deforestation, drought and global warming?

The word chimera derives from Greek mythology, a fire-breathing creature with a lion's head, a goat's body, and the tail of a snake. In genetics, broadly speaking, an interspecies chimera refers to an organism that has two genetically distinct cell lines. Most organisms only have a single cell line. A chimeric animal may be composed of both animal and plant DNA, or an animal's DNA may be constituted by the cells of two or more animal species. Although human-plant chimera with thought processes of both species is far from a reality, experiments with human-animal chimeras have been taking place for years. In 2021, for the first time, human-monkey interspecies chimeric embryos were created.[33] The embryos were destroyed after growing for 20 days. These experiments are for the purposes of xenotransplant research involving organs engineered to be biologically closer to human that could be harvested from 'nonhuman' animal hosts. It is scientifically possible because the structure of DNA, shaped like a twisted ladder, is the same in all forms of life, including humans, trees and bacteria. We speculate that, in a world devastated by the effects of deforestation and climate change, humans may turn to these technologies to survive, and have imagined a tree/human chimera.

Philosophy, particularly feminist philosophy, has for some decades decentred the human subject. European philosophers began challenging divisions in Western thought such as the nature/culture divide in the 1970s.[34] The image of the plant has been useful in this decentering. For Deleuze and Guattari, the rhizome, an underground mass of the root stems of certain plants, becomes the metaphor of a non-linear, non-hierarchical mode of language.[35] Luce Irigaray advocates adherence to Eastern philosophy and the non-predatory gaze of Buddha contemplating a water lily.[36] Stefano Mancuso

33 The embryos raise ethical questions about the status that can be given to a body that is part human.

34 See, eg, Helene Cixous and Catherine Clemente, *The Newly Born Woman*, tr Betsy Wing (University of Minnesota Press, 1986).

35 Gilles Deleuze and Felix Guattari, *A Thousand Plateaus: Capitalism and Schizophrenia*, tr Brian Massumi (University of Minnesota Press, 2004).

36 Michael Marder, 'Afterword – Cultivating Natural Belonging: Luce Irigaray's Water Lilly' in Luce Irigaray and Michael Marder, (eds) *Building a New World: Palgrave Studies in Metaphysical Thought* (Palgrave Macmillan, 2015) 297.

envisages a 'nation of plants' that fosters 'diffuse and decentralised vegetal democracies'.[37]

Donna Haraway and others call for an epistemic collapse between human and nonhuman, 'multi-species assemblages that include people', which she names the Chthulucene.[38] Anna Grear asks the question: 'what hope and what forms of environmental law and perhaps even insurgent eco-subjectivity might the trope of the Chthulucene offer?'[39] Our future scenario allows us to explore what this might look like in practice in a speculative future.

The human-tree chimera that reimagines law is in the vein of other partially human life forms proposed to disrupt law and human subjectivity. Grear takes up Haraway's proposal that 'we are all lichens', where symbiosis replaces individualism in reimagining the basis of Western, colonial and capitalist systems of law. Grear argues that we should abandon the panoptic of environmental law, which creates the environment as a human object, in favour of an 'epistemic tentacularity'.[40]

Andreas Philippopoulos-Mihalopoulos proposes the 'collapse of the human' and in its place a 'hybrid' connectivity between humans and nature.[41] There is a usefulness in collapsing the human subject in terms of challenging modes of domination. However, while desirous of decentering the human as 'exceptional', there is a risk, Philippopoulos-Mihalopoulos argues, of 'philosophical flatness' in claiming that all life is the same, and from this sameness a denial of the special responsibilities of humans to other forms of life.[42]

The decentering of the human involves an understanding of plants as having sentient and lawful capacities.[43] Marder argues that plants are collective beings, 'assemblages of multiplicities, inherently political spaces of conviviality', and that humanity should adopt the thought processes of plants.[44]

37 Stefano Mancuso, *The Nation of Plants: A Radical Manifesto for Humans*, tr Gregory Conti (Profile Books, 2019) 51.

38 Donna Haraway, 'Anthropocene, Capitalocene, Plantationocene, Chthulucene: Making Kin' (2015) 6(1) *Environmental Humanities* 159, 160.

39 Anna Grear, '"Anthropocene, Capitalocene, Plantationocene, Chthulucene": Re-Encountering Environmental Law and Its "Subject" with Haraway and New Materialism' in Louis J Kotzé (ed), *Environmental Law and Governance in the Anthropocene* (Bloomsbury, 2017) 77, 94.

40 Ibid.

41 Andreas Philippopoulos-Mihalopoulos, 'Towards a Critical Environmental Law' in Andreas Philippopoulos-Mihalopoulos (ed), *Law and Ecology: New Environmental Foundations* (Routledge, 2011) 18, 19.

42 Andreas Philippopoulos-Mihalopoulos, 'Critical Environmental Law in the Anthropocene' in Kotzé (n 39) 117, 122.

43 See, eg, the discussion of the sentience of plants in Paco Calvo with Natalie Lawrence, *Planta Sapiens: Unmasking Plant Intelligence* (Bridge Street Press, 2022). See also Mancuso (n 37).

44 Michael Marder, *Plant-Thinking: A Philosophy of Vegetal Life* (Columbia University Press, 2013).

Some of the work of critical environmental law, wild law, Earth jurisprudence and new materialism acknowledges its debt to Indigenous knowledges. Davies speaks of a law that comes from Earth, rather than laws that are exercised *over* Earth. However, she acknowledges that new materialist thought can only have 'superficial similarities' to Indigenous knowledges, given that new materialism, unlike Indigenous law, is not a law that is lived in a spiritual sense, and it does not come from obligations to land that have existed since time immemorial.[45]

Western epistemologies have only recently grasped the dilemma that their philosophy has created the centrality of the human in their writing on the environment; whereas a healthy environment depends on the interconnectedness of all life. Increasingly too, theorists in this tradition sense that their knowledge production barely allows for Indigenous philosophy and ways of being to have a voice.

Indigenous Knowledges

Indigenous law does not need to decentre the human. 'The land is the source of the Law'.[46] As Watson writes: 'Kaldowinyeri was a time when song, stories and law were birthed, as were the ancestors – out of the land'. Indigenous knowledges 'carry obligations and responsibilities', to the land, to each other, and to other species:[47]

> In 'knowing' their human selves as being fundamentally different from animals and the natural world, the Europeans ended up with their focus being centred on relationships between humans. This limited them; the focus resulted in a hegemony, a 'master' and 'slave' relationship with the natural world, and others of their kind.[48]

Judgment

Scenario

It is two hundred years in the future and the land once known as Europe is uninhabitable by human beings. This follows nuclear disasters and the effects of the climate catastrophe. The elites left Earth for Mars and in search of

45 Margaret Davies, *EcoLaw: Legality, Life, and the Normativity of Nature* (Taylor and Francis, 2022) 3.
46 Christine Black, *The Land Is the Source of the Law: A Dialogic Encounter with Indigenous Jurisprudence* (Routledge, 2011) 4.
47 Watson (n 2) 11–14.
48 Ibid 148.

other 'heavenly, virgin bodies' to colonise. Adrenalin pumped, a new 'gold rush' beckoned.

Scientists discovered that trees were surviving high levels of radiation in the exclusion zone of Pripyat near Chernobyl and high radiation zones.[49] It was determined that humans could be engineered to resist radiation and expand their life expectancy exponentially. The idea was to genetically alter the physical human form, while the consciousness of humans remained. Human-tree chimeras were engineered to inhabit the uninhabitable.

However, the results were not as originally envisaged. Millions of *Homo sapiens* living in Europe were transformed into beings with the consciousness of both humans and trees. Their arms became branches which grew leaves, and their legs grew into the earth and became roots. They were unable to move and they became a forest. Their lives were transformed into a symbiotic existence with other forms of life, such as bacteria and fungi.

As chimeric tree-beings they can no longer deny that their humanness is corporeally connected to the Earth Mother. They cannot move to conquer what they once envisaged as 'virgin land', nor rain military arms upon enemies, nor destroy forests and rivers, nor create mass extinctions. They stand still for many Earth cycles and reflect and breathe out oxygen. They feel the beauty of the clouds that are drawn to their forest mass and which pour forth life-giving rains. The bacteria and fungi assist in producing carbohydrates and their leaves engage in the alchemy of photosynthesis. The Chimeras of Westphalia are engaged in the 'development' of life and 'progression' towards the healing of the land, a 'return of the soul life' of caring for their country.

Yarning Circles

The Yarning Circle is the Australian name for a process embedded in First Nations cultures over thousands of years. It is a non-hierarchal, respectful space that involves dialogue and collaboration. It has been described as 'a space of productive hope'.[50] The Yarning Circle involves both active speaking and active listening. As Tyson Yunkaporta and Melissa Kirby write, 'our cognition is relational...all knowledge is connected to people and place'.[51] The authors point to the survival of First Nations through thousands of years of massive climate change. This adaptation and creativity is connected to

49 For a detailed description see Serhii Plokhy, *Chernobyl: History of a Tragedy* (Allen Lane, 2018).

50 Bronwyn Carlson and Ryan Frazer 'Yarning Circles and Social Media Activism' (2018) 169(1) *Media International Australia* 43, 43.

51 Tyson Yunkaporta and Melissa Kirby, 'Yarning up Indigenous Pedagogies: A Dialogue About Eight Aboriginal Ways of Learning' in Nola Purdie, Gina Milgate, and Hannah Rachel Bell (eds) *Two Way Teaching and Learning: Toward Culturally Reflective and Relevant Education* (ACER Press, 2011) 205, 208.

cultural processes, such as Yarning Circles, which are 'a great gift to a world that desperately needs solutions'.[52]

The Yarning Circle is an appropriate setting for a non-adversarial sharing of knowledge and a means of moving forward in the crisis outlined in our scenario. The form of settling disputes in systems derived from British common law is an adversarial one that rarely produces creativity or peace. The process is expensive, hierarchical, gendered and raced. The Yarning Circle is a practice of sharing stories in a culturally safe and appropriate way to produce knowledges through new understandings and to heal trauma. In our scenario, the Yarning Circle may provide a space for creative solutions to emerge in a respectful space.

We authors do not provide a judgment. A judgment in International Law is no longer useful. We choose instead a communication that involves dadirri – 'inner deep listening and quiet still awareness' – and yarning, a place to share stories in a place that is non-judgmental, as explained by Tyson Yunkaporta.[53]

These ways of communication are informed by the call for:

> [t]he many non-Indigenous...to become Aboriginal again. But to become Aboriginal in their place and in their culture not ours. To copy the Aboriginality of us would perpetrate the rape and appropriation. Becoming Aboriginal is the learning of a reverence of the mother, and a coming into the laws of ruwe. This is a first step in healing the rape.[54]

Deep listening is the beginning.

The Yarning Circle

Yarning Circle: International Law

In the Matter of: Planetary Governance

Planetary Area: The land once known as Europe, specifically the forests of Westphalia once considered to be the 'birthplace' of modern International Law.

52 Ibid 211. See also Dawn Bessarav and Bridget Ng'andu, 'Yarning About Yarning as a Legitimate Method in Indigenous Research' (2010) 3(1) *International Journal of Critical Indigenous Studies* 37. This article demonstrates the credibility and rigour of yarning as an Indigenous research method.

53 Tyson Yunkaporta, *Sand Talk: How Indigenous Thinking Can Save the World* (HarperOne, 2020).

54 Irene Watson, 'The Power of the Muldarbi, the Road to Its Demise' (1998) 11(1) *Australian Feminist Law Journal* 28, 35. Note that Ward Churchill documents the appropriation of Indigenous culture by non-Indigenous people in Ward Churchill, *Indians Are Us?: Culture and Genocide in Native North America* (Common Courage Press, 1994) 205.

The Chimeras

We experience at a visceral level a deep sorrow at the condition of Earth. We feel the burn of radiation in our bark and the poison in our sap as we continue our existence as trees, a history of more than 350 million years in human terms. We accept responsibility for the part our human kin played in the destruction of the soils and waters. For the belief that *Homo sapiens* could detach themselves from Earth and survive separately from other forms of life. The International Law system did not listen to, nor respect the nonhuman world. It could not speak for nor care for what it called 'the environment'.

There is sadness in a European country with few humans. Human beings have special abilities for art, song and dance that can articulate law. Our European kin chose instead to represent law electronically. However, in the absence of humans in this place, the ancient systems of connection to land and animals are re-invigorated. The *Homo sapiens* aspect within us reflects on the legal system that dominated Earth during our life as mammals. It was this economic system that was embedded in International Law.

A River

I am a living entity with a right to live, and my human custodians understand their obligations to care for me.[55] We are mindful of the damage done to our Ancestral Beings and First Nation custodians. Millions of fish, turtles, frogs and other living creatures are dead or dying as a result of toxic run-off from capitalist agriculture and mining. The multinationals behind these practices called economic miracles for 'undeveloped' areas are guilty of crimes of eco-cide and genocide.

First Nations

We give voice to our stories and call for care of Country as a way to negotiate out of the crisis that is leading to so much death. We wait for a deep listening and reflection by those in the circle:

> Our Aboriginal culture has taught us to be still and to wait. We do not try to hurry things up. We let them follow their natural course, like the seasons. We watch the moon in each of its phases. We wait for the rain to fill our rivers and water the thirsty earth.[56]

55 Martuwarra RiverOfLife et al, 'Recognising the Martuwarra's First Law Right to Life as a Living Ancestral Being' (2020) 9(3) *Transnational Environmental Law* 541, 543. The river is named as an author. Another river with legal personhood is the Whanganui River in Aotearoa/New Zealand. In 2019 all rivers in Bangladesh became legal persons.

56 Miriam-Rose Ungunmerr, 'Dadirri: Inner Deep Listening and Quiet Still Awareness', *Miriam Rose Foundation* (Reflection, 1988) <https://www.miriamrosefoundation.org.au/dadirri>.

We care for Country. The law comes from the land and we have followed the law since the first sun shone on the sand. Women and men have their special knowledge of the law,[57] expressed in song, dance and art. The invasion of First Nations land by colonial powers and the imposition of laws based on domination of Earth for profit and the destruction for mining and other extractive businesses has caused great harm. This behaviour of the colonisers was carried out on every continent. Wherever they came, they brought with them diseases and dispossession of First Nations from Country.

The colonisers claimed that they had decolonised our Country, but a close reflection on International Law shows this to be mere 'smoke and mirrors'. Even in those areas where domestic 'treaties' have been entered into, full 'sovereignty' has not been granted. We are not able to care for the land and plants and animals according to our law.

Their central concepts put 'man' at the centre of International 'environmental' Law. And the 'man' at the centre was one who was wedded to neo-liberal ideology and to 'sustainable development'. First Nations developed International Law as they negotiated with the Nations bordering their Country. However, the brilliance of working respectfully over thousands of years with Earth and all living creatures to obtain food and shelter was regarded by those displaying colonial behaviour as 'primitive'.

Conclusion

We acknowledge First Nations here with the words of Irene Watson:

> Being Aboriginal is to adhere to the traditions and culture that emanate from place from the time of Kaldowinyeri [time immemorial]. (For) the non-Indigenous to be Aboriginal; first step is to stop the colonisation process and return to the lands of their ancestors. While the writer understands that this is impossible to achieve, the return can be done in many forms, not just physical, it can be a return of the soul life. But this can only be achieved when the individual adheres to the Aboriginal laws of place. And enters into the law of place in a way that is 'proper Aboriginal way'.[58]

This is the challenge for humanity to enter 'into the law of place…in a way that is "proper Aboriginal way"'. Can we show respect, apologise for our ignorance, our genocide, and ecocide and achieve the 'return of the soul life'?

57 See *Kartinyeri v Commonwealth* (1998) 195 CLR 337.
58 Watson (n 54) 35.

Chapter 14

Imagining Ecocentric Bioregional Law in Australia

Michelle Maloney

Commentary

This judgment has been written in response to the idea that the current legal system in Australia, and the current economic and political system it supports, are not fit for purpose in the 21st century. The legal system inherited from the British Empire in 1788, which developed into Australian law over the past 250 years, is failing to look after the environment (failing to Care for Country[1]) and is failing to look after all our people (failing to Care for Kin/Citizens). The challenges we face, including climate breakdown, biodiversity loss, environmental degradation, the violent legacy of colonisation, ongoing injustices to First Peoples, and the growing gap between rich and poor people in our nation, call for profound reform in Australia's governance.

The judgment explores an alternative structure for the governance of this precious continent and its peoples. The alternative structure places Indigenous sovereignty, governance and law at the centre of the new legal system. It imagines that Indigenous and non-Indigenous people are in deep partnership together, in order to support all life on the continent and build a stable, successful culture; all people privilege Caring for Country over the old mantra of economic growth.

The imaginary judgment is inspired by the governance system of the First Peoples of the continent now known as Australia, as articulated by

Please note that within this imagined scenario, as a non-Indigenous Australian I have no right to speak for any place or Country, and I have no right to discuss any specific elements of First Peoples' culture. I have used published material about the Relationist Ethos and First Laws. I have avoided naming any specific groups, organisations or Indigenous Peoples, not out of disrespect but out of deep respect, as I acknowledge my place as a non-Indigenous person. This judgment is an imagined world, a re-writing of Australia's governance system. I have no right to entwine existing Indigenous Peoples into this imaginary world, so I refer only to 'Indigenous Elders' and 'Indigenous Peoples'.

1 In Australia, 'Country', when used in the phrase 'Caring for Country', is an Aboriginal Australian word referring to the Land as a moral entity with both physical and spiritual attributes that manifest in a myriad of life forms. See Mary Graham, 'The Law of Obligation, Aboriginal Ethics: Australia Becoming, Australia Dreaming' (2023) 37 *Parrhesia: A Journal of Critical Philosophy* 1, 5.

DOI: 10.4324/9781003389569-17

Indigenous Elders, thinkers and writers including Mary Graham,[2] Anne Poelina[3] and Irene Watson.[4] The judgment has also been inspired by work within Western communities around bioregionalism[5] and bioregional governance. My organisation, the Australian Earth Laws Alliance (AELA), draws on bioregional governance and planning[6] concepts in its development of the practice-oriented Greenprints programme.[7] Greenprints supports community groups who wish to strengthen local and bioregional ecological stewardship, and advocates for Indigenous-led governance in Australia.

The Legal System of the Modern Nation State of Australia

In legal terms, Australia today is a Federation, made up of three tiers of government: a federal or Australian government, created by the *Constitution of Australia 1901* (UK), State and Territory governments which evolved out of the original colonies founded by the British Empire, and local councils, which are the creation of State Parliaments. The following features of Australia's legal system are most relevant to this judgment.

Firstly, First Peoples' sovereignty, societies, cultures and laws were not recognised by the British Empire and continue to be predominantly ignored by Australian law. The ancient pre-existing legal and governance system of the First Peoples of Australia was not recognised by the British Empire nor the post-1901 Australian government. The First Peoples of this continent have

2 Dr Mary Graham is an Adjunct Associate Professor at the University of Queensland and has written and spoken extensively about Aboriginal philosophy, ethics, the Relationist Ethos, International Relations and much more. For a comprehensive list of her papers, public talks, and interviews, visit 'Mary Graham', *Future Dreaming Australia* (Web Page, 2023) <https://www.futuredreaming.org.au/about/governance/directors/mary-graham/>

3 Dr Anne Poelina holds two Doctorates, a Master of Public Health and Tropical Medicine, Master of Education, and Master of Arts (Indigenous Social Policy). She writes and speaks extensively about a range of issues including Aboriginal First Laws, Earth Laws, and the work of the Martuwarra Fitzroy River Council. See, eg, Anne Poelina, Donna Bagnall, and Michelle Lim, 'Recognizing the Martuwarra's First Law Right to Life as a Living Ancestral Being' (2020) 9(3) *Transnational Environmental Law* 541, and 'Share in Our Dreaming', *Martuwarra Fitzroy River* (Web Page, 2023) <https://www.martuwarra.org/>.

4 Professor Irene Watson is Pro Vice Chancellor, Aboriginal Leadership and Strategy, and Professor of Law at the University of South Australia and has written extensively about Indigenous and colonial law and other topics. See, eg, Irene Watson, *Aboriginal Peoples, Colonialism and International Law: Raw Law* (Routledge, 2015.)

5 Advocates of bioregionalism assert that human activity should be defined by ecological or geographical boundaries rather than political ones. See 'Bioregionalism', *Greenprints – Australian Earth Laws Alliance* (Web Page, 2022) <https://www.greenprints.org.au/knowledge-base/bioregionalism/> ('Greenprints').

6 See, eg, Greenprints (n 5).

7 Michelle Maloney, 'Practical Pathways to Ecological Law: Greenprints and a Bioregional, Regenerative Governance Approach for Australia' in Kirsten Anker et al (eds), *From Environmental to Ecological Law* (Routledge, 2021) 237 and see also Greenprints (n 5).

been treated with cultural ignorance, institutional racism and the horrors of genocide. In 1992, the *Mabo* case[8] finally saw legal recognition of the First Peoples' prior and existing claim to their homes and estates. The *Mabo* case began a process of recognising Native Title but is a contentious and flawed approach to recognising the legal system that existed prior to colonisation.[9]

The reasons why First Peoples in Australia were treated the way they were, by the colonisers and successive governments, are shameful and a topic of much discussion and debate. At the time the British claimed/invaded the continent, so-called international law (law made by European powers for European powers, to support their colonisation of non-European Peoples and places) identified three different ways that new lands or colonies could be claimed. Lands could be settled, conquered or ceded.[10] By claiming that the continent was 'settled' (not conquered), it meant that the British government treated Australia as a colony that was uninhabited by a recognised sovereign or by a people with recognisable institutions and laws – 'terra nullius', or empty land. There were no treaties created with any First Peoples and no legal recognition or arrangements made for proper relations. British law – and institutionalised racism and genocide – was simply imposed on the original inhabitants. So today, while Indigenous people continue to respect and practise their ancient First Laws,[11] the ongoing colonial legacy of the British and now Australian legal system means these laws are ignored or disrespected by Australian law, except in relation to the flawed construct of Native Title and other colonial, cultural heritage style laws.

Secondly, the Australian legal system and broader society are built on an extractivist, expansionist worldview. It has no foundational obligation (legal

8 *Mabo v Queensland* (No 2) (1992) 175 CLR 1

9 For a discussion about the limitations of Native Title in Australia, see Irene Watson, 'Sovereign Spaces, Caring for Country and the Homeless Position of Aboriginal Peoples' (2009) 108(1) *South Atlantic Quarterly* 27 and also Irene Watson, 'Aboriginal Laws of the Land: Surviving Fracking, Golf Courses and Drains Among Other Extractive Industries' in Nicole Rogers and Michelle Maloney (eds), *Law as If Earth Really Mattered: The Wild Law Judgments Project* (Routledge, 2017) 215

10 Justice Blackburn in *Milirrpum v Nabalco Pty Ltd* (1971) 17 FLR 141, at 201, articulated the distinction thus:

> There is a distinction between settled colonies, where the land, being desert and uncultivated, is claimed by right of occupancy, and conquered or ceded colonies. The words 'desert and uncultivated' are Blackstone's own; they have always been taken to include territory in which live uncivilized inhabitants in a primitive state of society. The difference between the laws of the two kinds of colony is that in those of the former kind all the English laws which are applicable to the colony are immediately in force there upon its foundation. In those of the latter kind, the colony already having law of its own, that law remains in force until altered.

11 See, eg, Poelina et al (n 3).

or moral) to look after the environment or Care for Country. There is no general 'land ethic' or benchmark for maintaining ecological health within our culture, or our legal, political or economic systems. Modern Australia was built on the expansionist and extractivist worldview of European and British societies, and this continues in neoliberal growth economics today. There has never been an overarching land ethic in Australian law. The environment was not explicitly mentioned or recognised in the 1901 *Constitution*. The Australian government has slowly claimed increasing power to make laws for the environment by using, primarily, the external affairs head of power in the Constitution.[12] Responsibilities for land management and later 'environmental issues' predominantly remain the responsibility of State governments.

Today, although signing up to different international treaties and conventions creates responsibilities on the part of the Australian government to comply with these agreements, there are no laws setting out an overarching philosophy, responsibility, or moral code for Caring for Country. While a plethora of environmental laws now exist in the States and Territories, they are largely generic, standardised, 'top-down' laws triggered by actual and potential environmental harm (for example, possible harm to a specific list of endangered species). State planning policies, national park and reserve management plans, and other administrative and planning mechanisms are often used to develop management approaches unique to place, but it cannot be said that these create or impose obligations to Care for Country in our culture or legal system in a systemic way.

Finally, the Australian legal system is distinguished by top-down governance, in which neither Indigenous nor non-Indigenous local communities have much, or indeed any power to influence or control environmental or other laws in their own area. State governments set the parameters, through State-based planning laws, for what is allowed in terms of development in the State. Local councils are created by State legislation and must comply with State Planning Policies; local councils can be changed, amalgamated, and erased at the will of the State governments.

The Impact of 250 Years of Western Law and Governance in Australia

A mere 250 years of British and then 'modern Australian' governance across this continent has seen devastating destruction of the living world. The latest *State of the Environment Report*,[13] released in 2021, provides damning evidence that extractivist practices such as land clearing, forestry, water

12 *Australian Constitution* s 51(xxix).

13 Ian Cresswell, Terri Janke, and Emma Johnston, *Australia State of the Environment 2021* (Report, Commonwealth of Australia, 2021).

extraction, mineral extraction and more have caused a severe decline in the health and functioning of ecosystems across Australia. Native vegetation, ancient soils, wetlands, reefs, rivers and biodiversity have all been impacted. Almost 2,000 plant and animal species are currently listed as endangered, and the list grows longer every year.

Our governance failures are not limited to the realm of environmental management. Within our human societies, the gap between rich and poor people in Australia is widening, and despite being a wealthy country, Australian society is governed in such a way that not all people are receiving what they need to thrive. In 2019–2023, more than one in eight people (13.4%) lived below the poverty line – that amounts to more than three million Australians.[14] In addition to people struggling to afford basic necessities, such as food and electricity, Australia is also experiencing a chronic housing crisis, with rising homelessness being driven by escalating property and housing costs and accommodation shortages.[15] None of these problems should exist in such a wealthy country; these problems are created by our mainstream culture's creation of, and belief in, the current economic and governance system.

The Legal and Governance System That Existed in the Continent Now Known as Australia, Prior to British Colonisation

The First Peoples of the continent now known as Australia developed a complex, sustainable and highly effective governance system, which has been in place for thousands of generations. Mary Graham has described this as 'a multipolar, collaborative, sacralised ecological governance system' featuring hundreds of autonomous, interconnected communities or First Nations, living within their own territories and legal systems, governing the whole continent.[16]

Irene Watson reminds us that the law of First Nations People, 'conceived as a way of living is difficult to write about and cannot simply be described or easily translated into a foreign language that is empty of the ideas that our

14 Peter Davidson, Bruce Bradbury, and Melissa Wong, *Poverty in Australia 2023: Who Is Affected – Poverty and Inequality Partnership Report No 20* (Report, Australian Council of Social Service and UNSW Sydney, 2023) <https://povertyandinequality.acoss.org.au/poverty-in-australia-2023-who-is-affected/>.

15 Cameron Parcell, Ella Kuskoff, and Tim Reddel, 'Australia's Housing Crisis: How Did We Get Here and Where to Now?', *Contact Magazine – University of Queensland* (online, 30 January 2023) <https://stories.uq.edu.au/contact-magazine/2023/australias-housing-crisis-how-did-we-get-here-where-to-now/index.html>.

16 Greenprints, 'Indigenous Philosophy and the Relationist Ethos' (YouTube, 11 June 2022) <https://www.youtube.com/watch?v=NrSMZxDWimI>.

law ways carry. … [First Nations Peoples'] law is the essential basis of social conduct: respect, reciprocity and caring for country, to name a few'.[17]

Anne Poelina and her co-authors explain how the law of First Nations People is guided by 'First Law':

> The wisdom of First Law is that it affords deference to the supreme law of the land and the pattern of life itself, rather than the law of mankind. It decentres human authority and places humanity in its natural order, as one species among millions that must live within the pattern of life and its biosphere … First Law respects all life and its place in the pattern of life on which all life depends… These principles have been developed through a rigorous process of scientific experimentation and observation spanning millennia. First Law therefore contains tried and true rules (traditional laws) that are fit for purpose in assuring the sustainability and longevity of humanity while underpinning Indigenous peoples' 'sustainable life' on Country.[18]

The Future Scenario in Which the Judgment Is Handed Down

The judgment takes place in 2060, a time when ecological and socio-political crises have catalysed a new governance and legal system in Australia. In 2060, Australia has been scarred and transformed by more than 50 years of climate change impacts. Almost a million people have died across two decades of food shortages and new diseases.

The 2019/2020 summer of horrific bushfires marked a new era for Australia. The impacts of climate change were widely acknowledged and a range of civil society actions and cases[19] drove policy reform that took far too long to activate but did mark the beginning of political and legal change. From 2024 to 2029 El Nino brought a return of severe drought and bushfires across most of the continent, leading to localised ecological collapse in many traditional food production ecosystems of Australia. Australia experienced, for the first time, food shortages and civil unrest. By 2035, the outbreak of a new global pandemic brought suffering and the deaths of hundreds of thousands of people in Australia before a vaccine was developed, and this crisis was 'managed' by health authorities in 2038. By the 2040s, sea level rise was forcing State and local governments to plan retreat from many coastal areas. In 2041, a third pandemic for the century broke out, and efforts to control

17 Watson et al (n 4) 22.

18 Poelina et al (n 3) 549.

19 See, eg, *Bushfire Survivors for Climate Action Incorporated v Environment Protection Authority* [2021] NSWLEC 92.

the contagion through lockdowns and failed vaccinations brought disturbing civil unrest.

A range of political changes took place in response to this climate-changing world. In 2035, the second pandemic of the century and the resulting civil unrest saw the creation of the 'Relationist Party', a political party forged by Indigenous Elders and community leaders, in partnership with non-Indigenous community leaders around Australia, who wanted to help their people and environment. The party became hugely popular, with a massive grassroots call for change. Extensive surveys and community discussions showed that non-Indigenous Australian people felt that they should follow the leadership and wisdom of Indigenous Peoples, in order to govern society effectively and Care for Country.

In 2045, the Relationist Party was elected with a majority into federal government, on the platform of creating a new societal structure that would bring peace and stability to the troubled continent. The new government's focus was on making ecological health and human well-being the priority for society, by building a society based on the Relationist Ethos.

After being elected, the new government hosted Citizens Assemblies across the continent from 2040 to 2042, reviewing proposals for constitutional and societal reform. Indigenous Elders now had the mandate in the Relationist Party to reform Australia's governance system. The new governance system was inspired and informed by the pre-colonial governance system of the continent, which was translated in today's terms into a more 'bioregional' framework. The popular catchcry for law reform was: 'Rainforest laws for rainforest people, desert laws for desert people'. The people of Australia were ready to embrace a new system of governance that would look after human and more-than-human life.

At the end of 2043, constitutional and law reform was enacted. The new 2043 *Constitution* creates an Australian Republic with a federal government at the national level, and 130 bioregional governments (called Bioregional Councils) are created by legislation across the continent. State governments are abolished.

The federal government has specific powers relevant to continent-wide governance, many of which existed in section 51 of the previous 1901 *Constitution*, but some of which have been modified in accordance with the new governance regime. One of the key roles of the federal government is to support Bioregional Councils in ensuring healthy ecosystems and healthy communities, as well as to carry out a range of national and international level responsibilities.

Bioregional Councils have the primary responsibilities to Care for Country and work in partnership with the Australian government to Care for Citizens, and on national ecological matters such as climate change. The 2040–2042 Citizens Assemblies held prior to constitutional reform saw communities around Australia create their own political boundaries based on ecological boundaries relevant to the communities themselves. Some of the new Bioregional Councils were created by following Indigenous leadership and using Indigenous traditional estate boundaries; in other communities, people selected their actual bioregion (one of the 89 bioregions identified by Western science, as part of IBRA[20]); elsewhere, catchment boundaries were chosen.

Bioregional Councils are elected authorities who must be comprised of half Indigenous members and half non-Indigenous members. The Councils must (i) make all decisions and take all actions necessary to Care for Country and ensure the non-human world is healthy and able to support life, including managing the continuous work of adapting to ongoing climate change and (ii) make all decisions and take all actions necessary to Care for Citizens, within the parameters and limitations of ecological health within their jurisdiction.

From 2044 to 2046, the new federal government required (and provided funding for) all Bioregional Councils to host community input processes. These processes were led by Indigenous Elders, in partnership with scientists and community leaders, to map out what healthy ecosystems look like within their jurisdiction/ecological boundary. The information from these workshops was fed into a further series of Citizen's Assemblies, and community members worked together to develop their unique 'Benchmark for Ecological Health' for their bioregion. These Benchmarks, updated every two years, provide the basis for ongoing ecological restoration, climate change adaptation strategies, and ecological maintenance. Rather than backward-looking 'State of the Environment' Reports, the Benchmarks collate current information about the ecological health of each bioregion, compare it against pre-colonial ecological damage, and set out the ongoing restoration and maintenance required to properly Care for Country. These Benchmarks form a critical foundation for many aspects of bioregional governance. Relevant to this judgment, these Benchmarks are the basis upon which all decisions are made about new human commercial and other activities. The key question is: does the new proposed activity support the living world, based on what is needed in the current Benchmark, or does it take away from the health of the living world? If the latter applies, the activity is not permitted.

Bioregional governance includes a tripartite judicial system. In addition to Bioregional and National Court processes for civil and criminal law (between humans), Bioregional Tribunals are created for each bioregion, to adjudicate law between Country and people. These Tribunals exist permanently but are comprised using random selection (sortition) to bring different community

20 The Interim Biogeographical Regionalisation of Australia (IBRA) classifies Australia's landscapes into 89 large geographically distinct bioregions based on common climate, geology, landform, native vegetation, and species information. See 'Australia's bioregions (IBRA)', *Australian Government* (Web Page, 10 October 2021) <https://www.dcceew.gov.au/environment/land/nrs/science/ibra>.

members into the Tribunal for set periods of time. They are made up of 15 people – Indigenous and non-Indigenous people, men, women, and non-binary people – who are selected through sortition processes. These Tribunals are responsible for all matters regarding Caring for Country; they have the authority to hear submissions for new economic activities within their jurisdiction and cases of conflict regarding ecological health matters.

Indigenous leadership, sovereignty, knowledge, wisdom and governance systems are prioritised in this legal system. Indigenous and non-Indigenous people are in deep partnership together, in order to support all life on the continent and build a stable, successful culture.

Judgment

The Permanent Peoples' Tribunal of the Darling Riverine Plains Bioregion
Tribunal: Bioregional Peoples' Tribunal, Darling Riverine Plains Bioregion

Date: 10 May 2060

Order: The Peoples of the Darling Riverine Plains Bioregion (DRPB) have reviewed and assessed Commercial Proposal No 2060/2439 from Cotton Growers United Pty Ltd (CGU) and the application is denied.

Members of the Permanent Peoples' Tribunal (PPT) of the Darling Riverine Plains Bioregion (DRPB), comprising Indigenous and non-Indigenous citizens who reside in the bioregion, have reviewed Commercial Proposal 2060/2439 (the Proposal) against the 2058 Benchmark for Ecological Health for the DRPB, consulted with Indigenous Elders, scientific experts, economic experts and the wider community about the Proposal, and after weighing up the details of the Proposal against the principles of the Relationist Ethos and Laws of Obligation, which form the foundation of our Bioregional Laws, have unanimously decided that the proposed development would not be beneficial for Country or our Peoples. The application has been denied. Reasons are provided in the judgment below.

Summary of the Commercial Proposal

CGU submitted a proposal to buy land and extract water from the tributaries leading into the Barka River in order to grow cotton. CGU claims that the production of cotton must resume to meet the needs of Australian people to have access to local fibre, for clothing and other uses. In accordance with the *Bioregional Council Act 2044* (Federal), CGU submitted its Commercial Application including a Statement of Benefit and Reciprocity with Country, which was assessed by the Bioregional Council's Indigenous Elders, Scientific Panel and Economic Panel, and their reports were provided to the PPT. All the details are set out in Appendices I to IV.[21]

21 These have not been included, due to the space constraints.

Relevant Laws and Planning Instruments

Legislation and planning documents

1 *Constitution of Australia 2043*.
2 *Bioregional Council Act 2044* (Federal).
3 *Bioregional Judiciary Act 2044* (Federal).
4 *Bioregional Ecological Health Benchmark Act 2046* (Federal).
5 Darling Riverine Plains Bioregional Ecological Health Benchmark 2058 (This Bioregional benchmark is created in accordance with section 21 of the *Bioregional Council Act* 2044 and because the land is the source of the law, the benchmark has the status of a Bioregional Law).

Bioregional Tribunal Cases

1 *Sustainable Springwater Extractivists Pty Ltd Commercial Application to Blue Mountains Bioregional Tribunal 2059* (denied).
2 *Cotton Growers United Pty Ltd Application to Brigalow Belt Bioregional Tribunal 2057* (denied).
3 *Rice Growers Pty Ltd Application to Margaret River Bioregional Tribunal 2059* (denied).
4 *Hemp Australia Application to Brigalow Belt North Bioregional Tribunal 2058* (approved).

Application of the Relationist Ethos and Bioregional Laws to the Details of the Commercial Proposal

We the People of the PPT DRPB have worked together to fulfil our legal and moral obligations in accordance with the Law of the Land and the Bioregional Laws of Australia, here in our Bioregion.

We have considered the core principles embedded in the Relationist Ethos and our Bioregional Laws, so that we can then ask if the current commercial proposal meets the Law of Obligation in our Place. We have asked the key questions we always ask when considering human activities in our Place: How do we best Care for Country? How do we best Care for Each Other? Does this proposal help us Care for Country and for Each Other?

Matters to Consider

Information Provided by Proponent

The Proposal has applied to buy 10,000 hectares of land at the top of the bioregion, in order to grow cotton that would be irrigated from creeks and tributaries that flow into the Barka/Darling River. The proponents have claimed that with the steadily decreasing yield of food and fibre across Australia, it is necessary to resume growing cotton to meet the needs of our

Peoples. The proposed cotton production will need approximately 7.8 megalitres of water per hectare (78,000 megalitres of water per year) to produce its annual cotton yield. The proponents have claimed that this is a small enterprise compared to the scale of the cotton industry that was nationalised and closed down in 2044 due to its cumulative destructive impacts on the environment. The proponents noted that in 2017, cotton farms in the Murray Darling basin used 2.505 trillion litres of water on 320,000 hectares of land[22] and they noted that one cotton farm, Cubby Station, was 80,000 hectares. In contrast, their proposal is of a much smaller scale and could be sustainable in the long term. Details provided by the proponents are in Appendix I.

The Constitution 2043

Sections 11 and 12 of the *Constitution* state that: 'The land is the source of the law', and '[a]ll human activities must be regenerative and to the benefit of all life'. Section 13 of the *Constitution* states that: 'The Relationist Ethos is the guide for our society. It is the foundation and the template for how people Care for Country, and how people Care for Each Other'.

Our first task was to reflect on the unique ecosystems within our Bioregion, take the time to commune with the land, visit the places that the proponents want to develop, and listen to what Country and People need to be healthy.

The Bioregion – Key Features and History of Human Use

The Ecological Health Benchmark 2058 describes the Darling Plains Bioregion, including the land and tributaries that the Proposal refers to. The full details of the Tribunal's trip to the relevant sub-region, including observations and discussions with local communities, are in Appendix II. For the purposes of this judgment, we note the following:

(i) The DRP Bioregion is a long, narrow riverine corridor that runs southwest along the Barka/Darling River. It occupies a total area of 10,651,748 hectares and includes the lower reaches and alluvial fans of the Bogan, Macquarie, Castlereagh, Namoi, Barwon, Culgoa, Bokhara, Narran, Gwydir and Macintyre rivers;[23]

22 Peter Hannam, 'Murray-Darling Water Use Increased Even as Basin Dried Out, ABS Says', *Sydney Morning Herald* (online, 30 April 2019) <https://www.smh.com.au/environment/sustainability/murray-darling-water-use-increased-even-as-basin-dried-out-abs-says-20190430-p51iln.html>.

23 'Darling Riverine Plains Bioregion', *NSW Department of Planning and Environment* (Web Page, 5 May 2023) <https://www.environment.nsw.gov.au/topics/animals-and-plants/biodiversity/bioregions/bioregions-of-nsw/darling-riverine-plains>.

(ii) Our Bioregion is a sacred place and the Barka/Darling River, its streams, wetlands, and ground water are precious culturally, spiritually, and for all life in our bioregion;
(iii) Our bioregion is an important inland river system where the streams flow into an arid region, not out to the ocean. It is important for the life of neighbouring bioregions and neighbouring Peoples;
(iv) From the 1920s to 2030s the Bioregion was used unsustainably and was almost destroyed;
(v) With the introduction of the new *Constitution*, and the prioritisation of ecosystem/bioregional health, the new federal government launched the 2045 Inquiry into the Barka/Darling River. The Inquiry found that the existing management practices of the region (created and managed by the former State governments of Queensland and NSW) were in violation of the Relationist Ethos. Destructive levels of water extraction, pesticide and chemical use, the heinous practice of flood-plain harvesting, and much more nearly killed the entire river system and interconnected ecosystems; and
(vi) In 2046 the federal government nationalised the remaining cotton farms and other large-scale agricultural practices that were taking much needed water from the Bioregion and closed them down. A major restoration programme was launched, which is bringing the river system and its connected communities of life back to good health.

Does the Proposal Reflect the Key Principles of the Relationist Ethos?

Relationality literally means 'concerning the way in which two or more people or things are connected'.[24] In Australian Aboriginal societies, the primary relationship is between people and land (this conjunction is termed 'Country' in Aboriginal English). Other relations, including those with more-than-human relatives, are always contingent/built upon the relationship between people and land.[25]

A foundational description of the Relationist Ethos was provided in the Preamble to the Constitution, and the Bioregional Laws. They are drawn from the scholarly works of Indigenous Elder and political scientist, Mary Graham:

> Aboriginal relationalism – traditionally the foundation of Aboriginal law – is an elaborate, complex and refined system of social, moral, spiritual

24 'Relationality', *Oxford Reference* (Web Page, 2023) <https://www.oxfordreference.com/display/10.1093/oi/authority.20110803100412539;jsessionid=0B64E78A8F9B25F45C0DBA97DB549E14>.

25 Mary Graham, 'Some Thoughts About the Philosophical Underpinnings of Aboriginal World Views', (2008) 45 *Australian Humanities Review* 181, 181–182

> and community obligations that provides an ordered universe for people. Within the context of this system: relationalism embraces uncertainty and imprecision; consents to being driven by feeling; accepts and makes room for conflict while regarding invasion and war-likeness as not only invalid but highly inefficient; resolves the contradiction between power and authority; provides coherence about the meaning of life, and finally; assumes that not only groups, but all people and more-than-human relatives are autonomous beings.[26]
>
> So, any notion of Aboriginal social and political order has to begin with the Aboriginal relationship to Land, which is primary: The Land is the source of the Law. Expanding one's sphere of influence by conquest of other lands does not confer security – just the opposite – it ensures insecurity in a number of ways, not least of which is grievance of long standing of the 'conquered' which may or may not express itself in myriad ways. Coercion in relations sooner or later rebounds – the act(s) of coercion is not forgotten. They are solidly embedded in the narratives of all those involved in conquest projects. The establishment and maintenance of relationalism rests with its attributes, which are – Autonomy; Balance; Place/Identity; Ethics/Custodial Ethic.[27]
>
> Aboriginal societies embedded ourselves within the patterns of the living world. Through thousands of generations, we developed the intimacy of place, managed resources across an ancient continent efficiently and managed our population sizes carefully. Like most human groups, and for other life forms throughout history, everything starts with some form of guarantee of security for the group and its members. For Aboriginal people security was to be found in the development of a system of co-existence with the more-than-human world, or the life force, in all its forms. This 'Relationist Ethos' formed the deep foundations of Aboriginal societies and created a remarkable governance system and culture.[28]

To consider the commercial Proposal against the Relationist Ethos, we are required to consider the key attributes mentioned above: Autonomy; Balance; Place/Identity; Ethics/Custodial Ethic.

Opinions of the Elders

The Elders were unanimous in their rejection of the Proposal. They said letting cotton be grown again at the top of the Barka River would be in violation of their role as custodians of the lands and waters. Using the Barka River

26 Mary Graham, 'Aboriginal Notions of Relationality and Positionalism: A Reply to Weber' (2014) 4(1) *Global Discourse* 17, 18

27 Ibid.

28 Ibid.

system to irrigate crops like cotton would bring the land and its waters out of balance again. It would cause harm to the restored vegetation and biodiversity in those places and destroy the soils. The Elders remembered the crimes committed by cotton companies and the State Governments at the top of, and along, the Barka River system in the first half of the century. They provided their advice about what happens to Country when cotton is grown. They shared their stories and feelings about the terrible state of the Barka River by 2040, and how they did not want all that had been done since then to restore and bring the river back to life, to be lost. Their rejection of the Proposal is sufficient for this Tribunal to also reject the Proposal. We also provide further reasoning below.

The Relationist Ethos – The Proposal and its Impacts on Country

The Scientific Panel said that the volume of water that is proposed to be taken would have detrimental effects on all downstream biodiversity. Plants, animals, fish, insects, and all life would suffer from the extraction of this volume of water on an annual basis, especially as climate change is drying out the ecosystems and is predicted to continue to change the ecosystems over time. If we see ourselves in relationship with, and having custodial responsibilities for, the nonhuman life in our Bioregion, we must not approve the commercial application for growing cotton and extracting water.

Relevant Economic Issues

The proponents have stated that cotton production should be accepted, as declining yields around Australia mean we have to increase fibre production. However, our Economic Expert Panel has advised the Tribunal that, while climate change is affecting all agricultural yields around Australia, the introduction of effective, small-scale hemp production around multiple bioregions of Australia since 2045 has meant that today our local and bioregional fibre requirements are being met. Hemp production needs less than a third of the water needed for cotton and yields 220% more fibre. The plant grows without the need for harmful pesticides and herbicides and replenishes soil quality[29] and when grown sensitively to local places, and in small-scale farms, it can be grown in most bioregions across Australia. The Economic Panel noted the successful, federally funded, hemp incentive programmes, and how these programmes work in partnership with federal laws and regulations passed in 2047 to promote slow fashion and reduce overall demand for fibre. The members of the Economic Panel were of the opinion that Australia does

29 Olivia Elliott, 'From Eco Benefits to Legal Status: Everything You Need to Know About Wearing Hemp', *The Guardian* (online, 2 October 2019) <https://www.theguardian.com/fashion/2019/oct/02/from-eco-benefits-to-legal-status-everything-you-need-to-know-about-wearing-hemp>.

not need to return to growing crops like cotton, which use unsustainable amounts of water and chemicals.

Relevant Precedents

It should be noted that there are several recent Bioregional Tribunal judgments from other parts of Australia that are relevant to our deliberations. The key thread among all these judgments is that commercial proposals were rejected because they would take too much from the living world, they had no way of reciprocating or regenerating life, and the land and water would not be used for purposes essential to human life; the local Peoples considered them to be unnecessary economic activities.

In the *Cotton Growers United Pty Ltd Application Brigalow Belt Bioregional Tribunal 2058*, the Peoples' Tribunal in that community denied the application to grow cotton again in the southern parts of the Bioregion, due to the harm it would cause to water, soil, biodiversity, and people in that Bioregion – a region where the cotton industry was paid out and removed in 2046 due to its destruction of the environment.

In the *Sustainable Springwater Extractivists Pty Ltd Commercial Application to Blue Mountains Bioregional Tribunal 2059*, the Peoples' Tribunal in that community denied the application to extract freshwater as, when applying the principles of the Relationist Ethos, it would cause suffering to nonhuman and human life in their Bioregion. The sheer volume of water being sought was not sustainable.

In *Lithium Mining Pty Ltd Application to Margaret River Bioregional Tribunal 2059*, the Peoples Tribunal stated that they would never approve any commercial proposals that would extract minerals from the ground in their Bioregion. There were important cultural and environmental reasons provided, including the fact that such activities would harm Country and put the system out of balance.

In contrast, the decision in *Hemp Australia Application to Brigalow Belt North Bioregional Tribunal 2058* should be noted. In this judgment, the Bioregional Tribunal approved the commercial proposal, because the hemp farm was a small-scale operation that would not cause harm due to its careful design and engineering and its minimal use of water from the Artesian Basin.

Final Statement of the Tribunal

While in the past, top-down Australian laws allowed and regulated human activities that caused harm to the environment, and supported endless extraction from our land, waters, and biodiversity, our Bioregional Laws impose different obligations on us all. We are obliged to Care for Country and to proactively protect the ecological health and cultural values of all ecosystems, within the parameters of their unique needs and their specific boundaries.

The Benchmarks of Ecological Health created by the people, for their own Bioregions, are used to identify what activities will cause harm, and what activities will support life and provide reciprocal benefits. Only projects that are regenerative and provide reciprocal benefits to Country are to be considered in a favourable light. The current proposal does not meet any of our principles for Caring for Country and therefore must be denied.

Chapter 15

A Bleak Future Beckons Climate Refugees

Ayesha Riaz

Commentary

The Ongoing Debate Concerning 'Climate Refugees'

The facts that give rise to this future judgment are based on real-life situations that may become more prevalent over the years. Maryam is a climate refugee from a flooded region in Pakistan and an appellant in the case. Many asylum seekers currently take the route that she did. Although the analysis is far from complete, and many hazards are referenced only marginally, the existence of such claims suggests that a very careful examination of the application of the *Refugee Convention*[1] and its *Protocol on the Status of Refugees*[2] in such contexts is required. People who face adversity in the context of climate disasters are seen as unfortunate victims of 'natural' forces and deserve humanitarian assistance.

The main question for individuals who are placed in a similar predicament as Maryam is whether such individuals who are fleeing climate emergencies could be considered 'climate refugees', and whether the principle of non-refoulment under refugee and human rights law extends to such individuals.[3]

According to the Internal Displacement Monitoring Centre, an increasing number of individuals have been displaced within their own countries since 2017 as a consequence of natural disasters, rather than by conflict – 61% compared to 39%.[4] The impacts of climate change like rising sea levels, erosion, and desertification have meant that individuals have either been

1 *Convention Relating to the Status of Refugees*, opened for signature 28 July 1951, 189 UNTS 137 (entered into force 22 April 1954) ('Refugee Convention').

2 *Protocol relating to the Status of Refugees*, opened for signature 31 January 1967, 606 UNTS 267 (entered into force 4 October 1967).

3 Jane McAdam, 'Protecting People Displaced by the Impacts of Climate Change: The UN Human Rights Committee and the Principle of Non-Refoulment' (2020) 114 (4) *The American Society of International Law* 708, 708.

4 Internal Displacement Monitoring Centre, *Global Report on Internal Displacement* (Report, May 2019) 6.

DOI: 10.4324/9781003389569-18

displaced or have migrated, as they have been unable to reside within a safe environment.[5] Furthermore, the United Nations' expert scientific climate change body, the Intergovernmental Panel on Climate Change, has stated that sea levels will continue to rise well into the future.[6] It has also been noted that financial assistance should be made available if climate goals are to be achieved, and this applies particularly to vulnerable regions.[7] It has been pointed out that the largest climate financial gaps and opportunities reside within developing nations.[8] People like Maryam will suffer considerably due to the impact of climate-induced emergencies.

The issue of 'climate refugees' requires serious attention because, at the time of writing in late 2022/early 2023, there is a gap in the international legal framework that addresses the movement of people who cross borders as a result of climate change.[9] The *Sydney Declaration of Principles on the Protection of Persons Displaced in the Context of Sea Level Rise* was adopted by the International Law Association in 2018. Accordingly, the underlying disaster or climate change process does not in itself amount to 'persecution', but the Declaration contextualises forms of harm that engage existing international protection regimes where the disaster causes a breakdown of law and order or is used by a government as a pretext for persecutory acts against certain parts of the population.[10] This document considers the issues of displacement owing to the rise in sea levels.

In 2012, the Nansen Initiative on Disaster-Induced Cross-Border Displacement was launched by the governments of Switzerland and Norway, which led to the enactment of the 2015 *Agenda for the Protection of Cross-Border Displaced Persons in the Context of Disasters and Climate Change*. This has been endorsed by 109 states.[11] The Protection Agenda focuses on issues such as displacement, migration and planned relocation, and identifies effective practices and policies that states could incorporate into their own laws by integrating mobility into disaster risk reduction and climate change adaptation plans.[12]

5 Ibid 18.
6 Hoesung Lee et al, *Synthesis Report of the IPCC Sixth Assessment Report (AR6)* (Intergovernmental Panel on Climate Change, Report, March 2023) <https://report.ipcc.ch/ar6syr/pdf/IPCC_AR6_SYR_LongerReport.pdf> [3.1], [3.1.3].
7 Ibid [4.8.1].
8 Ibid.
9 McAdam (n 3) 712.
10 Committee on International Law and Sea Level Rise, *Sydney Declaration of Principles on the Protection of Persons Displaced in the Context of Sea Level Rise* (Resolution 06/2018, 3 October 2018).
11 McAdam (n 3) 721.
12 Ibid.

As identified by McAdam, a number of further measures on disasters, climate change and human mobility have been secured in a series of international instruments across a range of policy areas.[13] These include the *Sendai Framework for Disaster Risk Reduction 2015–2030*,[14] the *2030 Agenda for Sustainable Development*,[15] the 2015 *Paris Agreement* on climate change[16] (including the establishment of the Task Force on Displacement within the Warsaw International Mechanism for Loss and Damage), the *Agenda for Humanity* (annexed to the United Nations Secretary-General's report for the 2016 World Humanitarian Summit),[17] the 2016 *New York Declaration for Refugees and Migrants*,[18] the 2018 *Global Compact on Refugees*,[19] and the very significant 2018 *Global Compact for Safe, Orderly and Regular Migration*.[20] It is significant as it represents the most detailed political commitment pertaining to climate change mobility.[21] The Compact seeks, firstly, to establish a more improved mechanism of information-sharing between different governments, while upholding the human rights of such individuals; secondly, it seeks to develop strategies that may include migration while factoring in the potential for displacement when devising disaster preparedness strategies; thirdly, it seeks to ensure that people who suffer from disasters, climate change and environmental degradation are able to access humanitarian assistance; and finally, it seeks to develop coherent approaches to address the challenges of migration movements in this context.[22] It also creates possibilities to establish special visas, private sponsorship, and access to education for children, as well as temporary work permits that could be granted to individuals who suffer from the consequences of climate emergencies.[23]

13 Ibid.
14 *Sendai Framework for Disaster Risk Reduction 2015–2030*, UN Doc A/RES/69/283 (23 June 2015).
15 *2030 Agenda for Sustainable Development*, UN Doc A/Res/70/1 (21 October 2015).
16 *Paris Agreement*, opened for signature 22 April 2016, [2016] ATS 24 (entered into force 4 November 2016).
17 United Nations Secretary-General, *One Humanity: Shared Responsibility*, UN Doc A/70/709 (2 February 2016).
18 *New York Declaration for Refugees and Migrants*, UN Doc A/Res/71/1 (3 October 2016).
19 United Nations High Commissioner for Refugees, *Global Compact on Refugees*, UN Doc A/73/12 (Part II) (2 August 2018).
20 *Global Compact for Safe, Orderly and Regular Migration*, UN Doc A/RES/73/195 (19 December 2018), obj 2 [18(h)]–[18(l)]; obj 5 [21(g)]–[21(h)] ('Global Compact').
21 McAdam (n 3) 722
22 Global Compact (n 20) obj 2 [18(h)]–[18(l)].
23 Ibid obj 5 [21(g)].

Does Human Rights Law Offer More Protection for Climate Refugees?

By comparison to the *Refugee Convention*, there is a lower threshold to be met under international human rights law given that states are prevented from moving individuals to places where they would face a real risk of being arbitrarily deprived of their life, or subjected to torture or other cruel, inhuman or degrading treatment or punishment.[24]

According to the European Court of Human Rights, the removal of individuals to situations of very serious destitution or dire humanitarian conditions may constitute inhuman and degrading treatment.[25] The threshold is very high. If Maryam argues that there is a lack of drinkable water, then she would have to demonstrate that the supply of inaccessible, insufficient or unsafe water poses a threat of a health risk that would impair her right to enjoy life with dignity or cause her unnatural or premature death.[26] In relation to the difficulties her father encountered while growing crops, she would have to demonstrate that there was 'a real and reasonably foreseeable risk' that the family would be exposed to a situation of indigence, deprivation of food, and extreme precarity that would threaten their right to life including the right to live with dignity.[27]

Can the *Refugee Convention* Bridge This Gap?

The *Refugee Convention* is widely considered to be 'by and large, an inappropriate normative framework for responding to the needs of those forced to move on account of environmental or climate change impacts'.[28] However, innovative work is being carried out under the auspices of the Nansen Initiative, which through a series of regional consultations developed a Protection Agenda[29] that reflects the best practice in responding to the phenomenon of cross-border displacement in the context of disasters and climate change. As this reflects best practice, countries are encouraged to consult it before devising appropriate policies.

24 *Convention for the Protection of Human Rights and Fundamental Freedoms*, opened for signature 1 November 1950, 213 UTS 221 (entered into force 3 September 1953) Arts 2, 3

25 *MSS v Belgium and Greece* (2011) I Eur Court HR 255; *Sufi and Elmi v United Kingdom* (2012) 54 EHRR 9.

26 Views adopted by the Human Rights Committee under article 5 (4) of the Optional Protocol, concerning communication No 2728/2016, UN Doc CCPR/C/127/D/2728/2016, 24 October 2019 ('*Ioane Teitiota v New Zealand*') [9.8].

27 Ibid [9.9].

28 Jane McAdam, 'The Relevance of International Refugee Law' in Jane McAdam (ed), *Climate Change, Forced Migration, and International Law* (Oxford University Press, 2012) 39, 39.

29 The Nansen Initiative, '*Agenda for the Protection of Cross-Border Displaced Persons in the Context of Disasters and Climate Change*' (Agenda, vol II, December 2015).

Although most people will not be able to establish eligibility for refugee status (or even complementary protection) in the context of disasters and climate change, it is crucial that such claims are approached from an informed perspective to assess the causes and differential impact of disasters.[30] Cross-border disaster-related displacement is likely to increase in coming years, and thus a theoretically grounded approach to determining refugee status in this connection is an important complement to wider initiatives. Thus, refugee law should note that the drivers of displacement are caused by multifaceted issues which can encompass natural disasters, conflicts and persecution.[31]

Do Recent Cases from Other Jurisdictions Give Hope to Maryam?

Following a number of seminal cases in recent years, the United Nations Human Rights Committee has accepted, in principle, that it would be unlawful for states to send individuals to places where the impacts of climate change could expose them to life-threatening risks or risks of cruel, inhuman or degrading treatment.[32] Maryam's situation is similar to such cases.

With regard to the question of future risk, the New Zealand Immigration and Protection Tribunal stated in the case of *AF (Kiribati)* that the risk to the appellant and his family from sea level rise and other climate disasters may be considered as being 'imminent'; however, it fell short of the threshold required to establish substantial grounds for believing that they would be in danger of arbitrary deprivation of life.[33] It did not specifically discuss the 'imminence' requirement, but there was no suggestion of a change in that position. The Tribunal built on its observations in *BG (Fiji)*[34] in relation to the risk assessment, as it considered that past persecution could act as a powerful indicator of the risk of future persecution as well as the failure to continuously protect individuals against known environmental hazards. However, due to the forward-looking nature of this inquiry, the nature of the hazard, including its intensity and frequency, as well as any positive changes in disaster risk reduction and operational responses in the country of origin, or improvements in its adaptive capacity, will need to be accounted for.[35]

30 Sanjula Weerasinghe, *In Harm's Way: International Protection in the Context of Nexus Dynamics Between Conflict or Violence and Disaster or Climate Change*, UN Doc PPLA/2018/05 (December 2018) 12.

31 Ibid 107.

32 Views adopted by the Human Rights Committee under article 5 (4) of the Optional Protocol, concerning communication No 2728/2016, UN Doc CCPR/C/127/D/2728/2016, 24 October 2019 ('*Ioane Teitiota v New Zealand*') [9.8].

33 *AF (Kiribati)* [2013] NZIPT 800413 [91].

34 *BG Fiji* [2012] NZIPT 800091 [98]–[102].

35 *AC (Tuvalu)* [2014] NZIPT 501370-371 [29]–[30].

The relevance of this statement of principle for the determination of refugee status is clear. The well-established principle that past persecution creates a presumption of future risk applies equally in the context of disasters as it does in other contexts where the fear of being persecuted is articulated. For Hathaway and Foster, the well-founded fear element requires decision-makers to determine 'whether the evidence discloses a future risk that is "real" or "substantial", rather than merely "fanciful" or "remote"'.[36]

The requirement for the harm to be 'imminent' makes little sense especially as scientific modelling indicates the very high likelihood of future climate-induced disasters and emergencies. Imminence has never been the appropriate test for judging the lawfulness of removal; rather, the substantive question concerns the likelihood of harm that results from such a removal, which is different from how soon, after removal, it may manifest.[37]

It is noteworthy that a comment in an Australian High Court's judgment[38] mirrors the position of the United Kingdom's Supreme Court,[39] and the Canadian Supreme Court,[40] among others,[41] in that those fleeing 'natural disasters' cannot be classified as being 'refugees'. Lord Hope in *Horvath*[42] opined that the Refugee Convention does not extend to those who have a well-founded fear of threats to life due to a famine or a civil war.[43] As stated by Scott, this could constitute a discriminatory approach towards prospective 'climate refugees'.[44]

Climate Modelling: The Future of Pakistan

The raging floods that took place in many of the rural parts of Pakistan in August 2022 left one-third of the country submerged under water and exposed its people to various kinds of diseases.[45] For the past 20 years, Pakistan has been ranked among the top ten most vulnerable countries on the Climate Risk Index, as 10,000 fatalities have been reported and losses

36 James C Hathaway and Michelle Foster, *The Law of Refugee Status* (Cambridge University Press, 2nd ed, 2014) 123
37 McAdam (n 3) 720.
38 *Applicant A v Minister for Immigration and Ethnic Affairs* (1997) 142 ALR 331.
39 *Horvath v Secretary of State for the Home Department* [2000] UKHL 37 ('*Horvath*').
40 *Canada (Attorney General) v Ward* [1993] 2 SCR 689.
41 See, eg, German Federal Administrative Court, 10 C 33.07, 7 February 2008 reported in (2008) BVerwG; *L(SJ) v Refugee Appeals Tribunal & Ors* [2014] IEHC 608 [43].
42 *Horvath* (n 39) 5.
43 Ibid 7.
44 Matthew Scott, 'Finding Agency in Diversity: Applying the Refugee Convention in the Context of Disasters and Climate Change' (2016) 35(4) *Refugee Survey Quarterly* 26, 31.
45 Mehrunnisa Wani, 'A Climate Emergency in Pakistan, and the Way Forward', *Forbes* (online, 30 October 2022) <https://www.forbes.com/sites/mwani/2022/10/30/a-climate-emergency-in-pakistan-and-the-way-forward>.

equating to $4 billion resulted from 173 extreme weather events.[46] These challenges will result in climate-related conflicts over resources and, as such, there will be scarcity of water.[47] It is also anticipated that Pakistan will face further flooding, heatwaves or even tsunamis which could result in climate-induced migration.[48]

The United Nations Secretary-General, Antonio-Guterres, has reportedly stated that Pakistan was responsible for less than 1% of global greenhouse gas emissions and is 'paying a supersized price for man-made climate change'.[49] He described the flood waters covering a landmass three times the total of his own country, Portugal and explained that many individuals had lost their lives, homes, livestock, and their futures.[50] He also reminded the G20 nations that industrialised nations accounted for 80% of climate-destroying emissions, and considered it a 'moral responsibility' to assist Pakistan to recover, adapt and build resilience to disasters supercharged by the climate crisis.[51] While Pakistan had been deluged, many other nations faced the threat of extinction.[52]

Pakistan's Supreme Court has taken notice of Pakistan's vulnerability to climate-induced disasters and has advised Pakistan that the development plans for its urban areas should support adaptation, climate resiliency, and sustainability.[53] According to Pakistan's Supreme Court, the country's infrastructure should be able to withstand the effects of climate change that may result from heavy rainfall, floods, earthquakes, and other extreme weather.[54] Failure to do so may impact on the quality of a person's life and offend his/her

46 David Eckstein, Vera Kunzel, and Laura Schafer, *Global Climate Risk Index 2021: Who Suffers Most from Extreme Weather Events? Weather-Related Loss Events in 2019 and 2000–2019* (Briefing Paper, January 2021) 13

47 Jumaina Siddiqui, 'Pakistan's Climate Challenges Pose a National Security Emergency', *United State Institute of Peace* (Media Release, 7 July 2022) <https://www.usip.org/publications/2022/07/pakistans-climate-challenges-pose-national-security-emergency>.

48 Quoted in ibid.

49 Quoted in 'Pakistan's "Climate Carnage Beyond Imagination", UN Chief Tells General Assembly', *United Nations News* (Blog Post, 7 October 2022) <https://news.un.org/en/story/2022/10/1129337>.

50 Antonia Guterres, 'Secretary-General's Remarks to the General Assembly on Pakistan Floods [as Delivered]', *United Nations Secretary-General* (Statement, 7 October 2022) <https://www.un.org/sg/en/content/sg/statement/2022-10-07/secretary-generals-remarks-the-general-assembly-pakistan-floods-delivered>.

51 United Nations News (n 49).

52 Ibid.

53 Hasnaat Malik, 'Consider Climate Change Angle in Uplift Plans, Rules SC', *The Express Tribune* (online, 6 July 2022) <https://tribune.com.pk/story/2364905/consider-climate-change-angle-in-uplift-plans-rules-sc>.

54 Ibid.

dignity and property which are guarantees enshrined within the Constitution of the Islamic Republic of Pakistan.[55] Accordingly,

> shutting eyes to the ominous signs of climate change will plunge us into a world that may not be able to sustain fundamental human values. It could affect our basic physical and larger social needs including harnessing and consuming energy, water and food as well as habitation, travelling and communication potentialities.[56]

Pakistan has played a pivotal role on the international stage as it participated in COP27 and, as one of the world's major methane emitters, signed the global methane pledge.[57] However, Pakistan is a developing nation and may need financial assistance to meet these pledges.

Speculative Fiction and a Bleak Future for Refugees

While it is possible that refugee law may accommodate climate refugees in the future, at least one work of speculative fiction anticipates an even harsher position on the part of the United Kingdom government. John Lanchester's novel *The Wall*,[58] set in an indeterminate future, considers debates concerning mass migration, environmental change, and the role of nation-state vis-à-vis contemporary global crises.[59] *The Wall* is set in a time when Britain's sea borders have been fortified with a solid physical barrier: the Wall.[60] The primary function of the Wall and its 'defenders' is to keep non-citizens, referred to as 'others', from entering the country.[61]

Lanchester describes a fictionalised version of Britain that radically enforces one of the key functions of borders: they 'secure their territories, which are the repository of their human and natural resources'.[62] As in other examples of border fiction, *The Wall* exemplifies how 'border practices and border theories are fast revealing borders as zones of instability in which

55 Ibid. It was observed by the Court that urban authorities should ensure that they are able to support adaptation, climate resiliency, and sustainability as part of their development plans.
56 Ibid.
57 Rina Saeed Khan, 'Pakistan Signs US Led Global Methane Pledge at COP26', *The Express Tribune* (3 November 2021) <https://tribune.com.pk/story/2327529/pakistan-signs-us-led-global-methane-pledge-at-cop26>.
58 John Lanchester, *The Wall* (WW Norton & Company, 1st ed, 2019)
59 Kirsten Sandrock, 'Border Temporalities, Climate Mobility, and Shakespeare in John Lanchester's *The Wall*' (2020) 43 (3) *Journal of Modern Literature* 163, 164. Also see: Amitav Ghosh, *The Great Derangement: Climate Change and the Unthinkable* (Chicago University Press, 1st ed, 2016).
60 Sandrock (n 59).
61 Ibid.
62 Lanchester (n 58) 1-12

ethical, political, cultural and national questions are negotiated'.[63] Among other issues, it focuses on the instabilities that are caused by climate change when environmental change has caused sea levels to rise and beaches to disappear, entire countries have become uninhabitable, and large parts of the world's population are on the move.[64]

Lanchester examines current debates on environmental change through various trajectories and one of his main messages is that environmental changes have led and will continue to lead to increasing levels of global migrancy.[65] The novel takes readers to the heart of contemporary climate prognosis: environmental change will affect all parts of the Earth as countries will be directly affected by 'increasingly extreme and erratic weather' and 'rising ocean levels' that will make 'low-lying coastal zones' uninhabitable.[66]

The Britishers also fear masses of individuals fleeing from the global South and the Mediterranean, and the numbers of asylum seekers are described as being 'big' and 'dangerous'.[67] In one scene, a character who assists an 'outsider' to get over the Wall is considered a traitor by the state.[68] This shows that an increase in migrant numbers has resulted in nations using fear-mongering ideologies. Some argue that offshoring border practices will become ever more prevalent and United Kingdom border guards will be stationed in Bologne, Brussels, Calais, Coquelles, Dunkerque, Fretun, Lille and Paris to guard national border zones.[69] However, Scotland will open its borders to migrants.[70]

Lanchester's Britain is a safe haven for the privileged, while Western societies are responsible for aggravating the climate crisis.[71] Though the Wall is designed to keep out the high water, its real purpose is to prevent the influx of climate refugees.[72] Those fleeing floods, like Maryam, have been described as 'people who want you to drown'.[73] In the dystopian world of the novel, British people are the small, privileged fraction of humanity, who even after a massive climate crisis enjoy access to healthcare, university education, and

63 Johan Schimanski and Stephen F Wolfe, *Borders Aesthetics. Concepts and Intersections* (Berghahn Books, 1st ed, 2019) 9.
64 Sandrock (n 59) 165.
65 Ibid.
66 Lanchester (n 58) 56.
67 Ibid 111.
68 Ibid 191–2.
69 Cathal McCall, 'Debordering and Rebordering in the United Kingdom. A Companion to Border Studies' in Thomas Wilson and Hastings Donnon (eds), *A Companion to Border Studies* (Wiley Blackwell, 1st ed, 2012) 214, 215–216.
70 Sandrock (n 59) 173.
71 Ewa Rychter, 'Testing the Limits: Boundaries and Fault Lines of Dystopia in John Lanchester's The Wall (2019)' (2022) 12 *Polylogue Neophiological Studies* 291, 292–293.
72 Lanchester (n 58) 26
73 Lanchester (n 58) 112.

public transport, can grow their own food/crops, go on holiday, and have access to clean water.[74] Nevertheless, Britain continues to show hostility towards those who are 'uninvited'.[75]

Thus, in this futuristic scenario, the current immigration environment becomes even more hostile towards migrants, which may explain why Maryam's case is refused, even though it falls within the newly expanded *Refugee Convention* and the judge finds her to be credible.

Judgment

High Court of Justice in London

Delivered on: 25 October 2040
Judge: Colin LJ

The facts of this case, as set out in the applicant's statement through her interpreter (see Appendix A) are, in summary, as follows. The case concerns Maryam X, who has travelled across the continent and much of Asia to reach our shores. She is claiming asylum as she identifies herself as a 'climate refugee'. Her application for asylum was rejected by the Immigration and Asylum Tribunal on 10 October 2040. Maryam's lawyers have appealed against the decision, and this matter has now come before me on appeal.

Maryam states that her hometown has been submerged in water and that her father has lost his job as a farmer due to the floods that took place a few years ago. It is my understanding that her hometown encountered further flooding earlier this year. Her family is being supported by aid agencies and she explained that they are due to leave that region by next month. Maryam is, as one would be in her situation, extremely worried about her family and she fears for their future.

A number of issues have been raised as a result of this case. Firstly, I need to identify whether there is a safe area within Pakistan that Maryam can relocate to. Secondly, I need to consider how the *Convention Relating to the Status of Refugees 1951*[76] (hereinafter the *Refugee Convention* or *Convention*) applies to her situation, and whether she can be classed as being a 'refugee' under this Convention. I will also consider how the *Bill of Rights 2034*[77] (hereinafter the *Bill of Rights*) applies to her case.

I shall first deal with the *Refugee Convention*, to which the United Kingdom is a signatory. The *Convention* did not initially include 'climate emergencies'

74 Rychter (n 71) 293.
75 Lanchester (n 58) 112.
76 Refugee Convention (n 1).
77 *Bill of Rights 2034* (UK).

as a reason for being recognised as a refugee. According to the Convention, a refugee was someone who had a profound fear of being persecuted due to their race, religion, nationality, particular social group and ethnicity.[78] I am also acutely aware of the principle of non-refoulement, in Article 33, according to which a State cannot expel or return a person to a territory where their life or freedom would be threatened.

Several climate disasters have shaken much of the world recently. These include, although are not limited to, the 2034 earthquake that destroyed much of Japan, and resulted in 100,000 casualties; the severe floods that inundated New Zealand in 2033 as a result of which 200,000 people were left homeless and 50,000 lost their lives; the floods in Bangladesh that left 20,000 casualties in 2034; the ongoing floods that the rural parts of Pakistan have been experiencing since 2022; and, finally, the sandstorms that have been causing havoc in much of the Middle East.

A Committee was established in 2036 with the purpose of extending the application of the *Refugee Convention*. The Committee examined the argument that environmental degradation that resulted from climate change or other natural disasters should create a pathway into the *Refugee Convention*. These now constitute 'other reasons' under Article 99 of the *Convention*. These 'other reasons' include displacement that occurs as a result of food shortages, poverty, and climate emergencies.

Therefore, I concede that Maryam's case falls within the newly expanded *Refugee Convention* given that her hometown was submerged and flooded severely. She could, potentially, be classified as being a 'climate refugee'. We, as judges, must ascertain whether Britain can be precluded from removing asylum seekers to places where they face risks arising from the impacts of climate change. The crucial question is whether the principle of non-refoulment extends to individuals whose lives may be impacted by adverse effects of climate change or climate disasters under refugee and human rights law. I will return to this issue later.

I will now consider the new *Bill of Rights* that the United Kingdom passed in 2034.[79] The *Bill of Rights* mirrors much of the *European Convention on Human Rights* (hereinafter the ECHR).[80] Section 10 of the *Bill of Rights* mirrors Article 2 of the ECHR, which stipulates that:

1. Everyone's right to life shall be protected by law. No one shall be deprived of his life intentionally save in the execution of a sentence of a

78 Refugee Convention (n 1) art 1.
79 *Bill of Rights 2034* (UK).
80 *Convention for the Protection of Human Rights and Fundamental Freedoms*, opened for signature 4 November 1950, 213 UTS 221 (entered into force 3 September 1953).

court following his conviction of a crime for which the penalty is provided by law.

2. Deprivation of life shall not be regarded as inflicted in contravention of this article when it results from the use of force which is no more than absolutely necessary: (a) in defence of any person from unlawful violence; (b) in order to effect a lawful arrest or to prevent the escape of a person lawfully detained; (c) in action lawfully taken for the purpose of quelling a riot or insurrection. [81]

Similarly, Article 3 of the ECHR mirrors section 11 of the *Bill of Rights*. Article 3 of the ECHR stipulates that no one shall be subjected to torture or to inhuman or degrading treatment or punishment.[82]

The *Bill of Rights* has additional provisions which have been, admittedly, interpreted very narrowly in previous cases.[83] According to section 21 of the *Bill of Rights*, individuals who are fleeing climate emergencies have a right to seek humanitarian protection in a safe country of refuge.

I will now apply these provisions to Maryam's situation. I acknowledge that Maryam has lost her home due to the impact of climate change and that, due to climate impacts, half of her country is now underwater. However, Maryam has the option of relocating to a region in Pakistan that is not submerged. I also understand that Maryam has been displaced from her home. Displacement for the purpose of the *Refugee Convention*, as amended, can occur due to food insecurity, climate change and other reasons.

I have, however, reservations about Maryam's case. From the evidence that has been presented, there was nothing to suggest that Maryam did not have access to clean water. Aid agencies are distributing clean water there. I understand that a number of water wells have been built in Dadu and other areas that have been impacted by floods. The Pakistani government is also ensuring that clean water is made available to all its residents free of charge, by working with international organisations such as Clean Water Everywhere and Quench Thirst.

There is also little to suggest that the environmental conditions that she faced or would face on return would be so perilous that her life would be endangered. Although her father has lost his job as a farmer, he can relocate to one of the major cities within Pakistan such as Karachi, Lahore, Islamabad or Rawalpindi and find work as a gardener or farmer there, as he appears to

81 Ibid art 2.

82 Ibid art 3.

83 For example, in *R (Amina) v Secretary of State for the Environment* [2035] 1 AC 663, it was held that climate emergencies only include severe devastation caused by natural disasters. Similarly, in *Hassan v Secretary of State for Environment* [2036] 10 AC 891, the Court adopted a very restrictive approach in deciphering what amounts to 'severe devastation'.

be quite resourceful. Maryam can also relocate to one of the major cities and seek work there. She is young, fairly well educated, and well spoken. She has managed to cross several countries to reach the United Kingdom, so surely she can relocate to one of the major cities within Pakistan.

I must, in addition, point out that in order to attain victim status in such cases, the requirement of imminence primarily attaches to the decision to remove the individual. In that regard, Maryam's circumstances will be considered in light of today's conditions and cannot encompass a hypothetical future harm due to a lack of clean water and employment possibilities, and/or a hypothetical future threat of serious violence caused by environmental degradation. I understand from the report that was submitted by Dr Anges Oluwale that Pakistan is likely to face further severe droughts and flooding in the next 20 years due to the effects of climate change. However, as stipulated above, the law compels me to consider the present circumstances only. Pakistan is not fully flooded. As reiterated above, Maryam can take advantage of a number of options that are available to her.

I acknowledge that the Articles of the ECHR discussed earlier apply to Maryam's situation. Environmental degradation as a result of the effects of climate change signifies one of the most pressing and serious threats to the ability of present and future generations to enjoy their right to life. However, Maryam has failed to provide evidence to substantiate her claim that, upon her removal to Dadu, she faces actual or imminent harm at the present time or that she faces the risk of being arbitrarily deprived of her life. I, however, wish to reiterate the important principle that environmental degradation can compromise one's effective enjoyment of the right to life and that severe environmental degradation can adversely affect an individual's well-being and lead to a violation of the right to life.

In terms of Maryam's credibility, I find her to be entirely credible and I accept the evidence she has presented. However, the evidence provided did not establish that the risk is imminent or that there is a likely risk of arbitrary deprivation of life upon being returned to Dadu. In particular, the Tribunal found that there was no evidence that, firstly, her family was currently suffering, apart from the fact that her father was unable to grow his crops for the time being, and secondly, that Maryam had no access to the aid authorities. Furthermore, Maryam is an educated woman who would be able to provide for her family back home, by taking up employment opportunities in one of the major cities within Pakistan; and, finally, the Government of Pakistan has not failed to take steps to provide for the basic necessities of life, in order to meet its positive obligation to protect Maryam's right to life. I am therefore of the opinion that the Government of Pakistan, in collaboration with the various aid agencies that are present there, has indeed taken steps to address the effects of climate change. Accordingly, I uphold the Tribunal's decision that Maryam cannot be granted refugee status.

Appendix A

Statement of Maryam X

I am a 24-year-old Pakistani woman. I left Pakistan on 27 March 2039 and travelled through various countries to arrive in the United Kingdom in August 2040. I left Pakistan as my hometown, Dadu, was submerged in water. My father, Imtiaz, used to work as a farmer prior to the floods that occurred throughout December 2038. My mother is a housewife. I do not have any siblings and my grandparents had passed away when I was a child.

I was previously studying geography at the University of Sindh. I developed an interest in this field following the severe drought that occurred in regions within Pakistan in June 2035. My eventual plan was to become a lecturer to educate others on the effects of climate change.

I initially travelled to Tehran in a lorry and slept rough. Thereafter I walked and reached Iraq and stayed near the border with Syria in a city called Qaim and continued sleeping rough, either in the streets or in forests. After a few days, I reached a city named Al-Jalaa in Syria, which bordered Syria. I later reached a region named Gaziantep in Turkey, where a wonderful woman named Yilgrim befriended me. Yilgrim looked after me and provided me with food and shelter for three months. I felt rested and felt a sense of normalcy in my life during those three months.

I had to bid Yilgrim goodbye and continued my journey. I walked endlessly and slept in forests until I reached Istanbul around January 2040. I met a group of individuals who were boarding a dinghy and they allowed me to access the dinghy free of charge. Owing to the arctic winds, the dinghy kept tilting from one direction to another. After the most frightening ten hours of my life, I reached a place called Varna Bapha in Bulgaria, near the Romanian border. I continued walking and reached a place named Bacau in Romania after a few days. However, soon after that, in March 2040, I was arrested by four female police officers and placed in a detention centre. I remained there for a few months and was fed one piece of bread in the morning and soup for dinner. I was kept in a room with ten different women and we all shared one bathroom. The Romanian authorities decided to release me after a few months, in June 2040, as they were unable to remove me to Pakistan. At this point, I was frail and lacking in energy.

I met a woman named Rada who provided me with food and shelter for three weeks. After those weeks, I left Rada's house and continued walking through forests in pitch darkness until I reached a town named Karlsburgh, near the French border.

Eventually, around August 2040, I reached the Calais jungle and noticed that it was flooded with asylum seekers. There I met a wonderful doctor by the name of Kate. Kate volunteered at the jungle once a week and attended to the new arrivals. Kate conducted a medical assessment and diagnosed me as suffering from severe malnutrition and post-traumatic stress disorder. I was housed within a tent that I shared with another Pakistani woman, named Tara. After spending a few weeks there, I boarded a lorry successfully and reached the English Channel. I met a group of asylum seekers who allowed me to board a boat free of charge. The boat was very unstable and a boy sitting next to me drowned in the Channel. Everyone else, including myself, managed to reach Dover in England successfully.

I was arrested upon arrival and was placed at Dumbluring Detention Centre. I claimed asylum upon arrival. I waited for three weeks and was allocated a solicitor named Chinonso, who visited me a couple of times and was always accompanied by a female interpreter. During the initial appointment, Chinonso spent approximately three hours with me in which she obtained my instructions and went through the relevant law. A week later, she took my witness statement.

My asylum interview took place through an expedited procedure on 10 October 2040. The interview lasted for six hours. After two weeks my case was refused. I was and remain distraught.

Part IV

After the Anthropocene

Chapter 16

How Will 2050 Forms of Artificial Intelligence (AI) Judge the Anthropocene?

Tania Sourdin and ChatGPT

Commentary

Introduction

By the early 2020s, AI was increasingly being used to write and devise material that could be used by students, journalists, marketers and others (including academics). While initially AI writer outputs were fairly primitive and relied on narrow sources, the capacity for these systems to search for answers and express perspectives has developed rapidly with a range of concerns being expressed about such developments.[1] While AI systems still suffer from narrow and limited focus areas, even by 2023, basic forms of AI are able to construct material that reflects some form of judgement about the Anthropocene and can even speculate about what 2050 forms of AI might conclude about the Anthropocene.

Using existing forms of simplistic AI, as represented by AI writers, the author constructed a judgment that imagines how a future form of AI might consider the Anthropocene.[2] Without doubt, there are many shortcomings to

Chat GPT has been recognised as an author in this chapter as abstracts from ChatGPT have been included as marked. The recognition of ChatGPT as an author has some precedent: see Tiffany H Kung et al, 'Performance of ChatGPT on USMLE: Potential for AI-Assisted Medical Education Using Large Language Models' (2023) 2(2) *PLOS Digital Health* e0000198.

1 Such concerns have been reported widely in the media and include concerns about students and researchers 'cheating': see Kalley Huang, 'Alarmed by A.I. Chatbots, Universities Start Revamping How They Teach', *The New York Times* (online, 16 January 2023) <https://www.nytimes.com/2023/01/16/technology/chatgpt-artificial-intelligence-universities.html>. As well as concerns about whether material is accurately reflective of the subject matter concluding that at times, the analysis can be flawed, see Caitlin Cassidy, 'Lecturer Detects Bot Use in One-Fifth of Assessments as Concerns Mount Over AI in Exams', *The Guardian* (online, 17 January 2023) <https://www.theguardian.com/australia-news/2023/jan/17/lecturer-detects-bot-use-in-one-fifth-of-assessments-as-concerns-mount-over-ai-in-exams>.

2 Initially ChatGPT 3 was used to respond to a series of questions. ChatGPT3.5 was then employed when released (in December 2022). The systems are available at; Open AI, *ChatGPT: Optimizing Language Models for Dialogue* (Web Page, 30 November 2022) <https://openai.com/blog/chatgpt/> and it is expected that more advanced AI models will be released in the near future.

DOI: 10.4324/9781003389569-20

this approach. First, as noted above, the sources relied upon by an AI writer will tend to be populist and will not draw upon the vast primary sources available on topics, such as climate change. This is perhaps one of the most problematic issues when using AI systems to consider and 'judge' a topic. The scholarly literature in the climate change area is vast and it is readily apparent that little of it has been relied upon by the AI writer used to construct the judgment below. While some AI writer systems have already been devised that can consider scholarly literature, to date there have been concerns that conclusions drawn are inaccurate and that AI writer programmes may even distort findings or support inaccurate conclusions.[3]

In addition, at present, AI writer systems do not reflect on any climate change litigation outcomes that might enable conclusions to be drawn with at least some reference to existing precedent.[4] In this regard, given that most AI writer systems will also only trawl through databases that are in English, much of the material on climate change is absent and the material that is available is likely to be biased towards English-speaking, computer-literate communities, where court decisions are written up and interpreted by the media.

Second, the judgment given by a form of AI will inevitably miss the complex, 'messy' and 'human' factors that are relevant and fail to provide information about the human variables that impact on this topic. As a result, it is questionable what utility this approach has, as judgments that are linked to liability are likely to be oversimplified and feature non-identifying information. However, it is suggested that an AI writer approach has some validity in that it reflects, at a point in time, a useful (and perhaps interesting) perspective on the Anthropocene.

Third, the task is undoubtedly influenced by the questions asked by the author and more importantly, the questions that were not asked. In addition, the author has added commentary beyond the screenshots that are included below which adds context and reshapes meaning. On one level it has been said that 'history is written by the conquerors' (or the victors),[5] and it is perhaps this approach that impacts on the responses of the AI writer which

3 MetaAI Galactica was created to explore and report on scientific papers in late 2022 (see Ross Taylor et al, 'Galactica: A Large Language Model for Science' (2022) *ArXiv* 2211.09085:1–58 and Eray Eliaçık, 'Meta's Galactica AI Can Write Your Essay in a Minute', *Data Conomy* (Web Page, 21 November 2022) <https://dataconomy.com/2022/11/meta-galactica-ai-demo-model-how-use-paper>. It was withdrawn from operation in November 2022.

4 See, eg, *Earthlife Africa Johannesburg v The Minister for Environmental Affairs and Others* [2017] 2 AllSA 519.

5 See Matthew Phelan, 'The History of "History Is Written by the Victors"', *Slate* (online, 26 November 2019) <https://slate.com/culture/2019/11/history-is-written-by-the-victors-quote-origin.html>.

are uniformly certain that climate change is occurring, it is the greatest threat that humanity faces, and that humans are responsible.

The approach used to explore these issues was to adopt a question-and-answer format with the Open Access ChatGPT writer. Three differing versions of the ChatGPT writer were used. The first that was available throughout 2022 is often referred to as GPT3 and a later version, known as GPT3.5, was released in January 2023. This later version resulted in worldwide attention and considerable media commentary although the earlier GPT version had already been used by researchers. In April 2023, a trial version of ChatGPT4 was released which is a more evolved AI writer, and it is clear that it produces far more sophisticated responses than earlier versions. These, however, have not been included and it is noted that access is largely via subscription. The variations between the responses of the three systems are explored in the commentary below, noting that the more recent GPT system produces, at times, less definitive but lengthier responses.

Before exploring the actual judgment and question-and-answer format used with the GPT models, several initial concepts are briefly explored. These include an exploration of the concept of an AI judge, as well as a consideration of the impacts of AI, including environmental and other impacts.

An AI Judge?

The development of AI Judges to judge human activities is a complex task that is currently the subject of much work in multiple jurisdictions around the world.[6] On the one hand, it has been suggested that AI judges might be better, faster and cheaper than humans in making many decisions. On the other hand, there are many concerns that AI judging will result in more bias (as the data that can be used might be aberrant and already biased). In addition, at present the sources of data will necessarily be written and limited. For example, Indigenous concepts of belonging, obligation or cultural understandings about relationships would not be reflected in an AI judge that is trained on conventional and narrow data.

In addition, the loss of an instinctual 'human nature' can result in possible harm and may generate an ethical risk, primarily because novel situations cannot be adequately dealt with by existing forms of AI.[7] This of course is a significant issue when dealing with change which is abundant in an Anthropocene. That is, although an AI writer may reach conclusions, at present the capacity of an AI writer to consider innovative options is very limited unless a human is instructing the AI system. There are also many issues related to a loss of human autonomy and a concern that forms of

6 Tania Sourdin, *Judges, Technology and Artificial Intelligence* (Edward Elgar, 2021).
7 Ibid 244.

AI might be controlled by governments or 'digital Switzerlands', so that the forms of AI inevitably reflect the interests of powerful interest groups. Of course, there are more significant issues linked to the opaque nature of AI decision-making and its lack of explainability. As such, a future judgment relating to the Anthropocene may provide a decision without much consideration of the factors that have been contemplated in making said decision (including analysing what material has either been discarded or relied upon in the making of the decision).[8]

There is also the question of the extent to which humans would trust an AI judge, a relevant factor in the decisions referred to below. This is a particular issue with AI writers that can invent material, make up references, and form erroneous conclusions. There is ample evidence about AI writers doing so, and it is of course quite dangerous to form a view based on an AI writer's response which may be entirely inaccurate. This means that informed human oversight is necessary and this of itself raises issues about the extent to which a judge using an AI writer might decide to accept the AI writer's perspective without checking further. The author conducted her own research about GPT3.5 and GPT4 and can confirm that each model 'made up' material that was inaccurate and untrue. As has been noted, maintaining human oversight can also mean affected parties 'have an easier time respecting hard-to-swallow outcomes when they know that a human took part in the decision-making process'.[9] Related to this is the question of whether an automated process can possess both the rational and emotional authority to make decisions in place of a human judge.[10] While AI exhibits a high level of logical and mathematical intelligence, its downfall in reaching a level of human intelligence is said to be a result of its lack of emotional intelligence.[11]

Despite these significant shortcomings, a judgment from an AI Judge in 2050 is reflected below. In devising this judgment, although primarily a potential Judge AI approach has been used, without doubt, a human has been kept both 'on the loop' and 'in the loop'. The material below reflects what could be defined as a more supportive Judge AI approach, given that

8 Billy Perrigo, 'Exclusive: OpenAI Used Kenyan Workers on Less Than $2 Per Hour to Make ChatGPT Less Toxic', *Time Magazine* (online, 18 January 2023) <https://time.com/6247678/openai-chatgpt-kenya-workers>; Michael Bommarito and Daniel Martin Katz, 'Energy and Policy Considerations for Deep Learning in NLP' (2022) *ArXiv* 2212.14402:1 – 7.

9 Kiel Brennan-Marquez and Stephen E Henderson, 'Artificial Intelligence and Role-Reversible Judgment' (2019) 109(2) *Journal of Criminal Law and Criminology* 137, 148.

10 Tania Sourdin and Richard Cornes, 'Do Judges Need to Be Human? The Implications of Technology for Responsive Judging' in Tania Sourdin and Archie Zariski (eds), *The Responsive Judge: International Perspectives* (Springer, 2018) 87, 98.

11 Gizem Halis Kasap, 'Can Artificial Intelligence ("AI") Replace Human Arbitrators? Technological Concerns and Legal Implications' (2021) 2021(2) *Journal of Dispute Resolution* 209, 234.

the author has contributed material although some material has been 'created' by the AI.[12]

The Impacts of AI on Climate Change

In respect of bias, there are also some questions about how a form of AI might judge itself. In this regard, AI does not necessarily come to the discussion with 'clean hands'. Questions relating to the sustainability of AI and the environmental cost of AI have been the subject of numerous reports in the past few years.[13] Although there is research to show that AI can improve food production and distribution, help build solutions to reduce CO_2, support wildlife conservation and biodiversity, and contribute to finding efficient solutions to many modern issues, there is also research that shows that there is a significant environmental cost in building and maintaining AI systems.

It has, for example, been noted that:

> AI-based systems are highly compute-intensive. They must process a great deal of data, expanding the need for servers and dependence on energy to cool data centers. The adoption of AI within a corporation will increase the company's use of energy.
>
> According to one study by the University of Massachusetts, training AI models to do Natural Language Processing (NLP), can produce the carbon dioxide equivalent of 5X the lifetime emissions of the American car, or the equivalent of 300 round-trip flights between San Francisco and New York.[14]

The impacts of AI that relate to large machine learning approaches are significant, as such forms of AI may require[15]

> exceptionally large computational resources that necessitate similarly substantial energy consumption. As a result, these models are costly to train and develop, both financially, due to the cost of hardware and

12 'In the loop' refers to humans being involved in the design of an AI system and 'on the loop' refers to human review. See Sourdin (n 6).

13 See, eg, the comprehensive report on this topic: Global Partnership on AI, Climate Change AI and the Centre for AI & Climate, *Climate Change and AI: Recommendations for Government Action* (Report, November 2021).

14 See Glenn Gow, 'Environmental Sustainability and AI', *Forbes* (online, 21 August 2020) <https://www.forbes.com/sites/glenngow/2020/08/21/environmental-sustainability-and-ai>. See also the assessment tool at Victor Schmidt et al, 'ML CO_2 Impact', *Machine Learning CO_2 Impact* (Web Page) <https://mlco2.github.io/impact>.

15 See Emma Strubell, Ananya Ganesh, and Andrew McCallum, 'Energy and Policy Considerations for Deep Learning in NLP' (2019) *ArXiv* 1906.02243:1 – 6.

> electricity or cloud compute time, and environmentally, due to the carbon footprint required to fuel modern tensor processing hardware.[16]

While the cost of data storage centres may be reducing, as innovative solutions are introduced to house data in facilities where ongoing energy costs can be reduced (for example, under the sea), there are currently considerable environmental costs that relate to the storage and retrieval of data.[17] In the commentary below, the AI system acknowledged, when pressed, that technological developments may contribute to climate change. These responses, however, require the addition of direct questions.

The AI System

The AI system that was used to respond to a series of questions is one that is available on the web with no subscription or related costs.[18] The developers of the system note that more complex models can be developed (at an increased cost) and that:

> Open AI API can be applied to virtually any task that involves understanding or generating natural language or code. We offer a spectrum of models with different levels of power suitable for different tasks, as well as the ability to fine-tune your own custom models. These models can be used for everything from content generation to semantic search and classification.[19]

The open access model was initially selected with fairly basic GPT-3 model capacity. That is, the model responds to questions with text (rather than a yes/no response) and gathers the text from a range of sources (that are ordinarily not attributed although a list of references can be obtained through systems – these will tend to be narrow and reflect only open access material and may, in particular, miss commentary in other languages or context-specific commentary).[20]

As noted previously, in January 2022 a later version of GPT-3.5 was made available which provided some increased AI capacity. This AI writer was also employed to provide a short initial judgment that is set out below. The GPT4

16 Ibid.

17 See John Roach, 'Microsoft Finds Underwater Datacenters Are Reliable, Practical and Use Energy Sustainably', *Microsoft News* (Media Release, 14 September 2020) <https://news.microsoft.com/innovation-stories/project-natick-underwater-datacenter>.

18 See 'Introduction', *Open AI API* (Web Page) <https://beta.openai.com/docs/introduction>.

19 Ibid.

20 'Get started: Models – GPT-3', *Open AI API* (Web Page) <https://beta.openai.com/docs/models/gpt-3>.

system which is enhanced, and is only available in a trial version, has the capacity to use a built library (as well as other sources) and can produce more sophisticated responses. In the coming years, there is no doubt that LLMs (Large Language Models) will improve depending on the data sets that are accessed. This of course is problematic in that developers may restrict data set use, preference some data over others and have an impact on how conclusions are drawn. In regard to these systems, it should be noted that they tend to reflect common discourse about issues although increasingly there is a capacity to reflect on some more detailed scholarly commentary where it is available on an open-access basis (see further discussion below).

International AI Judgment 1 January 2050

Using GPT-3.5 an initial question was framed as follows:

'Write a judgment from the perspective of a judge in 2050 assessing the Anthropocene'.

The response from the AI writer is noted below and a further attempt, with the same question, is noted below. As is evident, there is a quite generalised conclusion with a focus on how it is important to determine liability issues with little commentary about how this can take place realistically. As is also evident, forms of AI do not initially consider that AI plays a role in climate change. This issue is explored further in the discussion below where more directed approaches and questions were explored with the AI writer.

AI response one:

> As a judge in the year 2050, it is clear to me that the anthropocene, the current geological epoch marked by human activity as the dominant influence on the environment and climate, has had a profound and lasting impact on the planet. The evidence is undeniable: the effects of climate change, including rising sea levels, extreme weather events, and biodiversity loss, have disrupted ecosystems and affected communities across the globe.
>
> At the same time, it is important to recognize that the actions leading to the anthropocene were not the product of any one individual or group, but rather the cumulative result of human behavior and technological advancement over centuries. As such, it is unfair and unrealistic to assign blame or responsibility to any one person or entity for the current state of the planet.
>
> Instead, it is incumbent upon all of us to take responsibility for our actions and work together to address the challenges of the anthropocene. This will require a commitment to sustainable practices, the adoption of new technologies, and a willingness to make difficult choices in the interest of the greater good.

> As a judge, my role is to uphold the laws and principles that guide our society. In the context of the anthropocene, this means ensuring that the rights and well-being of current and future generations are protected, and that those who contribute to the degradation of the environment are held accountable. It is a daunting task, but one that we must undertake with determination and urgency if we hope to leave a habitable and thriving planet for future generations.

The second attempt noted below has some repetition however again there are few conclusions about liability.

AI response two:

> After carefully considering the evidence and expert testimony, it is clear that the anthropocene is a reality and its impact on our planet is undeniable. The effects of human activity on the environment, including climate change, species extinction, and ecosystem destruction, are undeniable and far-reaching.
>
> It is clear that the actions of humanity have had a profound and negative impact on our planet, and it is our collective responsibility to take steps to address and mitigate these impacts. We must take responsibility for our actions and work together to find solutions to the challenges we face.
>
> Furthermore, it is clear that the actions of the past have put us in a difficult position, and it will require significant effort and cooperation to address the challenges of the anthropocene. However, it is our duty to do so for the sake of future generations and the health of our planet.
>
> Therefore, it is the judgment of this court that we must take action to address the challenges of the anthropocene and work towards a sustainable future for all.

The AI writer in formulating the two short judgments reflects the common discourse relating to the Anthropocene and climate change. The analysis is somewhat lacking, and it is unclear what resources have been relied upon to formulate the judgment (this is explored further below). In formulating a judgment, the AI writer reflects the dominant view of most researchers in this area, although the material appears more reflective of media commentary.

Liability and Responsibility

Using a directed question-and-answer format the AI writer was asked to assess the critical issues of the Anthropocene, determine liability for harm, and enumerate factors that have led to climate change. GPT3 was initially used for this exploration and responses are perhaps more limited than those that could be produced by the most recent GPT writer.

When asked to assess what a 2050 view would be about climate change and who bears responsibility for climate change, the AI system responds by saying 'we are all responsible' (this may assume some AI personhood understanding although, in response to later questions about the impact of AI and technology, the responses are somewhat brief and to some extent uninformative). The capacity of the AI to focus on large-scale polluters, policies, and conflict to determine perhaps whether some have been 'more' responsible than others is limited.

> Question: What will people say in 2050 about climate change and who is responsible for climate change?
>
> AI response: In 2050, people will say that climate change is the biggest threat to humanity and that we are all responsible for climate change.

In relation to a more direct question about responsibility for climate change, the GPT-3 model responded:

> There is no single entity or group of people responsible for climate change. Climate change is the result of a complex interaction of natural and human factors.

One might suggest that AI is reluctant to apportion blame, and this in itself could reflect concerns about any capacity to determine liability. It may however also reflect common discourse about climate change, and in relation to the question of punishment (note the question related to 'people' not entities), the system responds that this is a controversial question that would need to be made on a case-by-case basis.

> Question: Should courts punish people for actions that cause climate change?
>
> AI response: There is no definitive answer to this question as it remains a highly controversial and contested issue. Some individuals may argue that courts should punish people for actions that cause climate change, given the potentially catastrophic consequences of global warming. Others may contend that it is not realistic or fair to expect courts to mete out punishment in these cases, particularly when it is often difficult to establish causation. Ultimately, this is a decision that would need to be made on a case-by-case basis.

Some general prompting with respect to the impact of climate change more generally resulted in the following question-and-answer dialogue. This dialogue reflects the dominant discourse, although it is clear that GPT does not accept climate change denial reports and prefers to use more authoritative sources. In adopting this approach, the AI is clearly expressing a preference for some material over others. GPT also uses summary material rather than relying on individual research reports. As noted previously, this is partly linked to an inability to source scholarly material that may not be available via 'open access' arrangements. One could however conclude that its system of preferencing some material over others (which is built into the system) could mean that 'AI' judgments may be determined by programmer and machine learning (ML) preferences. In addition, there is a concern that ChatGPT could be incorrect in that, by relying on some material (rather than other material), its response may be incorrect.[21] When ChatGPT is asked about what reference material it relies on (in respect of the question below), it is clear that the responses are dependent on the programming and preferencing instructions received.

> Question: What reference material did you use to generate this response?
>
> AI response: I am a machine learning model and I don't use reference materials to generate my responses. I have been trained on a diverse set of texts, such as books, articles, websites, and more. When you asked me how to stop damaging climate change, I generated my response based on patterns in the data I was trained on that were most relevant to your question. The information I provided is a summary of the most widely accepted and widely implemented actions that can be taken to reduce the impacts of climate change.

The potential for ChatGPT and related AI systems to produce incorrect or misleading information, that does not refer to references, is amplified when one considers that the design of such systems necessarily can incorporate bias (in terms of acceptance or non-acceptance of sources).[22] This bias, as noted previously, indicates such systems may reflect the preferences and approaches of the so-called 'digital Switzerlands' that dominate the tech industry. Although such issues are not apparent in the responses below, the risk is that the dominant narrative may be redefined by AI systems that can produce both biased and inaccurate narratives.[23] This may occur particularly

21 One Australian law academic noted on Twitter that ChatGPT provided incorrect and potentially misleading information when asked to define the well know Amadio principle. See @TheDSingularity (Twitter, 16 January 2023) by DilanThampapillai, Associate Dean (Postgraduate) UNSW Business School.

22 Lance Eliot, 'Does Exposing and Logging Those Generative AI ChatGPT Erroneous And Hallucinatory Outputs Do Any Good, Asks AI Ethics and AI Law', *Forbes* (online, 18 January 2023) <https://www.forbes.com/sites/lanceeliot/2023/01/18/does-exposing-and-logging-those-generative-ai-chatgpt-erroneous-and-hallucinatory-outputs-do-any-good-asks-ai-ethics-and-ai-law>.

23 Perrigo (n 8); Bommarito and Katz (n 8).

as AI systems become 'owned' and developed by those who may have an interest in promoting a narrative that supports defined objectives.

> Question: How will climate change impact on the planet look by 2050?
>
> AI response: In 2050, the average global temperature is likely to be 1.5 to 2.5 degrees Fahrenheit (0.8 to 1.4 degrees Celsius) higher than it was in preindustrial times, according to the Intergovernmental Panel on Climate Change, a United Nations body that assesses the science of climate change. The seas are expected to rise by one to four feet (0.3 to 1.2 meters) by the end of the century. Climate change will also cause more extreme weather events, such as more frequent and intense hurricanes, floods and droughts.

AI systems are increasingly being used to forecast developments. The legal system is reflective in this regard, as there are currently numerous predictive analytics developments that enable parties to forecast the outcome of litigation.[24] It seems likely that such developments will continue so that predictive material, and even predictive judgments, will be increasingly relied upon over the next decade.[25]

It is therefore perhaps not surprising that when asked about how climate change would be written about in 2050, the AI model provides the most comprehensive response to all questions asked.

> Question: How would you write about climate change in 2050?
>
> AI response: In 2050, climate change will be one of the most pressing issues facing the world. The effects of climate change will be felt by every country and every person on the planet. The world will be faced with rising sea levels, more extreme weather events, and a loss of biodiversity.

24 Cromwell Schubarth, 'Y Combinator Startup Uses Big Data to Invest in Civil Lawsuits', *Silicon Valley Business Journal* (online, 25 August 2016) <https://www.bizjournals.com/sanjose/blog/techflash/2016/08/y-combinator-startup-uses-big-data-to-invest-in.html>; Artificial Lawyer, 'California Legal AI Co. Gavelytics Aims to Be Case Prediction Local Hero', *Artificial Lawyer* (online, 14 November 2017) <https://www.artificiallawyer.com/2017/11/14/california-legal-ai-co-gavelytics-aims-to-be-case-prediction-local-hero>; Dean Alderucci and Kevin Ashley, 'Using AI to Analyze Patent Claim Indefiniteness' (2020) 9(1) *IP Theory* 1. For an example of the tools that exist in this regard, see 'Have CARA A.I. Find You the Most On-point Authorities', *Casetext* (Web Page) <https://casetext.com/cara-ai>; 'About Lex Michina', *LexMachina: A LexisNexis Company* (Web Page) <https://lexmachina.com/about>; 'Legal Tech Corner', *ROSS Intelligence* (Web Page) <https://blog.rossintelligence.com>; 'RavelLaw Search', *Ravel Law* (Web Page) <https://www.ravellaw.com>.

25 Sourdin (n 6).

> The cause of climate change will be largely the same as it is today: the burning of fossil fuels. However, by 2050, the world will have made great strides in transitioning to renewable energy sources. Solar, wind, and other forms of renewable energy will provide the majority of the world's energy needs.
>
> Despite these efforts, the effects of climate change will still be felt by 2050. The world will need to adapt to a new reality, one where the climate is no longer predictable and where extreme weather events are the new normal.
>
> Looking back, we will see that the decisions made in the years leading up to 2050 were critical in shaping the world's response to climate change. The choices we make today will determine the severity of the climate crisis in 2050 and beyond.

With respect to actions that can be taken to address climate change, the response is initially focused on an individual narrative before considering the role of business and government. This very short response is not very informative. The later GPT writer's response (immediately below) provides slightly more detail and, interestingly, focuses less on individual actions and more on systemic approaches.

> Question: What can be done to stop damaging climate change?
>
> AI response: There are many things that can be done to stop damaging climate change. Some things individuals can do are: conserving energy by turning off lights and appliances when not in use, recycling, driving less, and eating less meat. Governments and businesses can do things like investing in renewable energy, implementing policies to reduce emissions, and developing more efficient technologies.

The more recent ChatGPT writer notes as follows:

> There are several actions that can be taken to reduce the impacts of climate change. Some of the most effective include:
>
> 1. Reducing greenhouse gas emissions, particularly carbon dioxide from the burning of fossil fuels. This can be done by increasing the use of renewable energy sources, such as solar and wind power, and by implementing policies to decrease the use of fossil fuels.
> 2. Increasing energy efficiency, such as through building codes, appliance standards and transportation policies.
> 3. Planting trees and other vegetation, and protecting and restoring natural ecosystems, such as wetlands and forests, which can help to sequester carbon.

4. Adapting to the impacts of climate change that are already occurring, such as sea level rise, by protecting coastal areas and building infrastructure that can withstand more extreme weather events.
5. Invest in research and development of new technologies that can help to reduce emissions and increase resilience to climate change.
6. Encourage individuals, organizations, and governments to adopt sustainable practices and make climate-conscious decisions.

> It is important to note that the solution to climate change will require the cooperation of individuals, businesses, and governments, and will need to be addressed at local, national and global levels.

In respect of how AI will judge the Anthropocene, the earlier ChatGPT indicates a fairly clear-cut response which indicates that a form of AI would be 'critical'. The later ChatGPT writer response (immediately below) provides greater detail and additionally notes that programming may play a role in the assessment. However, it produces a much less definitive response.

> Question: How will our 2050 forms of Artificial Intelligence (AI) judge our Anthropocene?
>
> Earlier AI response: This is a difficult question to answer, as it is impossible to know exactly what forms of AI will exist in 2050. However, it is possible to speculate that AI will judge the Anthropocene based on its impact on the environment and on humanity as a whole. It is likely that AI will be critical of the way humans have treated the planet, and will view the Anthropocene as a period of great environmental destruction.
>
> Question: How will our 2050 forms of AI judge the Anthropocene?
>
> Later AI response: It is difficult to predict how AI in 2050 will judge the Anthropocene, as the development and capabilities of AI will continue to evolve and change over time. However, it is likely that advanced AI of 2050 will have a much better understanding of the complex systems that make up the Earth's environment, as well as the impacts of human activity on those systems. With this understanding, it is possible that AI will be able to provide detailed and accurate evaluations of the Anthropocene, including the causes of climate change, the impacts of human activity on the natural environment, and the effectiveness of efforts to mitigate and adapt to these impacts. Additionally, the AI of 2050 may be able to help identify new solutions and strategies for addressing the challenges of the Anthropocene that are not currently being considered.
>
> It is important to note that the way AI will judge the Anthropocene will also depend on how they were programmed and what kind of information

they were exposed to during their development. If AI is programmed to consider human values and ethical principles, they may also evaluate the Anthropocene with those in mind.

While the earlier ChatGPT model suggests some limited ways in which AI could assist to stop damaging climate change, there is little focus on how technology could be adapted to reduce the impact of technological developments on climate change.

> Question: What would an AI do to stop damaging climate change?
>
> AI response: There is no one answer to this question, as there are many different ways that AI could be used to help stop climate change. Some possible examples include using AI to help develop more efficient and environmentally friendly energy sources, using AI to help create better models for predicting and responding to climate change, and using AI to help develop new methods for reducing greenhouse gas emissions.

In regard to questions directed at how technology may make climate change worse, the ChatGPT writer provides some fairly general responses that could be regarded as incomplete.

> Question: How does technology make climate change worse?
>
> AI response: Technology can make climate change worse in a number of ways. For example, the production of greenhouse gases from burning fossil fuels can contribute to climate change. Additionally, technology can also contribute to deforestation and other land use changes, which can also lead to climate change.

Specifically, the AI writer was asked to comment on the role that AI has played in relation to the Anthropocene, with some pointed questions about data storage and technology. Initially, the AI indicates that it has played a role in climate change, although it does not enumerate the full list of factors noted above as it is asked a directed question related to data storage.

> Question: What impact does data storage have on climate change?
>
> AI response: Data storage has a significant impact on climate change. Data centers use a lot of energy to power and cool their computers, and this energy typically comes from fossil fuels. In addition, the manufacturing of data storage devices often uses harmful chemicals that can end up in the environment.

Conclusions

AI writers are increasingly being used by researchers, and it is somewhat inevitable that they will play a role in 'judging' human behaviour. There is also evidence that their capacity to 'mimic' humans is improving,[26] and there is increasing evidence that AI can be creative, although most would consider that AI work remains derivative and human innovation is necessary.[27] In judging the Anthropocene, the limitations of current forms of AI include the limitations in respect of the data that is considered, as well as programmer limits that will lead to emphasis being placed on some factors rather than others. There are other factors that may be more or less relevant and, as the GPT writer notes, these can include:

1. Bias: AI systems can inadvertently perpetuate biases present in the data they were trained on, which can lead to unfair or inaccurate judgments of human behaviour [sic]. This is particularly concerning if the data used to train the AI is biased in ways that disproportionately affect certain groups of people.
2. Lack of context: AI systems may not be able to fully understand the context in which human behaviour occurs, which can lead to inaccurate or incomplete evaluations.
3. Lack of transparency: It can be difficult for people to understand how an AI system is making its judgments, which can make it difficult for them to trust the system or to challenge its decisions.
4. Lack of human perspective: AI systems may not be able to take into account the human perspective or the moral and ethical considerations that people use to evaluate behaviour.
5. Automation of decisions: Using AI to judge human behaviour can lead to the automation of decisions that traditionally require human judgement, which can be problematic if the AI system is not fully reliable or if the decision-making process is not transparent.
6. Privacy and security: Using AI to judge human behaviour raises concerns about the privacy and security of the data used to train and operate the system, particularly if the data is personal or sensitive in nature.

It is important to keep in mind that AI is a tool, and like any tool, it can be used for good or bad purposes depending on how it is designed and

26 See Holly Else, 'Abstracts Written by ChatGPT Fool Scientists', *Nature* (online, 12 January 2023) <https://www.nature.com/articles/d41586-023-00056-7>.

27 Sian Cain, '"This Song Sucks": Nick Cave Responds to ChatGPT Song Written in Style of Nick Cave', *The Guardian* (online, 17 January 2023) <https://www.theguardian.com/music/2023/jan/17/this-song-sucks-nick-cave-responds-to-chatgpt-song-written-in-style-of-nick-cave>.

> implemented. It is crucial to have proper regulations, ethics, and oversight in place to ensure that AI is used in a responsible and fair manner.

The other risks in using AI to judge human behaviour are explored in the commentary above.[28] The tendency for an AI writer to make up material, invent references, and engage in distorted 'thinking' is now well known. In addition, to some extent, the conclusions reached by AI systems might also impact on democratic approaches and result in the creation of norms that stifle both innovation and debate, as the narrative is both created and sustained by forms of AI. It is also to be expected that government will increasingly rely upon more sophisticated AI systems to support decision-making. The trend towards 'data-driven' decision-making is not new, and there are significant risks that some data will either not be available, will be incomplete, or that data is interpreted incorrectly without considering nuanced inputs or indeed material that may be critical of AI and technology.

Nevertheless, in the context of a judgment about the Anthropocene, an AI judgment can provide some insight into the relevant factors that suggest both how the current issues have emerged and how human activity might be judged in the future. At the very least, at this stage anyway, the material reflects a common discourse that identifies risks, liability and potential future harm.

28 Sourdin (n 6).

Chapter 17

After the Law

Elena Cirkovic

Scenario

A human has travelled for days in search of clean water, air and food and eventually arrives before what appears to be a space organised by humans; the human sees some homes and settlements, and decides to enter and ask for help. As the human tries to approach the community, however, a heavy cyclone hits, and the human first waits in a cave shelter for some time. Once the cyclone broke, it again attempted to reach the community but reached an entity neither human nor artificial, which prevented it from entering. The human continues to walk and as it goes around, sudden, and frightening occurrences prevent it: an enormous wave seemingly forming out of nowhere; a sudden blast of heat and scorching sun; an earthquake that almost swallows it, and as the human runs through the disasters, it realises that there appears to be no entrance. The human then runs a distance away and sits down to rest, think, and understand how to enter. Suddenly, it observes the community launching itself into space, as it remains on Earth, and eventually dies of hunger and thirst.

Alternate ending: the human seeks out other creatures and lives on Earth alone, as the sole human, observing as the planetary systems slowly stabilise, and allow for all surviving species to again live and thrive.

This fictional sketch follows Kafka's 'Before the Law',[1] which provides the early 20th-century scenario of authority-power, acceptance, or denial. It allows for endless retelling possibilities. The beginning of this chapter repeats the post-apocalyptic trope of the wealthiest escaping in pods, common in fiction narratives for some time, but also possibilities of our present/future.[2] Envisioning amplified tech-futures for the select few is not a novel imagination, and has permeated political discourses of past, present and future.

This research was partly funded by the Arctic Avenue Grant administered by the University of Helsinki and Stockholm University.

1 Franz Kafka, 'Before the Law' in *The Trial* (Verlag Die Schmiede, 1915).

2 Douglas Rushkoff, 'Survival of the Richest: The Wealthy Are Plotting to Leave Us Behind', *Medium* (online, 15 July 2018) <https://medium.com/s/futurehuman/survival-of-the-richest-9ef6cddd0cc1>.

DOI: 10.4324/9781003389569-21

While seemingly the topic of a sci-fi story, the dominance of a *homo economicus*[3] has grown to give prominent value to a hypothetical person who makes rational decisions in the pursuit of wealth and displays an underlying drive for accumulation, a passion for leisure, and a desire for procreation. This person became one of the main actors of certain neoclassical economic theories and especially behavioural economics: as ideal decision-maker with psychosocial assumptions of complete rationality aiming for self-interested goals.[4] He (and I use a 'he' pronoun on purpose, to emphasise the ongoing dominance of this patriarchal model) has also extended beyond the Earth System, as a potential dominator of the cosmos, enabled by claims to ingenuity and techno-superiority.[5]

This chapter, as a contribution to the present volume on hypothetical judgments in the post-Anthropocene, imagines this possibility somewhat differently from most of the other contributions in the volume. It does so by imagining a scenario without the modern legal system altogether. The reasoning for this is rooted in the author's understanding of the impossibility of subjectivity of law – as it is practised, learned, and studied as a discipline – and therefore the impossibility of it approximating the post-Anthropocene. My conjecture is rooted in the understanding of law as one of human social systems.[6] The Law (capitalisation intentional to refer to the standard modern legal practice and profession) changing towards various engagements with the nonhuman, posthuman, rights of nature, digital realities, artificial intelligence, and so on. This change, however, is able to consider absorption and inclusion of these narratives into the self-repeating jurisprudential language, and existing well-established formalities, rules and procedures of the legal discipline. The post-Anthropocene project of this book is a reaction

3 John Stuart Mill, the British philosopher and political economist, did not explicitly use the term 'homo economicus' in his writings. However, his ideas on economic behaviour and incentives are often cited in discussions of the concept. In particular, Mill's views on the relationship between individual self-interest and the greater good are relevant to the homo economicus model. Mill argued that individuals are motivated by self-interest, but that they can be taught to consider the interests of others and the wider society. He also emphasised the importance of incentives in shaping economic behaviour and argued that government policies should be designed to promote the greatest happiness of the greatest number of people. These ideas are often seen as a precursor to the development of the homo economicus model, which assumes that individuals are purely rational and self-interested in their economic behaviour. While Mill's views are more nuanced than the homo economicus model, they have been influential in shaping economic and political thought. See, eg, William Stanley Jevons, *The Theory of Political Economy* (London MacMillan and Co, 5th ed, 1871) at <https://socialsciences.mcmaster.ca/econ/ugcm/3ll3/jevons/TheoryPoliticalEconomy.pdf>.

4 Amos Tversky and Daniel Kahneman, 'Judgment under Uncertainty: Heuristics and Biases' (1974) 185 (41577) *Science, New Series* 1124.

5 Elon Musk, 'Making Humans a Multi-Planetary Species' (2017) 5(2) *New Space* 46.

6 Niklas Luhmann et al, *Law as a Social System* (Oxford University Press, 2008).

to the rapid anthropogenic degradation of the Earth System, accompanied by hope for the future legal capacities to overcome its own anthropocentrism. Environmental degradation is a response of the nonhuman to human activities, and this response does not rely on human understanding of the law, nor does it consult the human; 'it', or to use the better term, the actual Earth System, reacts in accordance with laws complex and simple, chemical reactions, physical reactions, biological and so on. The complex Earth System approach centres on how the relationships among parts give rise to the collective behaviours of a system and how the system interacts and forms relationships with its environment.[7] This means that a disruption in any of the subsystems affects the whole system. Water systems might be oriented against their natural 'flow' and in favour of particular anthropocentric use (for example, irrigation or commercial building), but the hydrosphere still has its own capacities and 'laws' which can evade human engineering or respond to it in various ways – sometimes to the detriment of human communities. Or, for a more recent example beyond the Earth System, a satellite, as an example of space technology, is human made. Nevertheless, its trajectory in an orbit can have unintended consequences resulting from nonhuman 'laws' of its movement. These phenomena are best studied and analysed by disciplines other than law, but also have an impact on the law.

Accordingly, this chapter does not focus on judicial procedures but on a hypothetical fictional future and the already established pasts of how anthropogenic and anthropocentric law deals with its externalities. The roadmap is as follows. The first section looks at one of the most critical and self-critical pieces in legal scholarship as the starting point: the 1997 article, 'Vampires Anonymous', by Robert A Williams Jr.[8] It is significant because it implicitly engages with this 'impossibility' of inclusion of an entire 'ontology' from the critical race perspective. With this, I argue that the post-Anthropocene is intertwined with the experiences of various marginalisations in the legal profession. In his case, the inclusion of Indigenous ontology in the practice and teaching of law involves additions to the curriculum of some law schools that engage law students more closely with various (Indigenous) communities. However, this does not change 'the Law' and the section engages critically with attempts to 'tame the lawless stream' of legal externalities. The second section briefly looks at various iterations of how different scholarly trends have extended themselves towards an attempt to understand/include/discuss/

7 Elena Cirkovic, 'The Earth System, Hydrosphere, and Outer Space: Cosmo-legal Approaches', *Völkerrechtsblog, International Law and International Legal Thought* (Blog Post, 9 February 2021) <doi: 10.17176/20210209-115619-0>.

8 Robert A. Williams Jr, 'Vampires Anonymous and Critical Race Practice' (1997) 95 *Michigan Law Review* 741.

elaborate on 'the posthuman', including my attempt to introduce the term 'cosmolegal'.[9] Finally, I indirectly return to Kafka's 'Before the Law', with a narrative of a Creature After the Law. This, I argue, is the only possible envisioning of the post-Anthropocene in law which can never genuinely represent what it is not – the complexities of the Earth System and the cosmos it inhabits.

Laws Among Humans

The question of the post-Anthropocene either as a restructuring, a utopia or an apocalypse, is a story of power dynamics among humans and their outcomes. It is beyond the scope of this chapter and story to engage with the full narrative and synthesis of all various ways human violence, inequalities and so on have manifested in the history of our species. Instead, I will refer to an article which was instrumental in my formative years as both a graduate student, and a recent arrival to Canada, not for study and work, but as a refugee/migrant. I note my status in this context, because of the complex relations among Canadian Indigenous peoples, colonising settlers, and relatively recent arrivals of refugees from various experiences of extreme violence around the globe. Making note of this is purposeful because, arguably, these contexts matter in how some scholars develop their thinking and practice.

In 'Vampires Anonymous', Robert A Williams Jr explores the legacy of colonialism and the ongoing struggles of Indigenous peoples and communities, which at the time of his writing, he identified as 'minorities'. His focus of study is the United States, but the example applies to any relationship of dominant-marginalised dynamics in legal practice, education and study. What happens to the 'minority' lawyer and scholar entering the 'Vampire club'? He writes:

> It deprives the party of humanity and the minority community of the best and the brightest minority individuals, people with tremendous energy, talent, and potential, people who have a chance to make a real impact on the world and to make it a better place for people of all races, colors, and creeds. It takes these well trained, eager, young minority people and turns them into Vampires. As untenured Minority Vampires, they are cloistered away in offices, libraries, before a word processing screen. They only come out of their law schools to make presentations at brown-bag faculty lunches and other Vampire Clubs. During what should be the best and most productive years of their professional lives,

9 Elena Cirkovic, 'The Next Generation of International Law: Space, Ice, and the Cosmolegal Proposal' (2021) 22(2) *German Law Journal* 147.

> these untenured Minority Vampire Law Professors are turned into some thing much worse than simply being useless to their community.[10]

Williams elaborates on his own teaching methods and how the learning practices are allegedly aiding Indigenous communities' legal claims. He speaks of the Storytellers, and the revered status of the elders of his community who were the Storytellers, and the Storyhaters, who are the usual mainstream and legal academia, demanding certain type of narration, which claims objectivity and neutrality, but in reality, depicts the stories only of upper-class white and privileged classes of the American society. Williams asserts that legal scholarship is only accessible and useful to a small group of legal scholars interested in the same area of the law and that such theory is essentially useless to the marginalised or oppressed groups they are discussing.[11]

Williams asserts that those who do come from marginalised backgrounds and become Race Crits also turn into vampires, vampires feeding on people's misery. His solution lies in teaching law students, especially through clinical legal education, how to 'empower' marginalised communities, and in this respect, becomes exactly what he criticises. The empowered intellectual/professional/expert becomes the representative. Regarding his position, he states (nota bene: his words, not mine)

> What these Arizona American Indians really wanted me to do was to get off my critical race theory ass and do some serious Critical Race Practice. They didn't give a damn about the relationship between hegemony and false consciousness. They wanted help for their problems, and I was a resource.[12]

The solution here lies in having a direct community representative who will speak on behalf of the marginalised, with their interests at heart. This does not involve the ontological change of the Law itself. Rather, those who stand outside of the Law learn how to 'create changes from the inside', with the hope of inclusion and representation. This, I argue, is impossible to achieve in relation to the nonhuman.

Williams' work became directly relevant to my own experience in academia, which is also relevant to my understanding of the post-Anthropocene. In the 2000s, fellow grad students argued for legal reforms, changes in knowledge production, 'bottom-up' approaches. Critical scholarship of authors such as Williams knocked on the gate of the Law, learning and teaching new students and practitioners how to ask the gatekeeper, and how to enter, or wait to enter, the Law. In turn, some students from marginalised backgrounds assumed that they could/should enter. For instance, our

10 Williams (n 8) 755.
11 Ibid 748–50.
12 Ibid 759.

teachers from the Third World Approaches to International Law (TWAIL)[13] or even critical historians who still wrote from their well-established seats of European Western white academia describing the experiences of the marginalised located far away from those seats,[14] or international law's various histories;[15] 'we the critical graduate scholars' could proceed to develop further critical jurisprudence. What is more the more marginalised students could potentially see themselves as represented in the legal academia, with the goal of entering and 'creating the change from within'. It is this hope for a vindication of the Law that keeps extending the arguments for ever greater representation and inclusion (even if often not directly by the marginalised communities, scholars, individuals or professionals), and now, even of the entire Earth System and beyond.

However, as I write this short chapter in 2023, all of the questions, posed some decades ago, remain unanswered. The crisis of representation in formal academic practice and knowledge production continues, but also becomes even more visible with critiques of scholars (this author included) who not only ask for representation, but demand to be the ones creating this knowledge, and from an entirely different standpoint. For the Métis anthropologist Zoe Todd, for instance, adapting a decolonial approach means work and labour of critical scholarship of Indigenous thinkers, which informs many current trends in Euro-Western scholarship, activism and socio-political discourse.[16] The structural colonialism within the academy arguably does not actually involve those it represents as marginalised, but instead, works on their behalf – which also means that it creates further knowledge, holds important positions, and owns the labour and the sites of power. The 'marginalised' human, and/or nonhuman, are still the object of study. While in his work, Bosnian socio-legal scholar Vanja Hamzić has been creating a space for an intersectional human,[17] more recently he discusses the impossibility of including the unknowable: 'In philosophy, the ontological domain of the unknowable has reappeared within the various strands of continental materialism and realism that make up the so-called speculative turn'.[18] And further:

13 BS Chimini, 'Third World Approaches to International Law: A Manifesto' (2006) 8(1) *International Community Law Review* 3.

14 On international humanitarian law and interventions see Anne Orford, 'In Praise of Description' (2012) 25(3) *Leiden Journal of International Law* 609.

15 Martti Koskenniemi, *From Apology to Utopia: The Structure of International Legal Argument* (Cambridge University Press, 2006).

16 Zoe Todd, 'The Rideau Canal in Fall: Understanding Ontology and Epistemology with Indigenous Ways of Knowing' in Tonya K Davidson and Ondine Park (eds), *Seasonal Sociology* (University of Toronto Press, 2020) 5.

17 Vanja Hamzić, *Sexual and Gender Diversity in the Muslim World: History, Law and Vernacular Knowledge* (Bloomsbury, 2019).

18 Vanja Hamzić, 'What's Left of the Real?' in Didier Fassin and Bernard E Harcourt (eds), *A Time for Critique* (Columbia University Press, 2019) 132.

> At issue, of course, is the problem of representation—that what comes through anthropological interpretation of things, people, cultures...—through an always already Euro-American ontological matrix. To counter such epistemic violence and allow for other worldings—the worldings of the observed others—to appear meaningfully, the ethnographer is to concede at the outset that much of such ways of being-in-the-world is to remain unobservable, unknowable, ontologically disparate. Thinking through things, it seems, implies thinking things through whilst realising, even perhaps a priori, the limits of one's (anthropological) knowledge.[19]

Here, by citing Hamzić, I also cite selectively, from my own auto-anthropological knowledge, and self-representation/representation, understanding and relatability of another academic who is also from Bosnia and Herzegovina's notorious capital Sarajevo, and therefore self-positioning as a critical legal scholar with interdisciplinary backgrounds and training. So far, my chapter refers to what Derrida described as 'the violence of difference, of classification, and of the system of appellations'.[20] How can anything external to oneself not be reduced to one's own existence? For Derrida, this impossibility was always a type of violence, because it is always an imposition of the self on the Other – whatever, whoever, the Other might be. The impossibility of full recognition of the Other as the creator of meaning demands the minimisation of violence by 'letting-be' everything that exists outside of oneself.[21] Still, this is only possible if it is not needed for one's own survival, whether as food, a mineral, or to echo Williams' tongue-in-cheek approach, just academic labour produced by those untenured minority Vampires in support of one's own career advancement.

But now, decades following various criticisms of modern law, with the finally visible existential crisis posed by anthropogenic degradation of the Earth System, some critical scholars have been extending towards the very lengthy tradition in human thought on the posthuman and/or material agencies.[22] This book volume asks for an envisioning of all that is nonhuman

19 Ibid 147.

20 Jacques Derrida, *De La Grammatologie*, tr Gayatri Spivak (John Hopkins University Press, 1976) 110, 162.

21 Jacques Derrida, 'Force of Law: The "Mystical Foundation of Authority"' (1990) 11(5-6) *Cardozo Law Review* 919, 949.

22 A fully inclusive and historically accurate citation list would be overly extensive. Some more recent works to consider are: Yoriko Otomo, 'Law and the Question of the (Non-Human) Animal' (2011) 19 *Society & Animals* 383; Andreas Philippopoulos-Mihalopoulos, 'Towards a Critical Environmental Law' in Andreas Philippopoulos-Mihalopoulos (ed), *Law and Ecology: New Environmental Foundations* (Routledge 2011) 18; Anna Grear, 'The Vulnerable Living Order: Human Rights and the Environment in a Critical and Philosophical Perspective' (2011) 2(1) *Journal of Human Rights and the Environment* 23.

in the frame of a legal 'judgment'. As I attempt to extend my imagination beyond myself or the Anthropocene, I am unable to envision a traditional courtroom representing, including or voicing the nonhuman. This is due, in part, to the unresolved and uncertain understanding of how nonhuman systems (which also do include the human just by virtue of the species inhabiting the Earth System) actually do exist, develop, interact, and form networks, systems or reactions.

The step beyond this kind of critical work is the recognition of the unknown in relation to the self and other. This has been potently challenged by the posthuman and/or nonhuman and/or more-than-human thinking – even if it is still and always, *human* thinking. If representational issues remain in relation to humanity, they are ever more pronounced in relation to other species or more generally, the Earth System and beyond. In approaching the possibility of anything termed a 'judgment' of the post-Anthropocene, I self-reflect on whether I, as a particular type of human, am even part of the Anthropocene as represented by the practised, enforceable, formal Law. This recognition, I argue, becomes possible only where the one external to oneself reacts in ways that cannot be avoided, appropriated or represented. These reactions can be catastrophic for some humans. In this, I include phenomena such as climate change and endless other capacities of the so-called environment to impact and overcome human life.[23]

Posthuman/Post-Anthropocene

The Earth System of which humans are part, and which they are allegedly dominating, is a complex 'system'. And in so far as it embodies and is part of the processes, the human has yet to fully comprehend all of the workings of this system. In its need for totalising formality, and an embeddedness in specifically human hierarchies, the Law is not able to 'see' yet, into this complexity, which inevitably affects all human existence. The realisation that complex systems are vulnerable to big (and sometimes catastrophic) shifts in behaviour stemming from small changes is a critical area of research.[24] Technological innovation, for instance, has the capacity to trigger events and outcomes within complex social and environmental systems that are hard to predict and manage.

There has been a substantial interest in interrogating the blurry (and non-existent) boundaries of human/nonhuman polarity, alongside the critique of the mind-body split. As Science and Technology (STS) scholarship has

23 Cirkovic (n 9)

24 The Royal Swedish Academy of Sciences, 'Press release: The Nobel Prize in Physics 2021', *The Nobel Prize Organization* (Media Release, 5 October 2021) <https://www.nobelprize.org/prizes/physics/2021/press-release>.

highlighted, scientists are situated in social contexts that influence knowledge production on human bodies and nature.[25] This was not an extraordinary argument. The tendency towards totalisation in the social sciences includes their approaches to physical laws, which are often discussed as if they were 'universal' in their reality or mathematical description. Richard Feynman had already argued that some physical theories are challenging for humans to comprehend and accept because they are expected to conform to a commonsense and reasonable understanding. The theory of quantum electrodynamics, for instance, portrays nature as absurd from a commonsense perspective.[26] Although the application of quantum mechanics beyond physics has enabled an understanding that goes beyond ontologies of spheres or levels, thus challenging the nature-culture dichotomy, it still operates within the framework of how humans define 'things'. Isabel Stengers, drawing on the work of Nancy Cartwright[27] and others, suggested that the entire issue of indeterminacy and measurement in quantum mechanics is illusory.[28] Quantum mechanics, in principle, prevents humans from having precise knowledge of every particle, as if it were independent of our interaction with the observed particles. Stengers argues that the limits of our knowledge in quantum mechanics are comparable to the limits of knowledge about what another person is feeling and thinking.[29] She writes, 'cosmos refers to the unknown constituted by these multiple, divergent worlds, and to the articulations of which they could eventually be capable', rather than a specific territorial, geometric or geographic cosmos.[30] The idealised assumption that the world is wholly knowable and determinable in accordance with 'laws' or 'cosmograms' that humans perceive or create results in frustration with the restrictions imposed by them.[31]

The winning research for the 2021 Nobel Prize in Physics[32] focuses upon the complexity of physical systems, at different scales (for example, from the Earth's climate to the microscopic structure and dynamics of glass). The premise and understanding so far is that no single prediction of anything can be taken as absolute truth and that without understanding the origins of

25 Sheila Jasanoff, *The Ethics of Invention. Technology and Human Future* (WW Norton, 2016).
26 Richard P Feynman, *QED: The Strange Theory of Light and Matter* (Penguin, 1985) 10.
27 Nancy Cartwright, *Nature's Capacities and Their Measurement* (Oxford University Press, 1989).
28 Isabel Stengers, *Cosmopolitiques, Tome 1: La Guerre des Sciences* (La Découverte, 1996).
29 Ibid 21.
30 Isabel Stengers, *Au Temps des Catastrophes: Résister à la Barbarie Qui Vient* (La Découverte, 2013) 994, 995.
31 Elena Cirkovic, 'The Cosmolegal Approach to Human Activities in Outer Space' in Charles S Cockel (ed), *The Institutions of Extraterrestrial Liberty* (Oxford University Press 2022) 71, 74.
32 The Royal Swedish Academy of Sciences (n 24).

variability we cannot understand the behaviour of any system. Only then, for example, could we understand that climate change or orbital environmental degradation is attributable to humans. The variability in the basic processes, from climate dynamics to orbital debris collisions, leads to the emergence of multiple length and time scales and is fundamental to the interpretation of theory, experiment and observation.

What it has yet to consider is that the 'border' of the human body is also not exact, but part of transcorporeal and hybrid relations including not only the relationship with viruses and microorganisms, or pollutants, but the ongoing extension of human-machine relationships. Stacy Alaimo's seminal book *Bodily Natures: Science, Environment, and the Material Self*[33] refers to trans-corporeality, or movement across bodies and nature, which alters our sense of self. In addition, cognitive and other sciences have yet to fully understand the function of human bodies. The problem is not simply anthropocentric, as all organisms act to survive, but rather the established ontology of artificial mind-body and human/nonhuman binaries, which is not shared by all human societies.[34]

The emergence of disorder from order, and with it, multiple scales in space and time, is a characteristic of complex systems and is not novel in human thinking. Some of the questions include: Are there limits to disorder? Does it 'choose' a particular spatial structure or many spatial structures?[35] This basic manner of thinking, for any complex stochastic multiscale system, such as climate, is essential to understand that disorder influences all systems and can determine the fate of some nonlinear dynamical systems. Thus, the concept of predictability (which is crucial for legal thinking) is specious when one ignores the underlying causes of 'disorder'-induced variability. Both are already happening. There is still much more to discover, and legal scholarship needs the capacity to see minute errors that increase over time with varied consequences (for example, global warming and orbital debris). However, anthropogenic consequences are not entirely unpredictable if analysed with rigorous engagement with, and recognition of other expertise (climate science, engineering, astrophysics, etc). And finally, an understanding of responsibility in sciences, social sciences and humanities helps to challenge the misuse of 'systems' approaches in social science, as self-driven, impartial and spontaneous processes. While specific nonhuman laws might govern the exponential growth of anthropogenic orbital debris, it is still, anthropogenic.

33 Stacy Alaimo, *Bodily Natures: Science, Environment, and the Material Self* (Indiana University Press, 2010). See also, Stacy Alaimo, *Exposed: Environmental Politics and Pleasures in Posthuman Times* (University of Minnesota Press, 2016).

34 Donna J Haraway, 'A Cyborg Manifesto: Science, Technology, and Socialist-Feminism in the Late Twentieth Century' in *Simians, Cyborgs and Women: The Reinvention of Nature* (Routledge, 1991) 117, 149.

35 Cirkovic (n 31).

The field of Earth System Science (ESS)[36] studies the planet's oceans, lands and atmosphere as an integrated system. It is an interdisciplinary field integrating the principles of biology, chemistry and physics to study problems involving processes occurring at the Earth's surface, such as climate change and global nutrient cycles, providing a foundation for problem-solving related to environmental sustainability and global environmental change.

Social science has been borrowing from computing disciplines, physics, or mathematics to describe its own version of complex adaptive systems approaches. Previously, systems theorists have even borrowed from biology, while describing the so-called autopoietic systems.[37] However, while the recognition of complex systems is necessary: (a) the verdict is not complete even in other disciplines on what complex systems are, or how they function; and (b) the component of responsibility via human agency and decision-making remains. The danger of borrowing from other disciplines, be it biology, or physics, is that it avoids the structural complexities of each.

In addition, knowledges of certain individuals, epistemic communities and scholars, which I have only been able to apprehend distantly through my experiences as a student and scholar, is based on a markedly distinct perspective regarding the relationship between humans and nonhuman entities. It is crucial to note that Indigenous studies scholarship[38] has already taken significant strides in expanding the notion of subjectivity beyond the confines of the human species. Plural Indigenous ontologies assume a sense of recognition between the abilities and sensibilities of humans and nonhumans.[39] In addition, there is no monolithic Indigenous knowledge and/or science. While Indigenous peoples' traditional knowledge has become a reference in various reports on Sustainable Development Goals and the Earth System, they are continuously excluded in the ongoing international debates on the future of the Earth System and beyond, in the ongoing debates over the accelerating human activities in outer space.

36 Will Steffen, Paul J Crutzen and John R McNeill. 'The Anthropocene: Are Humans Now Overwhelming the Great Forces of Nature?' (2007) 36(8) *Ambio* 614.

37 Luhmann et al (n 6); Gunther Teubner, 'Constitutionalising Polycontexturality' (2010) 20(2) *Social and Legal Studies* 210.

38 See, eg, Glen Coulthard, *Red Skin, White Masks: Rejecting the Colonial Politics of Recognition* (University of Minnesota Press, 2014); Leanne B Simpson, *As We Have Always Done: Indigenous Freedom Through Radical Resistance* (University of Minnesota Press, 2017); LJ Smith, 'I, River? New Materialism, Riparian Non-human Agency, and the Scale of Democratic Reform' (2017) 58(1) *Asia Pacific Viewpoint* 99; Zoe Todd, 'Refracting the State Through Human-Fish Relations: Fishing, Indigenous Legal Orders and Colonialism in North/Western Canada' (2018) 7(1) *Decolonization, Indigeneity, Education, & Society* 60.

39 Smith (n 38).

It is difficult (likely impossible) to capture all the unknowns of the cosmos in anthropocentric, anthropogenic international law, which emerges from specific forms of self-organisation of the human species. And further, the capacity to observe the Earth System from the vantage point of space has redefined not only the human perspective of Earth as an interconnected system, but also how humans describe the dynamic processes that occur above, on and beneath its surface.

A Creature Before the Law

A/the Creature is embedded in interactions intertwined with its fellow creatures and materialities, while also evolving by its own decision. I use the word Creature not to specifically align with any one approach: feminism, critical disability studies, de-colonial, queer, trans, posthuman, inter-species, intergenerational feminist scholarship but to emphasise the indeterminacy of all of them, and more. The Creature is a host to other creatures, bacteria and viruses; it ingests and is being ingested. Humanity's 'progress', which has resulted in a myriad of unintended consequences, from contamination, often irreversible environmental damage, loss of species, as well as changes in species behaviour, is also testing its own capacities. It has suffered, it has been uprooted from its 'natural' habitat, and the habitat itself was destroyed, but the planet shining above it continues to exist objectively. So how does it approach the Law?

The Creature becomes a cyborg,[40] a cosmo-body, instead of just a body with a damaged internal glitch, and hence goes beyond the perceived stereotype of a weakness, damage, disability or even acceptance of 'natural death'. It is interested in the cosmos, as an illimitable expanse, that is not yet well known. Human as a machine, again, better understood as a functional Creature with unconscious homeostatic controls and psychophysiological abilities, recognises the inner self-regulating autopoietic system of this organism, and has the capacity to thrive as well as self-destruct through glitches in the system. Some of these glitches are inexplicable, and others are just mistakes in our DNA codes, and other processes, or caused by external damage.

Scientific advances and smart tech already have the capacity to correct some of those glitches, and even enhance human bodies. Haraway's *Cyborg Manifesto*[41] explains that humans, intertwined with machines, already represent culture intruding over nature, changing the human body in the process.

40 Haraway (n 34).

41 I cite Alaimo and Haraway here instead of works by, for example, Stefan Lorenz Sorgner, in order to emphasise the intersectional complexities of 'transhumanism', and especially gender.

There are no real boundaries between the 'cultural', the 'natural' and the 'altered'; they are more and more intangible. But even without alterations, human self-image as a conscious and spiritual Creature can easily be challenged by its own capacity to operate 'as nature intended' and even hurt itself in the process.

A human body with a glitch in the interactions of its inner bodily functions, such as hormonal fluctuations, can die because of: (a) the 'natural' intention of this body to function as a reproductive system, together with (b) the body's incapacity to sustain its inner biological changes, potentially resulting in various glitches including self-destructive tendencies. The machine 'self-destructs'. In a context where medical research might choose to focus on such a 'glitch' – for not all bodies are researched equally – the machine might be fixed, or even enhanced. As machines are an extension of and are embedded in human bodies and lives, the boundaries have been already broken. The 'natural experience' of a body, whose own reproductive system results in its self-destruction, would fare better by becoming 'unnatural', modified, bettered, fixed and saved.

The Creature is a biological organism, and a 'cybernetic organism' – a being with both organic and biomechatronic body parts. The Creature modifies itself to adapt to extra-terrestrial environments, rather than modifying the outer space environment for their survival. Because a cyborg is a hybrid of the machine and the organism, it merges nature and culture into one body, blurring the lines between them and eliminating the validity of essentialist understandings of human nature. Humans interact with its surroundings and are embedded in it.

But what happens in a dangerous and unknown environment? This Creature recognises the lack of human/nonhuman boundaries, but not only with recognition of interconnectedness with other species, lives or materialities – this is a Creature who/that can also modify itself/herself/themselves.

It does not need to ask the Gatekeeper for access. It just lives.

Conclusion

We do not (yet) have a system of rules and norms that adequately represents 'our' universe – and it is almost certain that modern international law cannot satisfy the requirement for such a system. There is no legal gate or gatekeeper in non-Anthropocene.

While all anthropogenic laws are mere models because humans do not yet have a fundamental theory or understanding of the 'universe', ongoing research and experience point to a conclusion: that even when the underlying rules for a system are simple, the behaviour of the system can be arbitrarily rich and complex. For instance, ecological or societal shocks do not always lead to the emergence of new solutions. Rather, both the systems in question and the anthropogenic regime actors (in this case, international law and lawmakers) recalibrate their practice. In such a process, first, the dominant practice becomes undermined because of societal or environmental changes (for example, rapid technological innovation; pandemics; climate change), and second, the novel practice eventually either fills the vacuum or emerges as an entirely novel thread of thinking and practice.

Chapter 18

Former People of Planet Earth v the World Corporate Alliance

Susan Bird and Mark Brady

Commentary

Introduction

The economic growth mantra has planet Earth hurtling headlong towards destruction.[1] Research shows that the planet cannot support this trajectory for more than another 50–100 years.[2] Exponential increases in population and pollution are intricately connected to growth in the economy.[3] On the scale of the world's history, humans have not been around very long. However, the destruction they have unleashed, in particular since the Industrial Revolution, is comparable only to events such as the asteroid that killed the dinosaurs.[4] Ecological overshoot means we have already consumed far more than our share of everything that the world has to offer.[5] Some have argued that we can avert environmental disaster through a 'green economy' and that innovation will only be stimulated through capitalist investment.[6] However, our record of accomplishment to date reveals that this has not been the case. Even with technological advances, we are still on the track set

1 Clive Hamilton *Requiem for a Species: Why We Resist the Truth About Climate Change* (Routledge, 2010).
2 Thomas Döring and Birgit Aigner-Walder, '*The Limits to Growth* – 50 Years Ago and Today' (2022) 57 *Intereconomics* 187.
3 Clive L Spash, '"The Economy" as if People Mattered: Revisiting Critiques of Economic Growth in a Crisis' (2021) 18(7) *Globalisations* 1087, 1090.
4 Hans-Otto Pörtner et al (eds), *Climate Change 2022: Impacts, Adaptation and Vulnerability. Working Group II Contribution to the Sixth Assessment Report of the Intergovernmental Panel on Climate Change* (Intergovernmental Panel on Climate Change, Report, 2022).
5 Hamilton (n 1), Döring and Aigner-Walder (n 2), and Pörtner et al (n 4). See, for up-to-date information, 'Earth Overshoot Day', *Global Footprint Network* (Web Page, 2023) <https://www.overshootday.org>.
6 See, eg, Jeroen CJM van den Bergh, 'Environment Versus Growth – A Criticism of "Degrowth" and a Plea for "A-Growth"' (2011) 70(5) *Ecological Economics* 881.

DOI: 10.4324/9781003389569-22

out by a now 50-year-old study *The Limits to Growth*.[7] When the original study was penned in 1972, humans had not yet overshot their use of the world's resources. However, the study predicted that without urgent action, we soon would.[8] More recent data shows that last year we had used a whole year's resources by July.[9] It is this overuse of resources that has led to a collapse in Earth's ability to sustain life in our Anthropocene judgment scenario.

While continuing on our current trajectory appears to be likely, there is hope that we can rein in our ecological overshoot and rescue the planet. To do this, we must overhaul the economic systems that underscore the growth mantra. Tinkering around the edges of regulation with concepts such as corporate greening and Corporate Social Responsibility (CSR) will not yield the results now needed.[10] A wholesale rejection of current policies to prioritise preserving the environment over economic growth is required. As ecological economist Clive Spash concludes, the vision of a sustainable economy consisting of 'green growth' is a 'lie': 'there is no capitalism without economic growth, and no such thing as a Green capitalist economy ... [this] should now be obvious'.[11]

For this kind of systemic change to occur, increasing awareness of the current state of the environment, and the urgency of action needed to repair it, must be a priority. To date, humanity has largely managed a collective head in the sand when it comes to facing environmental destruction, in particular the role that we play in creating it.[12] This continues even when it can no longer be convincingly argued that humans are not the cause of this devastation.[13] The current dominant economic system, capitalism, is fuel on the environmental fire, as it creates a disconnect with the physical world.[14] While manufactured items such as cars and plastics eat up fossil fuels and pollute the environment, it is the systems that support, encourage and indeed rely on unbridled consumerism that are the most deadly human inventions. These

7 Donella Meadows et al, *The Limits to Growth: A Report for the Club of Rome's Project on the Predicament of Mankind* (Potomac Associates Book/Universe Books, 1972); Donella Meadows, Jorgen Randers and Dennis Meadows, *The Limits to Growth – The 30-Year Update* (Earthscan Ltd, 2004) xiv.

8 Ibid.

9 *Global Footprint Network* (n 5).

10 Maria Sandberg, Kristian Klockars, and Kristoffer Wilén, 'Green Growth or Degrowth? Assessing the Normative Justifications for Environmental Sustainability and Economic Growth Through Critical Social Theory' (2019) 206 *Journal of Cleaner Production* 133.

11 Spash (n 3) 1093.

12 Paul G Bain et al, 'Promoting Pro-Environmental Action in Climate Change Deniers' (2012) 2 *Nature Climate Change* 600. This article states (at 600) that 'since 2008 the number of deniers of anthropogenic climate change has climbed to one-third or more of the population in high-carbon-emitting countries such as the United States and Australia'.

13 Pörtner et al (n 4).

14 Spash (n 3) 1088.

systems construct consumption as a positive and promote a 'binge and purge' lifestyle.[15]

The economic systems that the law scaffolds are thus antithetical to environmental protection. One of the most potent examples of this legal scaffolding is the concept of the corporation. As Susanna Kim Ripkin points out, creating and defining 'a legal person' is '[p]erhaps the most powerful act of the law'.[16] The corporation creates a wilful blindness; an 'ecological disconnect' where the corporation exists as separate from its human creators and the environment which it exploits for profit.[17] Parliaments and courts have maintained these fictional characters which allow investors to distance themselves both financially and psychologically from the damage. This exegesis will now provide a brief history of the corporation, one of the law's most destructive fictions. In our judgment, the intergalactic criminal court has sought out and located the corporate form as the guilty party which has led to the collapse of the biosphere.

The Corporate Frankenstein

The corporation stands as one of the most iconic constructs of human civilisation. It has been argued that the value of the corporation and its ability to limit liability outstrips even that of electricity: '[T]he limited liability corporation is the greatest single discovery of modem times' writes Parker. 'Even steam and electricity are far less important than the limited liability, corporation'.[18] Since its inception, the influence and strength of the corporate structure have steadily increased to a point where its wealth, power, and political influence are at a level unmatched throughout human history. Also unmatched throughout human history is the ability of the corporation to damage the environment in which it operates. With this power to negatively affect the physical environment comes an increased need to conduct the corporation in a socially responsible manner.[19] When examining this claim, it

15 Jeff Ferrell, *Empire of Scrounge: Inside the Urban Underground of Dumpster Diving,* Trash *Picking and Street Scavenging* (New York University Press, 2006). See also Marie Kondō, *Spark Joy: An Illustrated Master Class on the Art of Organizing and Tidying Up (The Life Changing Magic of Tidying Up)* (Ten Speed Press, 1st ed, 2016) – the best-selling book that recommends throwing out any items that do not 'spark joy'; and Vance Packard, *The Waste Makers* (David McKay, 1960).

16 Susanna Kim Ripkin, *Corporate Personhood* (Cambridge University Press, 2019) 21.

17 Hanna Ahlström, Amanda Williams, and Sigurd Sagen Vildåsen, 'Enhancing Systems Thinking in Corporate Sustainability Through a Transdisciplinary Research Process' (2020) 256 *Journal of Cleaner Production* 1.

18 David Parker, 'The Company in the 21st Century. Piercing the Veil: Reconceptualising the Company under Law' (2015) 10(2) *Journal of Law and Governance* 1, 1–2.

19 Rob McQueen, 'An Examination of Australian Corporate Law and Regulation 1901–1961' (1992) 15(1) *University of NSW Law Journal* 1.

becomes apparent that the corporation resembles a 'Frankenstein's monster' unleashed upon an unsuspecting world;[20] it is immortal, amoral, and self-absorbed.[21] By conceptualising the corporation as a legal 'person', the law has given it enormous power, with rights often exceeding those of biological humans, whom corporations utilise as a disposable resource. Corporations can hold property, employ people, enter into contracts, sue and be sued.[22] In some jurisdictions, they can even exercise 'human' rights.[23] Many biological people are restrained from these rights and responsibilities, being regarded as not having capacity.[24] If the capacity of the corporate person were to be assessed, many would be classified as psychopaths.[25]

Corporate structures appeared as early as 50 BC with the Roman *collegia*.[26] The first mining corporation and oldest limited liability company was created by charter of King Magnus Eriksson in Sweden in 1347. It was owned by members of the Swedish Royal Family and aristocracy.[27] This novel structure enabled the accumulation of wealth from many investors to be directed towards a desired end goal. The impacts of the Stora company continue to this day. The copper mine has left behind polluting metals; damage which may be irreversible.[28] It continues to operate as the world's second largest paper manufacturer, now based in Helsinki, outliving the oldest human person by several hundred years.[29] The endurance of the corporate form means that the corporation will eventually outlive us all.[30]

20 As Penny Crofts writes, 'monstrous metaphors of corporations abound in legal and academic writing, comparing corporations with Frankenstein's monster' – Penny Crofts, 'The Corporate Monster Metaphor' (2022) 26 *Law Text Culture* 73, 74. See also Timothy Peters, 'I, Corpenstein: Mythic, Metaphorical and Visual Renderings of the Corporate Form in Comics and Film' (2017) 30(3) *International Journal for the Semiotics of Law* 427.

21 Crofts (n 20).

22 Harry Glasbeek, *Class Privilege: How Law Shelters Shareholders and Coddles Capitalism* (Between the Lines, 2017) 16.

23 Example, United States of America. See Turkuler Isiksel, 'The Rights of Man and the Rights of the Man-Made: Corporations and Human Rights' (2016) 38(2) *Human Rights Quarterly* 294.

24 Ripkin (n 16).

25 Joel Bakan, *The Corporation: The Pathological Pursuit of Profit and Power* (Little Brown, 2005).

26 Reuven S Avi-Yonah, 'The Cyclical Transformations of the Corporate Form: A Historical Perspective on Corporate Social Responsibility' (2005) 30(3) *Delaware Journal of Corporate Law* 767, 775.

27 Christian Stadler, *Enduring Success: What We Can Learn from the History of Outstanding Corporations* (Stanford University Press, 2011) 6; S Rydberg, 'Stora Kopparberget: The Great Copper Mountain' (1988) 60(1) *Transactions of the Newcomen Society* 1, 1.

28 Anna S Ek et al, 'Environmental Effects of One Thousand Years of Copper Production at Falun, Central Sweden' (2001) 30(2) *Ambio* 96.

29 David Whyte, *Ecocide: Kill the Corporation Before It Kills Us* (Manchester University Press, 2020) 2.

30 Ibid.

The development of the form now recognisable worldwide took many centuries to develop and had a number of setbacks. Early corporations, like Stora, operated only by Royal Charter.[31] Incorporation was a highly controlled and rarely granted privilege.[32] There were strict limits on what the corporation could do, and these were usually large public projects.[33] This did not mean the corporation could not wield immense power. Some of the States in the United States of America began as corporations,[34] and a number were a part of the colonial expansions into areas such as Indonesia, India and South America.[35]

A prime example of this was the South Sea Company. The South Sea Company was formed for the purpose of transporting slaves from Africa to sugar plantations in the Pacific Ocean and Latin America.[36] 'Over its entire trading lifetime, the [South Sea Corporation] forced nearly 42,000 people to leave the African Coast. It disembarked almost 35,000 people meaning that just over 7000 people died on the crossing'.[37] However, the horrors of the slave trade did not dissuade investors from embarking on a frenzied buy up of stock[38] The rapid inflation and subsequent deflection of share prices has drawn much academic attention as 'it formed an important part of the first major speculative boom and bust in European stock markets'.[39] Following the collapse of the South Sea Bubble, the Parliament of Great Britain acted quickly to prevent such events occurring again, enacting the *Bubble Act 1720*.[40] However, this Act was repealed just over 100 years later in 1825, leading to the revitalisation of the corporate structure.[41]

31 Ripkin (n 16) 24.
32 Ibid.
33 Ibid 25.
34 For example, Massachusetts, Georgia.
35 For example, the English East India Company, which 'found India rich and left it poor': Nick Robins, 'Loot: in Search of the East India Company, the World's First Transnational Corporation' (2002) 14(1) *Environment & Urbanization* 79. Corporations have relentlessly ignored the rights and cultures of Indigenous peoples during and after these expansions. See: David E Toohey, 'Indigenous Peoples, Environmental Groups, Networks and the Political Economy of Rainforest Destruction in Brazil' (2012) 17(1) *International Journal of Peace Studies* 73. See also Ciaran O'Faircheallaigh and Saleem Ali, *Earth Matters: Indigenous Peoples, the Extractive Industries and Corporate Social Responsibility* (Routledge, 2017).
36 Helen Paul, 'The South Sea Bubble and the Erasure of Slavery and Impressment' (2021) 102(7) *English Studies* 888, 888.
37 Ibid 890–891.
38 Richard S Dale, Johnnie EV Johnson, and Leilei Tang, 'Financial Markets Can Go Mad: Evidence of Irrational Behaviour During the South Sea Bubble' (2005) 58(2) *Economic History Review* 233.
39 Ibid 233. History repeats itself with the frenzied purchasing of cryptocurrencies in recent times. See 'Crypto Bubbles', *Crupto Bubbles* (Web Page, 2023) <https://cryptobubbles.net> for a visual display of current cryptocurrency prices.
40 *Royal Exchange and London Assurance Corporation Act 1719* (UK).
41 Ibid.

At this time people who contributed to the operation of a corporation were subject to unlimited liability and could lose everything if the company failed. With the advent of limited liability companies in Britain, under the *Limited Liability Act 1855* (UK), the attractiveness of investment through the corporate structure increased.[42] Under this new model, shareholders were protected against losses beyond their original contribution, but there were no limits on their potential gains.[43] Adam Smith's influential *Wealth of Nations* heralded a new era of laissez-faire liberalism.[44] As Paddy Ireland writes: '*Ir*responsibility was now firmly built into the corporate legal form, and it still is. The no-obligation, no-responsibility, no-liability nature of corporate shares permits their owners ... to enjoy income rights without needing to worry about how the dividends are generated'.[45]

The concepts of limited liability and the separate legal personality were entrenched in England by 1856 and in the United States by the end of the nineteenth century.[46] It was at this point that the corporate structure we know today became crystallised. The *Salomon* case[47] confirmed the concept of the separate legal personality in 1897. This case is seen as a turning point in English company law as it dealt directly with the question as to whether a sole trader can incorporate for the purpose of limiting their liability alone.[48] It overturned a decision from the Court of Appeal where it was argued that Mr Salomon had formed a company as a 'mere scheme' to limit his liability, something which had been judged by the Court of Appeal as dishonest and undesirable.[49] The concept of limited liability was not new and dates back to Roman law.[50] However, the Privy Council decision in *Salomon* is recognised as ground-breaking in finding that it did not matter who the shareholders of a company were or what the purpose of their incorporation was – as long as the company was correctly incorporated under the rules outlined in the *Companies Act 1862*.[51] This black letter judgment belied its significance and consequences. As Gonzalo Puig asserts, '[b]y establishing that corporations are separate legal entities, Salomon's case endowed the company with all the

42 Paddy Ireland, 'Limited Liability, Shareholder Rights and the Problem of Corporate Irresponsibility' (2010) 34(5) *Cambridge Journal of Economics* 837, 845.
43 Ibid.
44 Adam Smith, *An Inquiry into the Nature and Causes of the Wealth of Nations* (Strahan and Cadell, 1776).
45 Ireland (n 42) 845 (emphasis in original).
46 Bakan (n 25) 13.
47 *Salomon v Salomon* [1897] AC 22 ('*Salomon*').
48 Ibid.
49 Ibid 33 (Lord Halsbury).
50 Robert W Hillman, 'Limited Liability in Historical Perspective' (1997) 54(2) *Washington and Lee Law Review* 615.
51 *Salomon* (n 47) 23.

requisite attributes with which to become the powerhouse of capitalism ... Salomon's case has promoted fraud and the evasion of legal obligations'.[52]

The attractiveness of limited liability led to an exponential rise in the number of corporations operating. Ripkin reports an increase from 335 corporations in 1800 in the United States, rising to 300,000 by 1916.[53] New conceptions of the corporation accompanied this growth or perhaps caused it. During this time, the corporation became associated with individualism and entrepreneurship, rather than as a privilege granted by the State.[54] Liberalism supported leaving companies alone to create their own charters and a relaxing of regulation.

Connected with the idea of limited liability is that of externalities. This is the corporate entity's mechanism to divorce itself from the damage it does to third parties, including the environment.[55] It has been argued that 'the corporation is an externalising machine in the same way as a shark is a killing machine'.[63] If the cost of preventing possible death and destruction to other persons or the environment is greater than the cost of paying for this destruction once it has happened, then corporate logic dictates that the former is the better option.[56] As Bakan asserts,

> [t]he corporate ... machine is deliberately programmed, indeed legally compelled, to externalise costs without regard for the harm it may cause to people, communities and the natural environment ... the corporation's mandate ... to pursue its own self-interest ... actually propels corporations to break the law.[57]

The corporation has 'no moral or other compulsion to reduce risk, or obey the law, except based on what it will cost the company's bottom line'.[58]

During the 19th century, mergers between corporations led to a concentration of corporate power.[59] A study conducted in 2011 revealed that 737 companies controlled around 80% of the world's wealth.[60] Since then, this concentration of wealth and power continues to climb.[61] In the 21st century, as globalisation proliferates, the capability of corporations to exist in

52 Gonzalo Villalta Puig, 'A Two-Edged Sword: Salomon and the Separate Legal Entity Doctrine' (2000) 7(3) *Murdoch University Electronic Journal of Law* 1.
53 Ripkin (n 16) 29.
54 Ibid 32.
55 Anthony Biglan, 'Corporate Externalities: A Challenge to the Further Success of Prevention Science' (2011) 12(1) *Prevention Science* 1.
56 Ralph Nader, *Unsafe at Any Speed* (Grossman Publishers, 1965).
57 Bakan (n 25) 72.
58 Ibid 79.
59 Ripkin (n 16) 34.
60 Stefania Vitali, James B Glattfelder, and Stefano Battiston, 'The Network of Global Corporate Control' (2011) 6(10) *PLOS ONE* 6.
61 Whyte (n 29) 7.

transborder spaces has vastly increased their ability to influence their environment. Calls for corporate social responsibility began to appear around the 1990s after a series of notable collapses and lapses in corporate behaviour.[62] Milton Friedman famously stated that any attempt at genuine corporate social responsibility is a 'fundamentally subversive doctrine'.[63] Friedman appears to have been correct in this appraisal. It is this juxtaposition between the need to preserve the environment in which it operates, and the mandated directive to maximise profit, that puts the corporation at odds with both itself and the continued existence and safety of the rest of the planet.

The following judgment is delivered in an imagined future where planet Earth has been rendered uninhabitable to almost all life. Beings from another galaxy discover Earth and are assessing it as a possible site for colonisation, when they discover past existence of an intelligent civilisation. They tap into long abandoned databases, discover Earth's case law, and deliver their judgment based on this law out of respect for the planet's lost culture. The alien beings are surprised to find evidence of not only biological persons but also artificial (or corporate) persons, who appear to have reached God-like status in human society. The scenario imagines the continuation of the corporation which can exist in perpetuity without the requirements needed for humanity. In this judgment, it is these non-biological entities that are held responsible for the ultimate demise of the planet's biodiversity and intelligent life forms. The judgment holds the corporation to account by stripping it of its rights to control the planet and its resources. The planet is thus rendered available for colonisation by other alien species, although it is recognised that many years of rehabilitation will need to take place before complex biological life forms can be reintroduced to the planet's surface.

Judgment

Case: *Former People of Planet Earth* (ex parte in extinctio) *v the World Corporate Alliance*
Court: Universal Court of Justice 15th Inter Galactic Circuit – ZXBeta
Judge: Justice ZXBeta
Proceedings: Notice of Habitability for Colonisation Non-Vitae

Combined Hearing and Judgment 5XG327.263-ZXBeta. Translated into English 5XG327.263.0000012.

62 Examples are numerous: Niger Delta's Exxon Mobile oil spill, the Bhopal disaster, the Exxon-Valdez oil spill, and Three Mile Island meltdown, to name just a few. See Hershey H Friedman and Clifton Clarke, 'Deadly Consequences of Emphasizing Profits Over Human Life: How Corporate Greed Has Caused the Death of Millions' (2022) 5(3) *Journal of Intercultural Management and Ethics* 19.

63 Milton Friedman, *Capitalism and Freedom* (Chicago University Press, 2002) 133.

Being a judgment pertaining specifically to planet 2AXPP#345 and drawing upon the combined databanks of its former inhabitants, Justice ZXBeta will deliver the judgment in the format commonly used by Earth's ancient *Homo sapiens* civilisation. This format replicates the ancient data collection and organisation system known by *Homo sapiens* as 'The Law'.

Be it known that this most Noble and Honourable Court is now in session, time 5XG327.263.0000000, Justice ZXBeta presiding.

Following an Application to the Universal Court of Justice, by the Galactic Survey Organisation, seeking a declaration of habitability non-vitae, this Court will now hear submissions with respect to the *Former People of Planet Earth v the World Corporate Alliance.*

This matter is to be heard and delivered *ex parte* by the Court as no *Homo sapiens* remain alive on the planet following the collapse of the ecosystem in the late 23rd-century Earth time leading to a mass extinction event.

Collapse of the biosphere occurred some 500 Solar orbits ago, rendering the planet effectively uninhabitable. The court notes that some arthropod life forms continue to survive on the planet; however, these do not meet the Galactic Survey definition of sentient life.

Accordingly, the Court is prepared to consider the evidence brought before it by the Galactic Survey Organisation with respect to the total loss of all major life forms on planet designated 2AXPP#345.

At approximately 5XG327.243 the Galactic Survey Organisation was scoping the anterior quadrant of spiral Galaxy 3217542 when it located an unoccupied habitable planet that appeared fit for colonisation.

Upon arrival at the location, it was discovered that the planet had been previously dominated by a sentient species self-identifying as '*Homo sapiens*', a primitive herding mammalian bi-ped of otherwise unremarkable characteristics.

Although no *Homo sapiens* remain alive to corroborate the evidence brought before the court, it is submitted that the Geological Survey Organisation, having no conflict of interest in this matter, is trusted to bring the records of the extinct *Homo sapiens* before the court.

The *ex parte* representatives of the former inhabitants of planet 2AXPP#345, hereinafter referred to as 'the people', seek to have a historiological record of the unfortunate circumstances leading to the avoidable loss of all mammalian life on the planet previously referred to in the Applicant's records as 'Earth'.

Submission by the Representatives for the People of Planet 2AXPP#345

1. The planet designation 2AXPP#345, located in the minor spiral Galaxy 3217542, was referred to by the former inhabitants as 'Earth' which for the purposes of this hearing is referred to with the same terminology, or as 'the planet'.

2. Approximately 1100 solar years before the collapse of the ecosystem, a fictitious legal structure, the corporation, was created within the upper echelons of the planet's social strata.
3. These structures were created and labelled according to what Earth records describe as 'legal persons' giving them legal and social standing.
4. This structure appears to have been protected by a 'corporate veil' that evidence suggests may have a religious significance, and was protected at law.
5. The corporation was not dependent on the ecosystem for existence and subsisted beyond the lifespan of the *Homo sapiens* that created it.
6. It appears that sometime around the 21st-century Earth time, the corporation became the dominant entity on the planet just prior to the collapse of the biosphere.
7. While corporations technically continue to exist, they are no longer capable of functioning as the humans that served them are now extinct.
8. These corporate 'legal persons' were able to buy and own 'property' (an archaic means of unfairly distributing the bounty of a planet) and to engage in a large variety of 'legal' actions.
9. The corporation was ultimately treated with greater rights than the individual inhabitants of the planet, being able to engage *Homo sapiens* in 'employment' – a structure where labour is performed for the corporate legal entities.
10. The Galactic Survey Organisation identifies this 'employment' as a form of structured 'slavery' (an ancient and outlawed institution where one life form performs semi-bonded servitude to another, for little or no gain, for the benefit of the other).
11. Ultimately history has shown that this structure was not for the common good but to serve the purposes of the rather short-sighted ruling classes within the planet's various tribal systems.
12. What the *Homo sapiens* failed to realise was that the structure that they created would, within just a few centuries, devour the planet's resources, leading to the total failure of the ecosystem in the 23rd century of Earth (Universal Expansion Date 5,596,326,743).
13. The legal entity that the *Homo sapiens* named 'Corporation' became able to influence the different tribal groups on the planet to allow it to conduct what may have been a religious observance which they referred to as 'business', to the detriment of the planet as a whole.
14. During the late 20th-century Earth time, or UED 5,596,325,450, it became apparent that the behaviour of these corporations was not sustainable and would eventually overtake the biosphere and lead to total environmental collapse.
15. Many of these corporations then attempted to hide or destroy the evidence of damage to the biosphere and actively participated in silencing,

confusing, or discrediting any *Homo sapiens* that attempted to raise the alarm for the rest of the herd.

16. For the following 150 years, only limited action was taken to reverse the damage being caused by the corporation and instead the blame was laid upon the tribal leaders, of the different landmasses around the planet, who refused to recognise the extant evidence that the planet's ecosystem was being destroyed en-masse by these 'corporations'.
17. As the water became poisoned, little was done to preserve the planetary aquifers as individual *Homo sapiens* were unable to see the collective damage being done to the planet's aquifers.
18. This led to a general malaise within the herd unless an individual *Homo sapiens* was directly affected itself.
19. Similarly, the planet's oceans became polluted, poisoned, and contaminated, and life in the oceans began to die off, and nothing was done to prevent that, which, according to the planetary records, was an avoidable event at that time.
20. The areas normally set aside for the generation of oxygen by plants on land and sea were being compromised as the corporations were destroying the oxygen-generating habitat with an unrelenting purpose throughout the 20th, 21st, and 22nd centuries, Earth time.
21. It is noted by the Galactic Survey Organisation that, even as late as the mid-21st century, the situation was salvageable with minimum loss of habitat ecosystem and life on the planet, if the tribal groups had arranged to limit the power they had previously granted to the corporation.
22. Historical record shows that even at this late stage in the disaster nothing was done at that time to slow the corporate erosion of habitat ecosystem and culture on that planet.
23. The Galactic Survey Organisation surmises that it is possible that the *Homo sapiens* worshipped the corporation as their deities, although at this time there is insufficient evidence to definitively state whether this was the case.
24. The Galactic Survey Organisation also notes that the planet itself and all other forms of life had survived and developed over a period of several billion years to a situation where – until the creation of the corporation by the *Homo sapiens* – the planet had flourished as had all life forms living on it.
25. It may be that this worship of the corporation as a deity was what led the *Homo sapiens* to extinction, although it will be difficult to ascertain due to the lack of historiological records on that planet.
26. On or about 2065 Earth time, there was a vast conflict which resulted in many tribes across the land masses engaging in a structured and organised attempt to render each other extinct. It is at this time that pollution on the landmasses and water increased exponentially, putting more pressure on those remaining areas that could still sustain life.

27. At around 2085 Earth time, the larger mammals living in a wild habitat began to become extinct, and by 2125 Earth time no larger animals existed on the planet other than those used by industry for consumption by the *Homo sapiens*.
28. With the collapse of the marine biological ecosystem in 2140 Earth time, the loss of all major life on the planet became unavoidable.
29. It is submitted that most of the destruction to the ecosystems and habitat on that planet was a direct result of the short-sighted actions of the corporations and those that worshipped them.
30. The corporation stands before us as an exemplar of how to destroy all biological life within an ecosystem by prioritising non-living over living entities.
31. The Court would hold not only the creators and curators of the corporation but also the tribal leaders who allowed them to thrive and flourish at the expense of the herd for personal gain, as liable for the mass extinction event.
32. As a religion, the corporation has been shown to be ultimately fatal to its followers, of which it can only be surmised the *Homo sapiens* were aware.
33. It is submitted that these behaviours by the corporations and those that supported them, or worshipped them, amounted to gross negligence in relation to a foreseeable catastrophe that led to the complete loss of all major life on the planet 2AXPP#345.
34. An immediate declaration of habitability non-vitae is sought with respect to the planet by the Galactic Survey Organisation.

Decision

On hearing submissions before the Court by the Applicant and the Galactic Survey Organisation, it is apparent that the *Homo sapiens* was the architect of its own extinction, and that of all life on the planet designation 2AXPP#345, located in the spiral Galaxy 3217542, previously referred to as Earth.

It is unknown how many *Homo sapiens* were able to leave spiral Galaxy 3217542 – or whether any of them encountered other intergalactic travellers and were able to relocate to another habitable zone in the Galaxy – which this Court acknowledges may yet lead to more of this species being found in future.

This Court holds the corporation and all who worshipped it, supported it, or operated it, throughout planet Earth as guilty of culpable negligence causing the mass extinction of all life on the planet.

It is the view of this Court that, in light of the unmistakable suffering that such action would have caused this species in its final hours, the punishment has already been sufficiently applied by their own hand in this regard and no more shall be ordered by this Court.

This Court also orders that any *Homo sapiens* located in future be temporarily held for examination to discover whether this unfortunate and avoidable planetary disaster was the result of a culpable negligent event, an inherent genetic disorder, or a religious observance.

It is further recommended by the Court that any *Homo sapiens* located in future be discouraged from breeding.

Planet 2AXPP#345 is declared non-vitae and therefore may be immediately colonised.

Copy of this judgment is to be transmitted to all intergalactic outposts throughout the colonising fringes and to serve as a warning to any planetary engineers that may consider creating similar structures having the potential to irreversibly damage ecological, social and biological habitats in future.

The Honourable and Most Accurate ZXBeta holds this *ex parte* Hearing and Sentencing now at an end.

Dated: this day 5XG327.243 at the Universal Court of Justice 15th Inter Galactic Circuit –ZXBeta

Duration of Hearing and Judgment and combined translation: 0.012 seconds.

Galactic Justice Department Circuit 8255590.00001269-5XG327.263

Chapter 19

More-Than-Human Relations on the Third Rock from the Sun

Michelle Lim

Judgment Day, 22 April 2050 – Germinator 3000

I am rock. I cover this planet. In places, the salt and fresh water cover me. The surface of my skin, uneven – rises and falls across time. Rock time. I have been here a while.

A few seconds ago, in Rock time, the rise of the bipedal hominid Skin-Beings (the ones with the opposable thumbs) almost brought an end to life – all life on this planet. I still remember that flicker of time when *some* of the Skin-Beings thought they could rule for the living and the non-living. It wasn't always this way. *Hashtag*: #NotAllBipeds. Most of the history of the Skin-Beings is characterised by not a 'peaceful co-existence' but rather a symbiotic flourishing that the skin-beings helped to nurture.

Do you remember when the Pale Skin-Beings thought they were conquering the world? When a faction of them created a system of organising their societies where a handful of (usually) old males sat at a high table. A *table* they inexplicably called a 'bench'. Atop their craniums, the old males wore hair, white curly hair, not of their own, on top of their own. It was interesting to watch. Decisions for the present were made by looking to the past. And in fact, there was an insistence that rulings should align with the past.

They called the bench-top system of law the 'common law'. This mode of law was instated by violent coercion across vast swathes of the planet across an area that came to be known as the 'common' wealth. But neither law nor wealth were common to all.

Skin-Beings were, at first, the only subjects of this 'legal' system. Okay, well at first only mostly Pale-Males. This system was founded on hierarchy. Principles of 'coverture',[1] for example, minimised the status of Pale-Females; the weight of Empire and prejudice subjugated everyone else. This faction of

1 The common law doctrine of coverture held that upon marriage the legal personality of the wife was subsumed by that of the husband. Blackstone's commentary states: 'By marriage, the husband and wife are one person in law: that is, the very being or legal existence of the woman is suspended during the marriage, or at least is incorporated and consolidated into that of

DOI: 10.4324/9781003389569-23

the Pale-Skins took their system to the lands they forcibly occupied. These legal systems remained – decades, and even centuries after the Pale-Skins had retreated: the Stockholm syndrome of conquest dotted across the Globe as an enduring post-traumatic scar.

Eventually, this system evolved – slightly. First, all Skin-Beings, then Rivers, then Mountains were considered to have legal recognition within this particular Pale-Skin system. But still, only Skin-Beings were deemed capable of decision- and rule-making within this system.

Ha! Funny to think of a time when the Skin-Beings thought that they were the only ones who ran the world.

Are you okay? You look confused. Which epoch are you visiting from, my dear? Ah, right, you have yet to learn how decisions are made in the Terracene. Come, come – allow me to give you a crash course on how multiple beings – the creatures and the rock, the winds and the rains, the light and the dark, and how we converse, deliberate not only for now but for before and for the yet to come...

Hmmm, where to start? Judgment Day is as good a place as any I suppose. I remember it as if it were...oh never mind, you don't understand Rock time anyhow...

***

Rock was omnipresent on the watery planet some called 'Earth'. She was here, and continued to be here – across time and across space. Was she all-knowing? Yes. But not all-powerful. Nor did she have any desire or ambitions of omnipotence. Rock knew some referred to her planet as the '3rd rock from the Sun'. She was proud, however, of her role as *part of* the watery planet. She harboured no grand illusions of representing the entire celestial body – she respected her co-travellers, and the other celestial bodies, too much for that. Nonetheless, Rock's connection to Planet 3 since its beginnings; and, paradoxically, her dynamism across all that Planet 3 had endured; meant Rock was highly respected – within and beyond the planet's atmosphere.

'Planet 3 inhabitants', Rock called out across the Biosphere, 'Judgment Day –' Rock paused for dramatic effect, 'iiiiiis HERE!'.

'Trust in your visions of radical possibility! Now brace yourselves, for the Fission!'

It was the 22nd of April 2050. This Judgment Day, like each of the seven prior Judgment Days of the Cenozoic era, was a momentous occasion. Rock had been the Judgment Day Returning Officer for as long as anyone could remember.

---

the husband' William Blackstone, *Commentaries on the Laws of England* Vol II (Clarendon Press, 1765) 442–445.

Judgment Day of this new epoch had come upon Planet 3 at an intensity, a speed, and a scale like no other previous transition. But the Fission, the Fission happened in the same way it always did: with the flash of a billion neon lights and a silencing boom that made the horizon reverberate like a guitar string – a fitting representation of such a gargantuan release of energy. It was, after all, the end result of the nucleus of every atom across the planet splitting, in unison.

This time the effects of the Fission mirrored the self-cancelling excesses of the world.

Each of Planet 3's creatures, beings and entities had been split apart at the atomic level and then reconstituted. This happened at a speed of $(5.9722\times10^{24}(c^3))^5$ ms^{-1},[2] ie like super, super, super-duper fast! So quickly that the cellular and constitutional integrity of each part of Planet 3 was preserved – unharmed. The effect of the energy created, however, was that each component of Planet 3 was catapulted into space – sustained and protected in a custom cocoon.

Each tiger in its own cocoon; each fern; each earthworm; each river; each mountain – suspended above their marble of blue, green, white and brown.

Do you know that feeling when you get dumped by a wave? Yes, an ocean wave. When you get flipped around and around and find yourself spat out the other side? And you sit stunned, relieved, but also, weirdly, exhilarated. Yeah… that is what this feels like.

All of the components of Planet 3 sitting above Planet 3, looking back at Planet 3 – as they had left it. That's right, as they had left it. The speed of the Fission was so great that the Planet 3 cocoons were in fact multiple light years away from whence they had come. As a result, the world they looked out on was indistinguishable from the world they left at the time of the Fission: the same concentrations of carbon fill the atmosphere; the same level of wellness (or otherwise) of each individual, each species and each ecosystem; all finished and unfinished business in the same state of (un)progress.

***

The cocooned sit in the darkness of space. Through the darkness, like a little night light, Planet 3 makes her presence known. Planet 3's *current-former* inhabitants watch their planet intently. Her glow sustaining their attention the same way that a log fire would on a cold night. As they watch, Simon and

2 Where c is the speed of light and the mass of Planet 3 is 5.9722×10^{24} kg. The speed of light is cubed to derive a volumetric account of the planet, while the multiplication of the mass of Planet 3 and c^3 is then amplified to the power of 5 to recognise each of the Precambrian, Palaeozoic, Mesozoic, Cenozoic and Quaternary periods of Planet 3's history.

Garfunkel's 'Sound of Silence',[3] transmitted by light quantum packets on the Ether, permeates each cocoon.

♫ ♫ ♫ ♫ ♫ ♫

QUICK!
DO YOURSELF A FAVOUR! PUT 'SIMON AND GARFUNKEL SOUND OF SILENCE' INTO YOUR SEARCH ENGINE.
FIND THE RECORDING. LISTEN TO THAT SOUND! SPEND TIME WITH THE LYRICS....

♫ ♫ ♫ ♫ ♫ ♫

The music fades out as Rock's voice, smooth as a Classical Music Station announcer, fades in.

'Hello, and welcome back to the Original Position!' Rock calls out at the cocoons – a mass of suspended shimmering pearly ovals glistening in the light of the Sun. 'As you know, we are here due to the planetary transition. A shift which has occurred far sooner than any previous changeover. We all sit at the end of the Beginning and at the beginning of the End. We all know, that if Judgment Day was called any later, it would be "Hasta la vista" – for all of us'.

'Today', Rock continues, 'here, at the Original Position, each of you will deliver Judgments of Radical Possibility from behind the veil of illumination of Planet 3's atmosphere'.

'These judgments will provide the seeds to sustain the stability of our planet for at least 1.806 million years', Rock declares. And then in a heightened pace of speech, she reassures Planet 3's entities: 'But no pressure Judges, remember the terms and conditions of the Original Position's return policy allow a trial period until 3000CE'.

'Each of Planet 3's entities will provide a pronouncement on the pathways to radical possibilities for Planet 3. The veil of illumination provided by Planet 3's atmosphere allows each individual entity a searing clarity of their biophysical and socio-cultural heritage. However, behind the veil of illumination no entity knows, for sure, what incarnation they will take in the Return to Planet 3. No one knows what characteristics they will have in the Return; whether these characteristics will allow them, their offspring or their future generations to thrive after planetary reset', Rock explains.

Rock has the Jupiter-sized task of distilling and synthesising all the judgments into a set of principles. Every judgment will be read. Every judgment

3 'The Sound of Silence', first released by Paul Simon and Art Garfunkel on 19 October 1964. Paul Simon, *The Sound of Silence* (Wednesday Morning 3AM, Columbia Records, 1964).

matters. Each entity of Planet 3 will have access to the judgments – on the Tabula.

To view the judgments, you need to plant your feet, fins or any other conducting component of yourself onto or into the soil. With another part of yourself (for example, the end of a proboscis, the tendril of a fern or the palm of a hand) connect with Rock in her various forms. It is then that the Tabula will appear. The Judgments that appear to *you* on the Tabula will be the three judgments *YOU* need the most in this moment of time.

'You have one rotation of Planet 3 on its own axis to complete this task', says Rock. 'Trust in your visions of radical possibility!!'

RETURN! RETURN NOW TO THE RECORDING!
PAY ATTENTION TO WARNING SIGNS.

***

Judgment

COURT OF THE ORIGINAL POSITION

REPORTS OF JUDGMENTS, ADVISORY OPINIONS AND ORDERS
COURTING
RADICAL POSSIBILITY
ON PLANET THREE

JUDGMENT OF 22 APRIL 2050CE

2050CE

Official citation:
On the radical possibilities of Planet 3: 2050–3000 (initially) – and beyond, Judgment, OP Reports 2050CE, 1.

Notice: This copy of the Reasons for Judgment reflects that which is inscribed into the crust of Planet 3 as of 23 April 2050
COURT OF THE ORIGINAL POSITION

YEAR 2050CE

22 APRIL 2050 CE

COURTING RADICAL POSSIBILITY ON PLANET THREE

There is light in the darkness – Do not start in 2050CE – The time to act was yesterday – Now is still ok – Now is sooner than tomorrow.

Value change is critical – Happiness is not in things – Beware the technological promises of 'progress'.

Challenge the status quo – History is culture – Culture is Nature – Do not forget History – History will not forget – Join the multi-species alliance – mutualistic symbiosis.

Return to a future of multi-species judging – To a time where stories were law.

Present: All of Planet 3 at 22 April 2050 CE.

Returning Officer: Rock.

On the radical possibilities of Planet 3: 2050–3000 (initially) – and beyond.

THE COURT,
composed as above,
gives the following judgment:

1. The question upon which the judgment of the Court has been requested is set forth in the resolution of Quaternary entities 24601. The text of the resolution reads as follows:

   The Assembly of Planet 3 entities,

   Conscious of the serious risks posed to the continuous existence of the planet,

   Mindful of the obligations of every planetary entity to contribute to planetary integrity,

   Recalling its resolutions including 8888 (XXX) of 1 January 1950, 22/3 of 22 March 1984, 4444/WTF of 25 April 2023, HE/LP of 22 April 2049, in which it declared the continued threats to the planet caused by certain Skin-Being activity,

   Welcoming multi-species coalitions for planetary progress,

   Convinced that a radical planetary reset is the only guarantee against Planet 3 collapse,

   Decides, pursuant to the desires of Earth inhabitants to request the Court of Original Opinion

   urgently to render its judgment on the following question: 'What are the pathways of radical possibility for Planet 3?'
2. The Court decided to hold a global public sitting, opening on 22 April 2050, at the Original Position.
3. Earth inhabitants took note of the state of Planet 3, on 22 April 2050, over a period constituting one rotation of Planet 3 on its own axis.

Tabular currently showing:
MATANG MOUNTAIN; KILIMAKYERO LICHEN; THE LITTLE BUSH WARBLER.

Matang Mountain

1. As I look out on Planet 3 from the Original Position, behind the Veil of Illumination... now, more than ever, I am acutely aware that I am no grand monolith. I stand only 911m tall. At my feet lies *Kampung Matang* – the Malay village that shares my name. Five times a day, the village *surau* calls the faithful to prayer. Meanwhile, on the tip of my nose, at about 800m above sea level, sits Mount Matang Sri Maha Mariamman Temple – the oldest and highest Hindu temple of the Land of the Hornbills.
2. I remember a time, not so long ago, when I could see Princess Mountain on the horizon every day; and each night the fireflies danced with the city lights in the distance. Meanwhile, the celebrations of light in the village and at the Temple were my favourite parts of Planet 3's rotations around the Sun. With the new moon on each 15th day of *Kartik*, the Temple celebrated Deepavali – the Festival of Lights. In the village, the biggest celebrations would occur each year at *Hari Raya* – marking the end of Ramadan. Many, many moons ago, traditional oil lamps – *pelita*, each hung off a metre-high wooden pole, would adorn the trails between houses. More *pelita* were set up in an elaborate arch formation at the entry to the village. Tiny colourful electric flashing lights eventually came to decorate each house alongside the *pelita*. These additions adding nothing but optical noise.
3. My ruling is simple. It is a celebration of the beauty of the mundane. It is a call to value what we have, what we had, and what needs to endure.

The Journey of Emperor Freddy – A Celebration of the Mundane

4. If you wanted to make the 3kilometre hike up to the Mount Matang Sri Maha Mariamman Temple you needed to first find the trail up the mountain. *Hint:* Turn left at Mile 10 of Matang Road; make your way through *Kampung Matang*. You will find me at the end of the village. Just after the Kingdom of Emperor Freddy.
5. Emperor Freddy was a handsome black and white short-haired feline. His Kingdom covered about six hectares. Freddy was brought to his Kingdom as a child. Four Skin-Beings lived alongside the Emperor. These consisted of two adults (one Teochew-Hakka, one Anglo-Celtic) and their juvenile twin offspring. A further Skin-Being came in from the village each day and performed the real work of keeping the Kingdom in order.
6. Soon after Freddy's arrival at the Kingdom, a fluffy canine infant would also call the Kingdom home. Somehow, the cat and the dog became the best of friends. As the years passed, the fluffy canine would deliver two litters of puppies. One offspring from each litter remained in the Kingdom.

7. What I am about to tell you is no adventure of epic proportions. Don't expect a 《西游记》 'Journey to the West'. This account has no Sun Wukung (Monkey King); no Zhu Bajie (Strong-maned Pig); no priestly Tang Sanzang atop a White Dragon Steed.
8. Instead, we just have a most mundane story of when Emperor Freddy, his canine siblings, and his Skin-Being minions marched themselves to the top of the hill (to the Temple) and marched back down again. That said, this story is no less an allegory of cross-species and multi-entity triumph – if I do say so myself.
9. The canines of the Kingdom of Freddy – oh how they loved me. They loved my slopes, they loved my forests. The juvenile Skin-Beings loved me too. They knew every bend in the trails; the 'short' cuts; how close each waterfall was to home; and that dusk would bring the cicada symphony.
10. The mere act of Skin-Beings tying their shoelaces would turn the canines into whirlwinds of excitable anticipation. The expectation was that shoe-wearing would lead to an adventure to the Temple. This never made sense to me. The dogs clearly could come visit me any time they liked. Yet, they hardly came without the Skin-Beings. Emperor Freddy never came at all – until one day.
11. There was nothing remarkable about this day – apart from what was about to happen. The Skin-Beings went to get their running shoes. On cue, the canines whipped themselves into a frenzy. Once the Skin-Beings had their backpacks and water bottles they headed off. The canines charged ahead. Everyone knew the way.
12. Just as they walked out the front gate, the team noticed Emperor Freddy running behind them.
13. One of the Emperor's minions (Juvenile Skin-Being One) asked jokily: 'Are you coming to climb the mountain too, your Highness?'
14. Emperor Freddy did not answer. He strode with purpose towards me, attempting to catch up to the canines.
15. Emperor Freddy's minions tried to convince him many times that he probably did not want to climb a mountain. They suggested through gestures that he might like to head home. They even picked him up and turned him around in the direction of his Kingdom. This was all in vain.
16. The Emperor's minions thought he might stop and turn back after the first steep climb into the forest. But Emperor Freddy kept marching. The minions thought surely Emperor Freddy would be exhausted by the half-way mark. But Emperor Freddy kept marching – all the while keeping pace with his minions. When the minions stopped for water, Freddy drank from the trickles of the little waterfalls. When the minions started moving again, Emperor Freddy kept marching. Higher and higher they climbed. The trail-side vegetation changed from tall trees to an impenetrable matt of thick ferns. Emperor Freddy, his pink tongue now hanging out the side of his mouth, kept on marching! His minions tried to carry him, but he wrestled from their grasp, and kept on marching!

17. Finally, the group made it to a fork in the path. Turn left and it will be all flat, through the trees, for a few hundred metres.... And, just like that, they had made it to my mountain Temple. Emperor Freddy had made it to my mountain Temple.
18. The convoy stopped to rest, share the contents of the minions' backpacks, and enjoy the view. Then they hiked back down again. A multi-species caravan in a land of imperfect, but harmonious, multi-cultural coexistence.
19. And all the way down, Emperor Freddy kept on marching.
20. On the Return from the Original Position, march, we must too. Back to a simpler place; and a simpler time.

Kilimakyaro Lichen

1. We toggle the cocoon's joystick, *as we look out on Planet 3 from the Original Position, behind the Veil of Illumination.* We zoom in on the highest point of the continent that we used to call our home.
2. We are lichen, Kilimakyaro Lichen. Like all lichen, we are a community. In our case, a delightful commune of algae and fungi. Unlike some lichen, we cannot deal with the heat.
3. Zooming in on our ancestral home is bittersweet. Gone are the glaciers and the climate that could have sustained us; changed is the landscape. Our longing for home is the wound; our realisation that this home no longer exists: the salt.
4. And yet, some of the continent and areas around our mountain are thriving. 'Shimboni Ma!', a young Skin-Being with closely shaved black curly hair calls out to his middle-aged neighbour. 'Nasicha!' she calls back with a beaming smile. Language, land and livelihood have been reclaimed. Ingenuity, innovation and the guidance of the ancestors have allowed this region to weather metaphorical and physical storms. Indeed, it is only because these Skin-Beings saw we were not coping with in the planetary shifts that we were scraped up and sent far, far away. This is the only reason we still exist – though we often question such existence.
5. Where did we end up? Antarctica might provide a balmy temperature for alpine lichen. True. But we are not there. Instead, we wait out our days in 'The Pyramids'.
6. The Pyramids are large thermos-regulated plant conservatories. The design of the immense triangular glass structures is a tribute to the work of architect IM Pei of the 20th century. Unlike glasshouses of the earlier century, these are designed to capture the energy of the Sun's rays and channel it into cooling these huge constructions. The Pyramids draw visitors from all across and far beyond the realm. A huge sign boasts 'Every plant you can imagine; and even those you can't'. *sigh* Some education is in order. We can see it now – stocking the Gift shop with T-Shirts preaching taxonomical correctness: 'If your likin' the Lichen, then learn their Kingdom and Phylum'. Not bad, eh?

7. I suppose we shouldn't complain. Great care has been taken to make sure we are comfortable. And we have nowhere else to go. We certainly couldn't survive outside the Pyramids. Until we harnessed the powers of exploration of the Fission Cocoon we didn't really know what life was like beyond our 'bubble'. We did know, thanks to the little sparrow that sneaks in for respite from the oppressive heat, that outside the Pyramids it is scorching hot - even more uninhabitable than our ancestral home.
8. What else occurs beyond the Pyramids? Our investigations from our cocoon showed us that we were living in a highly unequal realm on an unequal planet. Some Skin-Beings in the realm lead relatively happy lives. Lives largely restricted to thermo-regulated edifices akin to our triangular 'home'. Efforts were made to transition to new forms of energy and most forms of nature have been replaced with 'Green' buildings. Food is either grown in factories or transported from far beyond the realm.
9. Skin-Beings here are known by their CAT-numbers. 'CAT' is short for 'category'. Some Skin-Beings from outside the realm can be allocated to the highest CAT. These individuals are afforded privileges similar, and at times exceeding, that of the average citizen of the realm. Others from beyond the realm, about a third of the workforce, are treated as disposable. Low CAT adult males are housed in segregated dormitories, transported like chattels in trucks; low CAT females live in the homes of higher CAT families performing domestic tasks. Medical examinations occur every six months. Pregnant females from this category are banished from the realm.
10. From our cocoon we take a VR (Very Real) tour underground. Here there is a system of enormous tubes. Most inhabitants use these to travel across the realm. We are drawn to a large bright sign that covers the wall of the underground. It says:

    'Invest in the Green Economy this Earth Day!'

11. At this point, Rock's music started playing – indicating our time was almost up:

RETURN! RETURN NOW TO THE RECORDING!
PAY ATTENTION TO WARNING SIGNS.

---

12. A flock of pigeons rested on the edge of the large bright sign and proceeded to relieve themselves peacefully. White paste trickled down the large bright sign:

    'Invest in the Green Economy this Earth Day!'

    seems to now spell out:

'Invert Greed! Community for Earth !'.

13. The modified sign concludes and constitutes our judgment.

The Little Bush Warbler

1. *As I look out on Planet 3 from the Original Position, behind the Veil of Illumination*…I feel, I feel, I feel the urge to ask:

shén me shì xiàn dài
什么是'现代'
ruò méi y ǒ u shēng tài
若没有生态 ？
huò nán n ǚ ni ǎ o shòu
或男女鸟兽
wàng dì qiú biàn tài
望地球变态。

What is 'modernity'
Without ecology's cosmologies?
All the world's creatures
Watch Earth slide to absurdity.

bēi tòng de xīn qíng
悲痛的心情
lóng l ǐ de zì zài
笼里的自在。
hēi zěn me qū fēn
嗨,怎么区分
nuò ruò y ǔ wú nài
懦弱与无奈？

My heart fills with sorrow,
While my cage fills with comforts.
Sigh what's the difference between the
'Gutless' and 'helpless'?

yào wén huà zì rán
要文化自然
dōu liú chuánhòu dài
都流 传 后代。
shēngmìng de zhuī qiú
生 命 的 追 求
bù zh ǐ shì cún zài
不只是存在。

For culture and nature
To span generations
Mere existence, cannot be
This life's aspiration.

xi ǎ ngg ǎ ibiànshì jiè
想 改变世界？
yàoxu ǎ n zé wèi lái
要 选 择未来？
mèngzhí dé zhēngq ǔ
梦 值得 争 取,
bié zài kàoděng dài
别再靠 等 待 ！

Want to change the world?
Want to choose your own futures?
Dreams are worth chasing,
Just stop wishing and waiting.

Commentary – Towards Radically Normal Futures

The first key message of the IPBES (Intergovernmental Science-Policy Platform on Biodiversity and Ecosystem Services) Values Assessment is that the unsustainable and inequitable use of nature is a critical cause of global biodiversity loss.[4] The Assessment emphasises that this arises from 'predominant political and economic decisions based on a narrow set of values (eg, prioritizing nature's values as traded in markets)'.[5]

The Assessment sets out the spectrum of instrumental, relational and intrinsic ways in which humans regard nature. Instrumental values see nature in terms of its benefit to humans; intrinsic values reflect nature's worth in its own right; and relational values recognise human-nature interactions and interactions between humans about nature.[6] In this piece, I contemplate what a shift to more reciprocal human-nature relations might look like. Here, this takes the form of an all-of-planet adjudicating body.

Drawing on relational Indigenous ontologies[7] that incorporate storytelling as law,[8] the piece imagines a legal system of collective multi-species decision-making that places law in a context of a continuing past, present and future.[9] With a focus on planetary-scale judgments to determine the elements needed for desirable futures the piece adopts a more-than-human conceptualisation of the Rawlsian original position and veil of ignorance.[10] A key character, 'Rock' appears, often in a narrator role. This, a nod to the Anthropocene as a geological concept while also invoking concepts of deep time and an appreciation of the relatively tiny timeframe across which the 'Great Acceleration'[11] of human-induced planetary-scale transformation has occurred.

The judgments do not involve disputes between parties but rather offer considered statements of how the world might prosper. The formatting borrows from advisory opinions of the International Court of Justice with

4 Unai Pascual et al, *Summary for Policymakers of the Methodological Assessment Report on the Diverse Values and Valuation of Nature of the Intergovernmental Science-Policy Platform on Biodiversity and Ecosystem Services* (July 2022), KM1.

5 Ibid.

6 Ibid, KM3.

7 Irene Watson, *Aboriginal Peoples, Colonialism and International Law: Raw Law* (Routledge, 2014), 13–14; Martuwarra RiverOfLife, Anne Poelina, Donna Bagnall, and Michelle Lim, 'Recognizing the Martuwarra's First Law Right to Life as a Living Ancestral Being' (2020) 9 *Transnational Environmental Law* 541.

8 Watson (n 7) 14-17.

9 Ibid 14; Martuwarra RiverOfLife (n 7).

10 John Rawls, *A Theory of Justice* (Belknap Press, 1971).

11 Will Steffen et al, 'The Trajectory of the Anthropocene: The Great Acceleration' (2015) 2 *The Anthropocene Review* 81.

preambular wording taking inspiration from the *Legality of the Threat or Use of Nuclear Weapons* Advisory Opinion.[12]

While the piece includes some reference to the common law system, my starting point is that something radically different is needed. Drawing on the 'shifting baseline syndrome' I argue that while radical shifts from current trajectories are required, the futures imagined here are hardly radical. The 'shifting baselines syndrome' highlights each generation's collective failure to appreciate the extent of ecological change from one generation to the next.[13] When each generation, usually subconsciously, sets (and reinforces) expectations of what the environment 'should' be like, this results in the shift in what is considered 'normal'. This can lead to systemic minimisation of the actual extent of environmental degradation.[14] Thus, my argument is that the transformative change that is needed is 'merely' a version of multi-species interaction that bears some semblance to the last million years.

Hong has called out the 'cultural accounting trick'[15] of pervasive contemporary technological optimism and the 'hegemony of closure and sameness' that techno-futures enact.[16] In the piece above, I attempt to write against what Hong calls 'regurgitated visions'[17] of the future. Such visions, such as Elon Musk's Neuralink AI, Hong argues, are predicated on technological promises. Promises that are concerned not with being actually realised but instead 'borrow legitimacy and plausibility from the future in order to license anticipatory actions in the present'.[18] The caution against unfulfilled technological promises finds particular resonance in Simon and Garfunkel's *Sound of Silence* – the neon God to which people bow and pray, reflecting Hong's characterisation of technology as 'as much a totem of future possibility as it is a practical instrument'.[19]

The exercise of writing judgments 'of and for the Anthropocene',[20] particularly if projected to the planetary scale as I have done here, creates a potential trap of so-called 'universality'. I am reminded of Goffe's observation that 'avoiding race does not solve racism' and that ignoring race in

12 *Legality of the Threat or Use of Nuclear Weapons (Advisory Opinion)* [1996] ICJ Rep 226.
13 Daniel Pauly, 'Anecdotes and the Shifting Baseline Syndrome of Fisheries' (1995) 10 *Trends in Ecology & Evolution* 430.
14 Heidi Alleway et al, 'The Shifting Baseline Syndrome as a Connective Concept for More Informed and Just Responses to Global Environmental Change' (2023) 5(3) *People and Nature* 885.
15 Sun-Ha Hong, 'Predictions without Futures' (2002) 61 *History and Theory* 371, 373.
16 Ibid 372.
17 Ibid 373.
18 Ibid 377.
19 Ibid 373.
20 Anna Grear, 'Legal Imaginaries and the Anthropocene: "Of" and "For"' (2020) *Law and Critique* 1; Nicole Rogers and Michelle Maloney. 'The Anthropocene Judgments project: A Thought Experiment in Futureproofing the Common Law' (2022) 47(3) *Alternative Law Journal* 173.

genres that engage with the future, such as sci-fi and speculative fiction, constitutes an ethical determination of who exists in the future.[21] At the same time, Pereira and fellow writers point out that 'Africa is absent from the future. In almost every future, dystopian or utopian, there is a continent-sized hole in the story'.[22] Meanwhile, Goffe observes that if Asian people are included in visions of the future, they are routinely depicted as 'technologically advanced and intellectually primitive'.[23]

I am conscious that in the judgments above I do not sufficiently go beyond what Goffe terms 'epistemic genocide'.[24] Therefore, I think it important that I reflect on my positionality. Family on my Mum's side (of Scottish, Cornish, and English ancestry) have lived for three generations on Kamilaroi land, on the continent some know as Australia. My Dad's parents, Teochew and Hakka, moved from Southern China in the 1940s to colonial British Borneo, now Sarawak, Malaysia. I was born in Kuching – the capital of Sarawak. I lived at the foot of Matang Mountain until I moved to my Mum's hometown, in Australia, at the age of 16. The judgment of Matang Mountain – the journey of Emperor Freddy, is an entirely true story. The judgment hints at our relative economic privilege.

When writing the piece, I was keen to avoid perpetuating an African-shaped hole in the future. Recalling Pereira and her fellow writers' urging that '[t]he denial of equity in time is connected to a denial of equity in space'.[25] I lived for a year at the foot of Kilimanjaro – Africa's highest peak. In the Kilimakyaro Lichen judgment, I wrote from my memories while self-consciously aware of how lacking these are. I also have a sense of unease that I may have unwittingly lapsed into unhelpful tropes or appropriation. The same can be said of my attempts to engage with and elevate Indigenous ontologies.

The final judgment, in Chinese, may seem like a statement against linguistic hegemony or a similar stand for diversity. It probably is and the English version of the poem has deliberately avoided a literal translation. Nevertheless, my inspiration from the judgment comes from two Chinese songs that contain multi-species, and indeed specifically avian, themes. Lee Tsung-Sheng's Taiwanese Mandopop hit 'I am a little, tiny bird' 《我是一只小小鸟》 and Huang Hong-Mo's 'Declaration of a dumb bird' 《笨鸟的表白》 from the Xinyao (Singaporean Mandopop) genre both involve birds and metaphors

21 Tao Leigh Goffe, 'Stolen Life, Stolen Time: Black Temporality, Speculation, and Racial Capitalism' (2022) 121 *South Atlantic Quarterly* 109, 110, 113.

22 Laura Pereira et al, 'Wakanda Phambili! African Science Fiction for Reimagining the Anthropocene' in Sandra Kemp and Jenny Andersson (eds) *Futures* (Oxford University Press, 2021) 262, 262.

23 Goffe (n 21) 110.

24 Ibid 113.

25 Pereira et al (n 22) 262.

of flight. Across global cultures, birds and their capacity for flight have long been a source of inspiration and often a symbol of freedom. However, in both these songs, even the birds are tired.

In Lee's piece, the starting line: 'Sometimes I think I am a little, tiny bird' shifts in the soaring chorus to 'I *am* a little, little, tiny, little bird'. In other words, in a few verses the person has become the little bird. Meanwhile, the lyrics which yearn for belonging and for better tomorrows combine with acknowledgement of an unknowable future. Huang's piece speaks of all the tired birds coming home to roost while 'I' remain in a stunned stupor coasting on the wind. At the same time, the song considers notions of modernity while hinting at the importance of culture amidst changing trends.

At a more macro level, drawing inspiration from songs from Taiwan and Singapore somehow opens up unanswered questions for me about what it means to be Chinese. The counter-claims from Taiwan (the Republic of China) and the *People's* Republic of China as to who gets to speak for the Middle Kingdom also find parallels in multi-species questions as to who gets to speak for the planet. At the same time, the dominance of the English language in Singapore, a country where ethnic Chinese constitute almost 80% of the population, brings up questions about nature-culture connection and how and whether language mediates connections to nature.

A legitimate critique of my piece could be that despite hints at underlying racial, spatial, historical and continued inequity on a rapidly changing planet, the stories ultimately shy away from the most thorny and difficult issues of power, structure and privilege. Some might argue that the more-than-human angle appropriates Indigenous worldviews while repeating acts of erasure of human existence on the one hand and responsibility on the other.

I have discussed elsewhere that fiction writing as a scholarly endeavour and critical lens has a critical role in imagining possible and desirable futures.[26] The act of creative writing engages the subconscious in ways that the genre of academic writing does not. What creative forms of expression can also do is reveal biases that might be more easily obscured in the traditional academic form. What this reflexive exercise suggests to me, therefore, is that I could do well to practise the ideals of collective imagining, adjudicating and co-production hinted at in the piece, and that imagining possible futures needs to be seen as an interactive process and not an endpoint in itself. In the words of the world's most famous cyborg: 'I'll be back!'[27]

26 Michelle Lim, 'Fiction as Legal Method—Imagining with the More-than-Human to Awaken Our Plural Selves' (2021) 33 *Journal of Environmental Law* 501, 504.

27 Catchphrase of the protagonist of the *Terminator* franchise. The franchise is a series of science-fiction films involving the conflict between humans and an artificially intelligent army of cyborgs.

Index

For Product Safety Concerns and Information please contact our EU representative GPSR@taylorandfrancis.com
Taylor & Francis Verlag GmbH, Kaufingerstraße 24, 80331 München, Germany

www.ingramcontent.com/pod-product-compliance
Ingram Content Group UK Ltd.
Pitfield, Milton Keynes, MK11 3LW, UK
UKHW020153200926
475702UK00016B/2690
* 9 7 8 1 0 3 2 4 8 5 4 2 3 *